T0348475

Toxicogenomics-Based Cellular Models

Toxicogenomics-Based Cellular Models

Alternatives to Animal Testing for Safety Assessment

Edited by

Jos Kleinjans, PhD
Professor and Chair
Department of Toxicogenomics
Maastricht University, Maastricht, The Netherlands

AMSTERDAM • BOSTON • HEIDELBERG • LONDON
NEW YORK • OXFORD • PARIS • SAN DIEGO
SAN FRANCISCO • SINGAPORE • SYDNEY • TOKYO

Academic Press is an imprint of Elsevier

Academic Press is an imprint of Elsevier
The Boulevard, Langford Lane, Kidlington, Oxford, OX5 1GB, UK
225 Wyman Street, Waltham, MA 02451, USA

First published 2014

British Library Cataloguing in Publication Data
A catalogue record for this book is available from the British Library

Library of Congress Cataloguing in Publication Data
A catalogue record for this book is available from the Library of Congress

ISBN: 978-0-12-397862-2

For information on all Academic Press publications
visit our website at elsevierdirect.com

Typeset by MPS Limited, Chennai, India
www.adi-mps.com

Printed and bound by CPI Group (UK) Ltd, Croydon, CR0 4YY
10 9 8 7 6 5 4 3 2 1

Contents

SECTION 4 REPRODUCTION TOXICITY

SECTION 6 TOXICOINFORMATICS

SECTION 7 SELECTION AND VALIDATION OF TOXICOGENOMICS ASSAYS AS ALTERNATIVES TO ANIMAL TESTS

SECTION 8 TOXICOGENOMICS IMPLEMENTATION STRATEGIES

List of Contributors

Marjam Alloul-Ramdhani
Department of Dermatology, Leiden University Medical Center, Leiden, The Netherlands

Scott S. Auerbach
Biomolecular Screening Branch, Division of the National Toxicology Program, NIEHS, Research Triangle Park, North Carolina

Giulia Benedetti
Division of Toxicology, LACDR, Leiden University, Leiden, The Netherlands

Harrie Besselink
BioDetection Systems b.v., Amsterdam, The Netherlands

André Boorsma
The Netherlands Organisation for Applied Scientific Research, Microbiology, and Systems Biology, Zeist, The Netherlands

Bram Brouwer
BioDetection Systems b.v., Amsterdam, The Netherlands

Ines Chaves
Department of Genetics, Erasmus University Medical Center, Rotterdam, The Netherlands

Maarten Coonen
Department of Toxicogenomics, Maastricht University, Maastricht, The Netherlands

Emanuela Corsini
Department of Pharmacological Sciences, University of Milan, Milano, Italy

Erik H.J. Danen
Division of Toxicology, LACDR, Leiden University, Leiden, The Netherlands

Marjo de Graauw
Division of Toxicology, LACDR, Leiden University, Leiden, The Netherlands

Eugin Destici
Department of Genetics, Erasmus University Medical Center, Rotterdam, The Netherlands; Department of Medicine, School of Medicine, University of California, San Diego, California

Marja Driessen
RIVM—National Institute for Public Health and the Environment, Bilthoven, The Netherlands

Jennifer Fostel
National Toxicology Program, National Institute of Environmental Health Sciences, Research Triangle Park, North Carolina

Abdoelwaheb El Ghalbzouri
Department of Dermatology, Leiden University Medical Center, Leiden, The Netherlands

Sanne A.B. Hermsen
Center for Health Protection Research (GZB), National Institute for Public Health and the Environment (RIVM), Bilthoven, The Netherlands

Kristina M. Hettne
Department of Medical Informatics, Biosemantics Group, Erasmus University Medical Center, Rotterdam, The Netherlands

Wouter T.M. Jansen
PricewaterhouseCoopers Advisory N. V., The Hague, The Netherlands

Barae Jomaa
Division of Toxicology, Wageningen University, Wageningen, The Netherlands

Jos Kleinjans
Department of Toxicogenomics, Maastricht University, Maastricht, The Netherlands

Jan A. Kors
Department of Medical Informatics, Biosemantics Group, Erasmus University Medical Center, Rotterdam, The Netherlands

Dinant Kroese
The Netherlands Organisation for Applied Scientific Research, Risk Analysis for Products In Development, Zeist, The Netherlands

Robert Luebke
Cardiopulmonary and Immunotoxicology Branch, US Environmental Protection Agency, Research Triangle Park, North Carolina, USA

Karen Mathijs
Department of Toxicogenomics, Maastricht University, Maastricht, The Netherlands

Romana Nijman
Department of Genetics, Erasmus University Medical Center, Rotterdam, The Netherlands

Richard S. Paules
Toxicology and Pharmacology Laboratory, Division of Intramural Research, NIEHS, Research Triangle Park, North Carolina

Jeroen Pennings
National Institute for Public Health and the Environment (RIVM), Centre for Health Protection, Bilthoven, The Netherlands

Aldert H. Piersma
Center for Health Protection Research (GZB), National Institute for Public Health and the Environment (RIVM), Bilthoven, The Netherlands; Department of Toxicogenomics (TGX), Maastricht University, Maastricht, The Netherlands

Jan Polman
Department of Toxicogenomics, Maastricht University, Maastricht, The Netherlands

Leo S. Price
Division of Toxicology, Leiden Amsterdam Center for Drug Research, Leiden University, Leiden, The Netherlands

Tessa Pronk
National Institute for Public Health and the Environment (RIVM), Center for Health Protection, Bilthoven, The Netherlands

Sreenivasa Ramaiahgari
Division of Toxicology, Leiden Amsterdam Center for Drug Research, Leiden University, Leiden, The Netherlands

Erwin L. Roggen
3Rs Management and Consultancy, Kongens Lyngby, Denmark

Peter Schmeits
RIKILT—Institute of Food Safety, Wageningen University and Research Centre, BU Toxicology and Bioassays, Wageningen, The Netherlands

Jan Hendrik R.H.M. Schretlen
PricewaterhouseCoopers Advisory N. V., The Hague, The Netherlands

Jia Shao
RIKILT—Institute of Food Safety, Wageningen University and Research Centre, BU Toxicology and Bioassays, Wageningen, The Netherlands

Rob H. Stierum
The Netherlands Organisation for Applied Scientific Research, Microbiology, and Systems Biology, Zeist, The Netherlands

Laura Suter-Dick
University of Applied Sciences and Art, Northwestern Switzerland (FHNW), School for Life Sciences, Institute of Chemistry and Bioanalytics, Muttenz, Switzerland

Ewa Szalowska
RIKILT—Institute of Food Safety, Wageningen University and Research Centre, Wageningen, The Netherlands

Cornelis P. Tensen
Department of Dermatology, Leiden University Medical Center, Leiden, The Netherlands

Elisa C.M. Tonk
Department of Toxicogenomics (TGX), Maastricht University, Maastricht, The Netherlands

Geert R. Verheyen
Drug Safety Sciences, Janssen Research and Development, Beerse, Belgium

Oscar L. Volger
RIKILT Institute of Food Safety, Wageningen University and Research Center, Wageningen, The Netherlands

Annelieke S. de Wit
Department of Genetics, Erasmus University Medical Center, Rotterdam, The Netherlands

Bart van der Burg
BioDetection Systems b.v., Amsterdam, The Netherlands

Joost van Delft
Department of Toxicogenomics, Maastricht University, Maastricht, The Netherlands

Freddy van Goethem
Drug Safety Sciences, Janssen Research and Development, Beerse, Belgium

Henk van Loveren
Center for Health Protection Research, National Institute of Public Health and the Environment, Bilthoven, The Netherlands

Eugene van Someren
The Netherlands Organisation for Applied Scientific Research, Microbiology, and Systems Biology, Zeist, The Netherlands

Leo van de Ven
RIVM—National Institute for Public Health and the Environment, Bilthoven, The Netherlands

Bob van de Water
Division of Toxicology, LACDR, Leiden University, Leiden, The Netherlands

Gijsbertus T.J. van der Horst
Department of Genetics, Erasmus University Medical Center, Rotterdam, The Netherlands

Louise von Stechow
Division of Toxicology, LACDR, Leiden University, Leiden, The Netherlands

SECTION 1

Introduction to Toxicogenomics-Based Cellular Models

CHAPTER

1.1 Introduction to Toxicogenomics-Based Cellular Models

Jos Kleinjans

Department of Toxicogenomics, Maastricht University, Maastricht, the Netherlands

We live in an era where we witness increased needs for alternative models to ultimately replace the "gold standard" in repeated-dose toxicity testing: the rodent bioassay. In demand are tests that are more reliably predicting human health risks, are less costly and time consuming, and are preferably non-animal based, also to meet with ethical concerns on animal welfare. Since the turn of the millennium, so-called 'omics technologies, coming from the endeavor of unraveling the human genome, have been increasingly applied to these challenges in chemical safety assessment, in an approach which is generally referred to as "toxicogenomics." In addition, a wide range of cellular models, exploiting human cell lines, human primary cells, and human embryonic stem cells, have been put to the test. While initial results are promising, the upgrading of such cell assays, the absorbing of the newest 'omics technologies, and in particular the managing of the tsunami of 'omics data, still pose a major challenge to the research community of toxicogenomics.

Multiple user groups have high expectations of such toxicogenomics-based approaches towards developing novel test systems for chemical safety assessment. However, a thorough overview that will simultaneously introduce toxicogenomics to a wider readership is still lacking.

The goal of this book is to describe the state of the art in developing toxicogenomics-based cellular models for chemical-induced carcinogenicity, immunotoxicity, and reproduction toxicity, all important endpoints of toxicity, the evaluation of which to date costs large numbers of animal lives. Also, where 'omics technologies tend to generate "big data" requiring extensive bioinformatics and biostatistics efforts for actually retrieving toxicologically meaningful results, the field of toxicoinformatics will be thoroughly introduced. The book will also address how to validate toxicogenomics-based alternative test models, and will provide an outlook to societal and economic implementation of these novel assays.

Toxicogenomics-Based Cellular Models.

1.1.1 The demands for alternatives to current animal test models for chemical safety

The current default model for assessing repeated-dose toxicity of novel or existent chemicals is the rodent bioassay involving mice or rats. Sometimes, however, also guinea pigs and rabbits, and on rare occasions dogs or monkeys, are used. Options here are the 28-day oral toxicity test, the 28-day dermal toxicity test, and the 28-day inhalation toxicity test. For sub-chronic toxicity testing, 90-day oral, dermal, or inhalation toxicity studies are available. Lastly, chronic rodent assays have been put in place, such as the 2-year treatment protocol for carcinogenicity testing. All protocols involve the daily administration of the compound of interest. These assays aim to quantitatively analyze whether and to what extent toxicity, a persistent or progressively deteriorating dysfunction of cells, organs, or multiple organ systems, is present upon repeated administration to the animal of the chemical under investigation. Repeated-dose testing in vivo enables evaluating particular molecular and histopathological endpoints of toxicity in organs, but also provides information on perturbations of more complex (e.g. hormonal, immunological, neurological) systems. Focus is on establishing dose–response relationships, from which a NOAEL (no observable adverse effect level) is derived that forms the basis for setting safety standards for human health in relation to daily lifetime exposure to the chemical. For formalized safety testing, international regulatory authorities such as the Organization of Economic Cooperation and Development (OECD) have developed a series of dedicated guidelines. It should be noted that toxicity testing protocols may differ to some extent, depending on the domain of ultimate application—for example, pharmaceuticals, cosmetics, or food. Current estimates calculate that approximately 14% of all animals annually tested within the EU for the purpose of chemical safety assessment are used in sub-chronic and chronic safety assessments, e.g. repeated-dose toxicity.

Obviously, the underlying assumption for using animals in safety testing is that results can be extrapolated to humans relatively reliably, because of resemblances at the molecular and physiological levels: basically, we are all mammals. It may be convincingly argued that the repeated-dose animal test has probably prevented mankind in the past from major chemical disasters. By contrast, it is worthwhile to note that your average Shakespearian king, under constant threat of being poisoned, would not have trusted an animal for his private chemical safety testing but would have required a serf for pre-tasting his food. In more civilized terms, over the last decade or so, we have learned to ask critical questions concerning the actual relevance of the rodent bioassay for assessing chemical safety to human health. A short overview follows.

Over the last decade, the pharmaceutical industry has suffered from high attrition rates of novel candidate drugs, in particular because of adverse findings in the last research and development stages—for example, during clinical trials. These related to disappointing efficacies, or inadequate adsorption, distribution, metabolism and excretion (ADME) properties, but in 30–40% of cases also to overt toxicity, in particular for the liver and the heart, despite the fact that animal tests had reported these novel compounds to be safe [1]. Unexpected toxicity in humans may even occur after market introduction. Each year, about two million patients in the United States experience a serious adverse drug reaction when using marketed drugs, resulting in 100,000 deaths, thus representing the fourth leading cause of death [2]. Similar percentages have been estimated for other Western countries such as The Netherlands [3]. Here is a case where the "gold standard" animal model for chemical safety clearly lacks sufficient sensitivity. Failure in the last phases of drug development obviously is to the disadvantage of patients, but also, in view of the extreme costs of developing new drugs, involves huge economic losses [4].

Simultaneously, other examples demonstrate that animal models for repeated-dose toxicity may also over-report human health risks: the US *Physicians' Desk Reference* has reported that out of 241 pharmaceutical agents used for chronic treatment, 101 agents were demonstrated to be carcinogenic to rodents. However, epidemiological studies among chronically treated patients as reviewed by the World Health Organization (WHO) International Agency on Research on Cancer have identified only 19 pharmaceuticals, mostly intended for anticancer treatment or hormone therapy, to be actually carcinogenic to man.

This apparent lack of sufficient specificity and sensitivity of the rodent bioassay for repeated-dose toxicity is underscored by a report indicating that only 43% of toxic effects of pharmaceuticals in humans were correctly predicted by tests in rodents [5].

These examples demonstrate that better tests for predicting human drug safety are in demand.

Next, there also are ethical concerns with animal toxicity testing. In general, the way animal welfare is considered has often be claimed to represent a sign of civilization. In the words of the twentieth century Indian civil rights activist and political leader Mahatma Gandhi, "The greatness of a nation and its moral progress can be judged by the way its animals are treated." Immanuel Kant, the great eighteenth century German moral philosopher, had already stated that "We can judge the heart of a man by his treatment of animals." With regard to animal experimentation, also implying the use of animal tests for chemical safety evaluations, these concepts were successfully adopted by William Russell and Rex Burch in their 1959 book *The Principles of Humane Experimental Technique*, in which they presented the 3R principle, referring to *replacement*, *refinement*, and *reduction* of animal testing. Since then, the 3R principle has also found political recognition, for instance within the EU where the *Protocol on Protection and Welfare of Animals* annexed to the European Community (EC) Treaty aims at ensuring improved protection and respect for the welfare of animals as sentient beings. It is stated that in formulating and implementing the Community's policies, the Community and the member states shall pay full regard to the welfare requirements of animals. All industry sectors, including pharmaceuticals, chemicals, cosmetics, agrochemicals, and foods manufacturers, are consequently obliged to apply available methods to replace, reduce, and refine animal use (three Rs) in safety and efficacy evaluations under the existing animal protection legislation (Directive 86/609/EEC) [6]. The most prominent political action within the EU in this context, concerning safety testing of cosmetic ingredients, undoubtedly is the 7th Amendment (Directive 2003/15/EC) to the Cosmetics Directive, which requires the full replacement of animals in safety testing, and which sets a timetable for the availability of alternative testing methods for assessing safety of cosmetics ingredients and products, with a deadline of 2013.

All in all, we now see the need for better, reliable predictive tests for assessing chemical safety for human health, which come less costly and less time consuming, and preferably are no longer animal based.

1.1.2 The toxicogenomics approach

With the advent of genomics technologies to the domain of toxicology, hopes have been set high that the so-called toxicogenomics approach may actually deliver the desired alternative test systems to the current animal models for chemical safety. This is, for instance, expressed in the 2006 EU Regulation on the Registration, Evaluation, Authorisation and Restriction of Chemicals (REACH), which addresses the production and use of chemical substances, and their potential impacts on both human health and the environment. The REACH legislation states:

> *The Commission, Member States, industry and other stakeholders should continue to contribute to the promotion of alternative test methods on an international and national level including computer supported methodologies, in vitro methodologies, as appropriate, those based on toxicogenomics, and other relevant methodologies.*

The principle approach for developing toxicogenomics-based predictive assays for chemical safety, and in particular for the purpose of hazard identification, implies that genomic data are derived from exposure of bioassays to known toxicants. Bioassays may refer to animal models, but for developing non-animal-based tests, the exploration of human cellular models in vitro is quite obvious. Per endpoint (class) of toxicity, prototypical compounds are derived from available toxicological databases such as the US National Toxicology Program, and the bioassay of choice is challenged by such compounds for training a classifying gene set, predicting the particular toxic phenotype. Subsequently, to strengthen predictivity, this classifier may be validated by a second set of model compounds. It is obvious that such 'omics-based gene profiles for toxic mode of action gain more predictive value as statistical power increases, implying that genomic profiles should be generated from as many model compounds as possible, while mechanistic specificity of such profiles increases with the availability of specific model compounds. The selection of as many prototypical compounds for respective classes of toxicity as possible from available databases, thus, is crucial to improve predictivity [7].

These gene profiles are compared to a set of genomic changes induced by a suspected toxicant. If the characteristics match, a certain toxic mode of action can be assigned to the unknown agent, thus identifying a potential hazard of that compound to human health.

To date, whole-genome gene-expression analysis by applying microarray technology has been the dominant technique in toxicogenomics. A few years ago, standardized procedures were developed for this. The Toxicogenomics Research Consortium, a consortium of 10 US-based laboratories, was formed to compare data obtained from three widely used platforms using identical RNA samples. The outcome was that there were relatively large differences in data obtained from different laboratories using the same platform, but that the results from the best-performing laboratories agreed rather well. It was concluded that reproducibility for most platforms within any laboratory was typically good. Microarray results can be comparable across multiple laboratories, especially when a common platform and set of procedures are used [8]. Next, the MicroArray Quality Control (MAQC) project, pursuing similar aims by analyzing 36 RNA samples from rats treated with three prototypical chemicals, each sample being hybridized to four microarray platforms, showed intra-platform consistency across test sites as well as a high level of inter-platform concordance in terms of genes identified as differentially expressed [9].

MAQC demonstrated that different bioinformatics/biostatistics approaches to analyze 'omics data caused the major differences between participating laboratories. So, in the subsequent MAQC-II project, multiple independent teams analyzed microarray data sets to generate predictive models for classifying a sample with respect to endpoints indicative of lung or liver toxicity in rodents. It was demonstrated that, indeed, with current methods commonly used to develop and assess multivariate gene-expression-based predictors of toxic outcome, differences in proficiency emerged, and this underscores the importance of proper implementation of otherwise robust data analytical methods [10].

Over recent years, more 'omics platforms have been put to use for investigating molecular mechanisms of toxicity and for developing better gene classifiers for predicting chemical hazards for humans; proteomics and metabolomics being first explored for this purpose. Where such a combination of technologies seems of added value [11], insights into the increasing complexity of gene regulation networks have forced the toxicogenomics research community to rapidly absorb even more upfront techniques, such as microRNA analysis [12] and whole-genome DNA methylation analysis [13]. The most challenging 'omics technology to incorporate into the toxicogenomics approach in the near future is undoubtedly next-generation sequencing, which is capable of generating big data on molecular events induced by toxicants while simultaneously providing extensive information on yet unexplored responses, such as alternative splicing.

It may be rightfully argued that the previously described toxicogenomics approach for retrieving classifying gene sets from relevant bioassays simply provides statistical results, not necessarily generating mechanistic insights in toxic modes of action. For instance, within such gene sets, genes may be present (and actually are) with still unknown functionalities. It stands to reason to anticipate that toxicogenomics-based alternatives to current animal test models will be better accepted if predictive gene classifiers plausibly represent relevant molecular mechanisms for inducing toxic phenotypes. Also, in this respect, toxicogenomics is considered promising.

In view of the capacities of these toxicogenomics technologies, it is foreseen that they will leverage a so-called systems toxicology concept, which refers to the 'omics-based evaluation of biological systems upon perturbation by chemical stressors by monitoring molecular pathways and toxicological endpoints and iteratively integrating these response data to ultimately model the toxicological system [14]. This was followed up in 2008 by Francis Collins et al., who proposed the bypassing of animal-based human safety testing by shifting toxicology from a predominantly observational science at the level of disease-specific animal models in vivo to a predominantly predictive science focused on broad inclusion of target-specific mechanism-based biological observations in vitro. Key is to develop bioactivity profiles that are predictive of human disease phenotypes, by identifying signaling pathways that, when perturbed, lead to toxicities. This mechanistic information is then to be used for iteratively developing computational models that can simulate the kinetics and dynamics of toxic perturbations of pivotal signaling pathways, ultimately leading to systems models that can be applied as in silico predictors for human drug safety [15].

1.1.3 Upgrading cellular models

The appropriate non-animal bioassay for generating 'omics-based gene profiles for toxicological hazard identification in man will generally be a cellular model, preferably consisting of organotypical human cells, that can be operated in vitro. This may consist of immortalized cell lines, primary cells obtained, for instance, in the course of surgery, or stem-cell-derived organ-like cells. Obviously, to date, such cellular models can hardly copy full organ complexities in vivo; however, the relevant question is to what extent they are actually capable of mimicking organ function in situ, in order to achieve a reliable prediction of toxicant-induced risks of organ damage (at the molecular, functional, and/or morphological level) in vivo. Issues considered key in answering this question relate to whether these models are metabolically competent, whether they express critical checkpoints for inducing functional toxicity, whether they remain phenotypically stable when being cultured for a longer period, and whether they are robust, e.g. return reproducible results. It has to be noted, though, that currently such cellular models are only available for a limited range of possible target organs for repeated-dose toxicity in vivo [16].

In addition to critical factors for cell function in in vitro toxicological assays already mentioned, it has been argued that these models do not represent sufficiently well cell–cell and cell-matrix–environment interactions as occurring in intact tissue, whilst simultaneously may have profoundly different transport characteristics. To cope with these deficiencies, organ slices, in particular from the liver, are explored. But recently, more advanced three dimensional (3D) culture techniques, in particular multicellular spheroid culturing, have also been introduced. First results on toxicological studies in 3D cellular models, in particular for the skin and for the liver, are now becoming available.

In addition, the zebrafish embryo (*Danio rerio*) has been proposed as an alternative to current rodent test models. The reasons for this are manifold. First, with the zebrafish genome sequencing nearing completion now, and once DNA microarrays were manufactured, this model became suitable for toxicogenomics approaches. Secondly, the minute size (2–3 mm in length) of zebrafish embryos allows for culturing in a 96-well format, so actually enabling in vitro investigations of toxicological endpoints in the whole organism, implying analysis of target organ functions and physiological processes, as well as of absorption, distribution, metabolism, and excretion of toxic agents, in a model vertebrate. Given its embryonic stage, this model is obviously best suited for the purpose of developmental toxicological studies in particular. Within this context, the zebrafish embryo model has already been extensively explored for studying teratogenicity, multiple organ toxicities, immunotoxicity, and carcinogenicity [17]. Obviously, the big challenge lies in extrapolating findings on zebrafish embryo toxicity to higher organisms, including human health risks.

With respect to the goal to develop alternative non-animal-based toxicity tests, the zebrafish embryo model has yet another advantage: it is not considered an animal, at least not by EU Directive 2010/63/EU on the protection of animals used for scientific purposes. Where independent feeding is considered as the stage from which free-living larvae are subject to regulations for animal experimentation, here, the earliest life stages of animals are not defined as protected, and, thus, do not fall under these regulatory frameworks. Basically, as long as the zebrafish embryo is attached to its yolk sac, it is not considered to independently feed itself. It has been suggested that, taking into account yolk consumption, the development of digestive organs, free active swimming, and the ability to incorporate food, this covers a period until 120 hours after fertilization [18].

1.1.4 Regulatory aspects

From the foregoing, it may be already obvious that toxicology in general, and consequently the toxicogenomics approach with respect to its endeavor to develop alternatives to current animal models for assessing chemical safety, operates within a regulatory environment. Societal acceptance of, in particular, 'omics-based alternative test models is thus dependent on whether important stakeholders will become convinced of their relevance and reliability. For this, their biological, technical, and formal validation is key. To advance regulatory acceptance, molecular alterations and mechanistic insight derived from human cellular models need to be correlated with injury or potential for injury in humans [19]. It has been demonstrated that, for this purpose, it is indeed feasible to integrate gene-expression transcriptomics data from human cells with chemical and drug information together with disease information into what has been referred to as a "connectivity map" [20].

In addition, toxicogenomics applications require further technological standardization as well as biological standardization, especially with respect to the annotation of genes and pathways related to toxicologically relevant endpoints.

As early as 2003, a workshop on "Validation of Toxicogenomics-Based Test Systems," organized jointly by the European Centre for the Validation of Alternative Methods (ECVAM), the US Interagency Coordinating Committee on the Validation of Alternative Methods (ICCVAM), and the National Toxicology Program (NTP) Interagency Center for the Evaluation of Alternative Toxicological Methods (NICEATM), aimed at defining principles applicable to the validation of toxicogenomics platforms as well as validation of specific toxicological test methods that incorporate toxicogenomics technologies [21]. The three focus areas included: (1) biological validation of toxicogenomics-based test methods for regulatory decision making, (2) technical and bioinformatics aspects related to validation, and (3) validation issues as they relate to regulatory acceptance and use of toxicogenomics-based test methods. Some important recommendations from this workshop, which still stand today, are:

- Conduct toxicogenomics-based tests and the associated conventional toxicological tests in parallel, to (1) generate comparative data supportive of the use of the former in place of the latter, or (2) provide relevant mechanistic data to help define the biological relevance of responses within a toxicological context.
- Determine and understand the range of biological and technical variability between experiments and between laboratories and ways to bring about greater reproducibility.
- In the short term, favor defined biomarkers that are independent from technology platforms, and therefore are easier to validate; in the longer term, focus on pathway analysis (i.e. a system biology approach) rather than just on individual genes.
- Harmonize reference materials, quality-control (QC) measures, and data standards and develop compatible databases and informatics platforms that are key components of any validation strategy for a toxicological method.
- Determine performance standards for toxicogenomics-based test methods that will serve as the yardsticks for comparable test methods that are based on similar operational properties.

It has to be noted that the legal and institutional organizational environment in regard to worldwide regulations for the impact on alternatives to animal testing involves multiple regulatory stakeholders. The regulatory and legislative background of alternatives is governed by worldwide agreements, with consensus entities and global institutions such as the OECD directing, which are then transferred into supranational and national laws; for example, pan-European guidelines and directives [22]. It is also important to consider that industrial stakeholders play a major role, where early acceptance of particular alternatives to animal testing by the chemical manufacturing industries may stimulate their ultimate acceptance by the international regulatory authorities. And, last but not least, nongovernmental organizations are strong opinion leaders and definitely capable of advancing the acceptance of novel alternatives to current animal toxicity models.

Within this context, it is of utmost importance that the toxicogenomics research community reaches out to these stakeholders.

1.1.5 This book

Toxicogenomics-Based Cellular Models sets out to present an overview of research efforts within the domain of developing in vitro alternatives to the current rodent models for hazard identification of toxic agents by taking the toxicogenomics approach. A range of leading toxicogenomics scientists has been found willing to describe the state of the art in their research on seeking common 'omics-based denominators in toxic perturbations. Endpoints of toxicity under consideration are genotoxicity/chemical carcinogenesis, immunotoxicity, reproduction toxicity, and liver and kidney toxicity, all representing major targets in repeated-dose toxicity and requiring the largest turnover of animal experimentation for human hazard identification. Well-defined cellular models representing these targets in vitro are described. Some examples of the systems toxicology approach are presented, which demonstrates how multiplex 'omics, applying advanced transcriptomics (including analysis of non-coding RNAs), proteomics, and metabonomics in combination with concomitant analysis of relevant phenotypic endpoints of toxic modes of action, is used for identifying central hubs in signaling pathway responses, by using elaborate bioinformatics and biostatistics techniques. Issues concerning validation, facilitating societal implementation and regulatory embedding, are discussed. Lastly, venues for commercializing toxicogenomics-based assays for predicting human toxicity are indicated.

References

[1] Kola I, Landis J. Can the pharmaceutical industry reduce attrition rates? Nat Rev Drug Discov 2004:711–5.
[2] Giacomini KM, Krauss RM, Roden DM, Eichelbaum M, Hayden MR, Nakamura Y. When good drugs go bad. Nature 2007;446(7139):975–7.
[3] van der Hooft CS, Sturkenboom MC, van Grootheest K, Kingma HJ, Stricker BH. Adverse drug reaction-related hospitalisations: a nationwide study in The Netherlands. Drug Saf 2006;29(2):161–8.
[4] Paul SM, Mytelka DS, Dunwiddie CT, Persinger CC, Munos BH, Lindborg SR, et al. How to improve R&D productivity: the pharmaceutical industry's grand challenge. Nat Rev Drug Discov 2010;9(3):203–14.
[5] Hartung T. Toxicology for the twenty-first century. Nature 2009;460:208–12.
[6] http://ec.europa.eu/enterprise/epaa/1_2_3r_declaration.htm.
[7] Vinken M, Doktorova T, Ellinger-Ziegelbauer H, et al. The carcinoGENOMICS project: Critical selection of model compounds for the development of omics-based in vitro carcinogenicity screening assays. Mutat Res 2008;659:202–10.
[8] Toxicogenomics Research Consortium. Multiple-laboratory comparison of microarray platforms. Nat Methods 2005;2:345–50.
[9] MicroArray Quality Control project. Rat toxicogenomic study reveals analytical consistency across microarray platforms. Nat Biotechnol 2006;24(9):1162–9.
[10] MAQC Consortium. The MicroArray Quality control (MAQc)-ii study of common practices for the development and validation of microarray-based predictive models. Nat Biotechnol 2010;28(8):827–38.
[11] Ellinger-Ziegelbauer H, Adler M, Amberg A, et al. The enhanced value of combining conventional and "omics" analyses in early assessment of drug-induced hepatobiliary injury. Toxicol Appl Pharmacol 2011;252(2):97–111.
[12] Lizarraga D, Gaj S, Brauers KJ, Timmermans L, Kleinjans JC, van Delft JH. Benzo[a]pyrene-induced changes in microRNA-mRNA networks. Chem Res Toxicol 2012;25(4):838–49.
[13] Smeester L, Rager JE, Bailey KA, et al. Epigenetic changes in individuals with arsenicosis. Chem Res Toxicol 2011;24(2):165–7.

[14] Waters MD, Fostel JM. Toxicogenomics and systems toxicology: Aims and prospects. Nat Rev Genet 2004;5:936–48.
[15] Collins FS, Gray GM, Bucher JR. Transforming environmental health protection. Science 2008;319:906–7.
[16] Adler S, Basketter D, Creton S, et al. Alternative (non-animal) methods for cosmetics testing: Current status and future prospects-2010. Arch Toxicol 2011;85(5):367–485.
[17] Peterson RT, Macrae CA. Systematic approaches to toxicology in the zebrafish. Annu Rev Pharmacol Toxicol 2012;52:433–53.
[18] Strähle U, Scholz S, Geisler R, et al. Zebrafish embryos as an alternative to animal experiments: A commentary on the definition of the onset of protected life stages in animal welfare regulations. Reprod Toxicol 2012;33(2):128–32.
[19] Paules RS, Aubrecht J, Corvi R, Garthoff B, Kleinjans JC. Moving forward in human cancer risk assessment. Environ Health Perspect 2011;119(6):739–43.
[20] Lamb J, Crawford ED, Peck D, et al. The Connectivity Map: Using gene-expression signatures to connect small molecules, genes, and disease. Science 2006;313:1929–35.
[21] Corvi R, Ahr HJ, Albertini S, et al. Meeting report: Validation of toxicogenomics-based test systems—ECVAM-ICCVAM/NICEATM considerations for regulatory use. Environ Health Perspect 2006;114(3):420–9.
[22] Garthoff B. Alternatives to animal experimentation: The regulatory background. Toxicol Appl Pharmacol 2005;207(Suppl. 1):388–92.

SECTION

Genotoxicity and Carcinogenesis

2

CHAPTER

2.1 Application of In Vivo Genomics to the Prediction of Chemical-Induced (hepato) Carcinogenesis

Scott S. Auerbach[*] and Richard S. Paules[†]

[*]Biomolecular Screening Branch, Division of the National Toxicology Program, NIEHS, Research Triangle Park, North Carolina, [†]Toxicology and Pharmacology Laboratory, Division of Intramural Research, NIEHS, Research Triangle Park, North Carolina

2.1.1 Introduction

Prediction of human cancer hazard is one of the major goals of a regulatory-level toxicity assessment. The assay currently used by regulatory agencies to predict human cancer hazard is the rodent cancer bioassay [1]. As it is currently formulated, the bioassay is a 2-year study that provides a comprehensive characterization of the cancer hazard in both sexes of two rodent species. A chemical that causes cancer in a rodent bioassay is presumed to be a cancer hazard in humans until proven otherwise. The basic premise that a rodent cancer hazard extrapolates to humans is rooted in sound scientific observations that the biochemical and physiological processes that determine toxicity (and its often secondary byproduct cancer) susceptibility generally are conserved between rodents and humans. Data-driven reviews of animal testing results indicate it has significant predictive power when it comes to human toxicity [2,3]. Further, the rodent bioassay has been reported to have nearly 100% sensitivity when it comes to detection of human carcinogens [4], although this number has been challenged [5]. Despite its use as the regulatory standard for identification of human cancer hazard, the bioassay certainly is not perfect. There are species-specific modes of toxicological action that are rooted in the divergence of genes and pathways involved in mediating toxicity [6–8] that can lead to carcinogenic processes that are arguably of little relevance to humans [9]. Another criticism of the rodent bioassay is the use of dose levels that are far in excess of the true human exposure doses, thereby creating a significant challenge for risk extrapolation. It should be noted that the criticisms of dose levels are not always valid [10]. In addition to the scientific challenges that surround the bioassay, there are a number of ethical issues associated with the use of large numbers of animals as well as a significant cost in terms of time and money associated with performing each 2-year rodent bioassay. Couple these challenges with recent legislation mandating toxicity data for a large number of chemicals in commerce [11,12], along with the fact that only a small fraction of these chemicals have actually been assessed for carcinogenicity [13], and it becomes clear that alternative, more efficient methods for the identification of chemical cancer hazards for humans are needed.

Toxicogenomics-Based Cellular Models.

Considering the challenges related to the bioassay, it is not surprising that there has been a range of efforts to identify more efficient methods to discover potential human carcinogens. Examples include structure activity alerts [14,15], bacterial mutagenesis assays [16], in vitro mammalian-cell-based genotoxicity assays [17], in vivo micronucleus assays [18], in vivo single cell gel electrophoresis assay [19], mammalian cell transformation assays [20,21], histological changes in sub-chronic toxicity studies [22–26], initiation/promotion models [27], combined weight-of-evidence models [28], and accelerated bioassays [29,30]. While some of these approaches are still under consideration or are accepted as supportive data by regulatory authorities, none have effectively replaced the 2-year bioassay. There are a number of reasons for this [24], but one of the main reasons is that the bioassay is the most comprehensive preclinical assessment of apical toxicity and therefore provides confidence to regulatory decision makers that an undetected/unanticipated toxic liability will not manifest in a human population upon chronic exposure. It should be noted that the veracity of this line of thought has been debated extensively and the reader is referred elsewhere for further discussion of this issue [4,31–35]. The uncertainty introduced by any new approach, particularly when those approaches are black box in nature or seemingly cover limited toxicological space as it relates to carcinogenicity (e.g. the Ames mutagenesis assay does not detect non-genotoxic carcinogens), gives regulators significant pause and therefore change is slow to occur. It has been suggested that toxicogenomics, assuming technological and comfort barriers can be overcome [36,37], may be a means of overcoming the uncertainty associated with adopting new approaches. This is primarily due the very broad assessment of biological space that can be achieved by toxicogenomic approaches.

Toxicogenomics is the study of whole-genome responses following exposure to toxic agents [38] and most frequently refers to genome-wide alterations in messenger RNA levels, or transcriptomics, although it also encompasses other genome-wide studies including global changes in protein levels (proteomics) and protein function (metabonomics) (recently reviewed by Afshari et al. [39]). When a toxicogenomic experiment is performed at a genome-wide level on protein-coding mRNA using microarray platforms, there are approximately 21,000 data points captured per sample, which cover (at the RNA level) nearly all of characterized (i.e. annotated into a pathway/biological process) and uncharacterized pathway/molecular biological space. In other words, the technology provides a large net that allows a researcher to capture perturbations to a system in the absence of prior knowledge of a chemical's toxicological mechanism. In addition to being able to query a wide swath of biological space, transcriptomics also enables a toxicologist to perform objective, standardized assessment of descriptive features, which allows for creation of generalizable toxicity signatures [40], and to perform comparative genomic analysis of responses between test species and humans [41]. In toxicology, transcriptomic technology has been used in four primary ways: (1) to identify pathways/biological networks altered by chemical treatment, (2) to classify compounds by mode of toxicological action, (3) to derive signatures of pathology that can be used for diagnosis or obtaining a molecular-level understanding of pathogenesis [42], and (4) to derive predictive signatures that indicate a likelihood of eventual toxicological outcome (e.g. cancer). Although these uses represent a continuum of toxicogenomic application space, the last of the four will be the focus of this chapter. There have been a number of predictive toxicogenomic studies that have focused on a variety of organ systems: reproductive/developmental [43–46], pulmonary [47], neurological [48], hematological [49–51], renal [52–55], immunological [56], cardiovascular [57–59], and dermatological [60]. This review will focus on how toxicogenomics has been used to predict hepatocarcinogenic hazard in rodents. The reason for the focus on liver carcinogenesis is that it is also the most common cancer target organ in the rodent cancer bioassay [61,62] and therefore, not surprisingly, it has undergone extensive study using toxicogenomic-based approaches. It should be noted that the toxicogenomic-based prediction of hepatocarcinogenic activity has been reviewed previously (Guan et al., 2008; [63]; Paules et al., 2011; Waters et al., 2011) and readers are encouraged to refer to these publications for additional perspectives on this topic.

2.1.2 Toxicogenomics-based prediction of hepatocarcinogenic hazard

For some time it has been recognized that hepatocarcinogens (and carcinogens in general) can be grouped into two general classes (or modes of action), genotoxic carcinogens (GTCs) and non-genotoxic carcinogens (NGTCs) [64]. Within these two classes, there are agents that act by a variety of different mechanisms to yield cancer in rodents. In the case of genotoxic carcinogens there are direct- and indirect-acting genotoxicants, and the distinction between the two is critical for regulatory purposes. Mechanisms that underlie indirect-acting genotoxicity include inhibition of DNA repair/topoisomerases, formation of reactive oxygen species, and mitotic interference due to tubulin binding [65]. In the case of non-genotoxic carcinogenic chemicals, and in particular liver NGTCs, there are four well-characterized mechanisms of action, including activation of the ligand-regulated transcription factors AhR (aryl hydrocarbon receptor), CAR/PXR (constitutive androstane receptor/pregnane X receptor), and PPARA (peroxisome proliferator alpha receptor), in addition to chronic cytotoxicity [9,66,67]. It should be noted that chemicals often don't fit into one predicted mode of action and may, depending on the dose, act by both genotoxic and non-genotoxic mechanisms [63]. At the outset of the toxicogenomics era, there were significant efforts focused on the identification of mechanism-of-action signatures. A noteworthy fraction of these studies, using both in vitro and in vivo systems, has focused on the liver (or liver-derived cell lines), primarily because the toxic/carcinogenic mechanisms of action noted above are relatively well understood [68–78]. In addition, other studies attempted to differentiate mechanisms of genotoxicity [79–88]. The logic behind this avenue of research as it relates to predicting cancer outcomes is that chemicals that act by well-documented mechanisms of action linked to carcinogenesis will produce genomic signatures that are predictive of hepatocarcinogenic outcomes and can therefore be used to predict cancer hazard in rodents and, based on whether the mode of action can be extrapolated to humans, may be useful in predicting the cancer hazard posed to humans. Although this approach is likely to identify a sizable fraction of chemicals that will go on to produce liver cancer in rodents, it is well documented that not all chemicals that act by certain modes of action turn out to produce cancer in a 2-year bioassay [22,89]. The chemicals that would be incorrectly identified by mechanism-of-action-based approaches suggest a deeper inspection of the data is necessary, to identify secondary effect patterns that produce more reliable prediction of carcinogenic hazard. Such an inspection of the data would have to focus on the distinctions between cancer-discordant chemicals that activate the same pathways. The studies reviewed below (Table 2.1.1) attempt to identify such cancer-outcome-anchored signatures while also keeping in mind mode of carcinogenic action (i.e. genotoxic vs non-genotoxic), which is critical for evaluating human cancer risk (see www.epa.gov/cancerguidelines).

In vivo studies in rats

In one of the first attempts to identify genomic biomarkers that predict rodent liver carcinogenesis, Kramer and colleagues performed a correlation analysis that evaluated the relationship between

Table 2.1.1 In Vivo Toxicogenomics Studies Focused on Predicting Hepatocarcinogenicity

Primary Study	Independent Validation Studies	Related Studies with Overlapping Datasets	Model	Toxicity Study Details	Number of Chemicals	Platform	Gene ID/ Modeling Approach	Genes and Effect of Carcinogen Treatment	Data Repository
[90]			Male Sprague-Dawley rats	5-day repeat dose; single dose level typically exceeding a sub-chronic MTD; oral route	9 (1 GC, 6 NGC, 2 NC)	Incyte RatGEM1.0	Correlation with carcinogenic activity	**Up**: *Por*, *Aldh1a*, *Ugt1a7c* **Down**: *Tsc22d1*, *Bhmt*, *Hspb1*, *Hao2*, *Gulo*	CEBS
[91]		[92,93]	Male Wistar Hanover rats	1-, 3-, 7-, or 14-day repeat dose; single dose level; oral route	Training set = 13 (5 GHC, 5 NGHC, 3 NC) Test set = 16 (4 GHC, 6 NGHC, 6 NHC)	Affymetrix RG_U34A Affymetrix RAE230A	Variety of gene-selection and machine-learning methods	256 to 2048 probe sets depending on modeling approach	ArrayExpress database: E-TOXM-16 (Partial dataset)
[94]		[95]	B6C3F1 mice	14-day repeat dose; single dose level; oral route	6 (2 GHC, 1 NGHC, 3 NC)		Pooled results of individual t-tests that compared treated to control	421 genes	Array Express database: E-TOXM-18
[96]	[97]	[98]	Male Sprague–Dawley rats	5-day repeat dose; single dose level at a 5-day MTD; multiple routes	Training set = 100 (25 NGHC, 75 NHC) Test set = 47 (21 NGHC, 26 NHC)	Codelink Uniset 1 Bioarray (5443 probes were used for modeling)	Adjusted sparse linear programming algorithm	37 genes	GEO database: GSE8251 (primary dataset); GSE8858; GSE2409

[99]	[97]	[71,72]	Male Sprague–Dawley rats	1-day single dose; single dose level; multiple routes	Training/test set = 52 (24 NGHC, 28 NHC); 18 GC were also studied but not used for modeling	Custom array (1471 genes represented)	Semi-exhaustive, nonredundant gene-selection algorithm	6 genes **Up**: *Nutf2*, *Pgrmc1*, *Ugt2b1* **Down**: *Mt1a*, *Sel1l*, *Mat1a*	CEBS database
[100]	[51]	[101–105]	Male Sprague–Dawley rats	3-, 6-, 9-, or 24-h (single dose)/ 3-, 7-, 14-, or 28-day repeat dose; single dose level at a 7-day MTD; oral route	Training set = 8 (2 NHGC, 6 NC) Test set = 22	Affymetrix RAE230A	ANOVA; Prediction Analysis of Microarrays (PAM) with external validation set	112 probe sets	TG-Gates database
[106]		[107]	Male F344 rats	3-, 14-, and 28-day repeat dose; single dose level with exception 18 chemicals administered at 4 dose levels; dose level approximated a sub-chronic MTD and lower; oral route	Training set = 73 (21 GC, 26 NGC, 12 GNC; 12 NGNC) Test set = 12	NEDO-ToxArray III (6709 genes represented)	Welch's approximate t-test; hierarchical clustering	140 genes (high-stringency 28-day model)	GEO database: GSE16743; GSE16752; GSE16394; GSE16340
[108]			Male F344 rats	3-, 14-, and 90-day repeat dose; one or two levels; dose level approximated a sub-chronic MTD; oral route	Training set = 10 (3 GHC, 1 NGHC, 3 NC, 3 vehicle treatments) Test set = 10 (3 NGHC, 3 NC, 1 vehicle, 1 untreated, 2 untested)	Agilent Whole-Rat-Genome Microarray 4 × 44K	SVM-RFE with a variety of data combinations; independent validation was performed	3 to 59 probes depending the data set that was modeled	CEBS database

(Continued)

Table 2.1.1 In Vivo Toxicogenomics Studies Focused on Predicting Hepatocarcinogenicity (Continued)

Primary Study	Independent Validation Studies	Related Studies with Overlapping Datasets	Model	Toxicity Study Details	Number of Chemicals	Platform	Gene ID/ Modeling Approach	Genes and Effect of Carcinogen Treatment	Data Repository
[109]			Male Sprague–Dawley rats	5-day repeat dose; single dose level at a 5-day MTD; multiple routes	Training set = 100 (25 NGHC, 75 NHC) Test set = 47 (21 NGHC, 26 NHC)	CodeLink Uniset 1 Bioarray	Nearest-centroid method with a minimum-redundancy- and maximum-relevancy-based feature selection	90 genes	GEO database: GSE8251
[110]		[111]	Female B6C3F1 mice	90-day repeat dose study; up to 5 dose levels/ chemical; dose level approximated a sub-chronic MTD; oral or inhalation routes	2	Affymetrix Mouse Genome 430 2.0 Array	Correlation of genomic-based dose response to cancer benchmark dose		GEO database: GSE18858

CEBS, chemical effects in biological systems; GC, genotoxic carcinogen; GEO, gene expression omnibus; GHC, genotoxic hepatocarcinogen; GNC, ; MTD, maximum tolerated dose; NC, noncarcinogen; NGC, non-genotoxic carcinogen; NGHC, non-genotoxic hepatocarcinogen; NGNC, non-genotoxic, non-carcinogen; NHC, non-hepatocarcinogen; RFE, recursive feature; SVM, support-vector machine.

carcinogenic activity (i.e., 2-year liver cancer burden) and rat liver gene-expression changes elicited by three different doses of two non-carcinogens (NCs), one GTC, and six NGTCs after 5 days of repeat dosing [90]. They reported expression of three genes that were positively correlated with carcinogenic potency (*Por*, *Aldh1a*, and *Ugt1a7c*) and five genes that were inversely correlated (*Tsc22d1*, *Bhmt*, *Hspb1*, *Hao2*, and *Gulo*). The positively correlated genes are consistent with microsomal-enzyme-inducer compounds that are commonly found to be hepatocarcinogenic in rodents. Considering that a number of the down-regulated genes are also liver enriched genes, their down-regulation may be indicative of partial dedifferentiation that is a byproduct of the tissue regeneration, a process consistent with hepatocarcinogenic activity [112].

A series of manuscripts [71,72], culminating in one describing a cancer-prediction model [99], was published by a group based at Johnson and Johnson Pharmaceuticals. In the study describing the carcinogenesis-prediction model, data were obtained from male rats treated for 24 hours (single dose) with one of 24 NGTCs and 28 NCs. A semi-exhaustive, non-redundant gene selection algorithm yielded six genes, which by cross-validation classified chemical treatments with 88.5% accuracy. In an independent validation, the six-gene signature was notably effective at identifying data from rat peroxisome proliferators or chemicals that induce oxidative stress. However, the signature largely failed to identify a number of compounds known to activate liver macrophages (a presumed inflammatory mode of carcinogenic action in the liver). A simple comparison of expression data from NGTC- and NC-treated rats identified 152 genes that were differentially expressed. A network-based analysis using these genes indicated a pronounced role of *Myc* in the gene-expression alterations mediated by the NGTC. *Myc* plays a role in cell cycle progression, apoptosis, and cellular transformation [113], and genomic alterations in this gene have been associated with a number of human cancers, including liver cancers. In a follow-up independent validation study [97], the six-gene signature had a 55 and 64% accuracy classifying liver microarray data derived from rat exposed to over 100 chemicals [96]. In still another independent validation, using a reverse transcriptase polymerase chain reaction (RT-PCR)-based approach, the six-gene signature accurately classified samples taken from rats treated with three liver NGTC and two NC [114].

A group in Japan supported by the New Energy and Industrial Technology Development Organization has pursued the use of liver gene expression derived from 1-, 3-, 7-, 14- or 28-day studies in F344 rats to identify signatures indicative of hepatocarcinogenic activity. In their first study, published in 2006, they evaluated gene expression elicited by four pairs of carcinogenic and NC chemical isomers. They reasoned that evaluating paired isomers would allow for the differentiation of cancer-related expression from that produced by the basic skeleton of the isomers [107]. All eight of these chemicals have been shown to have mutagenic properties. Using a ratio filtering and Welch's approximate t-test analysis, 54 and 28 genes were identified based on comparisons of the 28-day expression data from three or four sets of isomers, respectively. Hierarchical clustering of the 28-day data clearly differentiated the carcinogenic from the NC isomers and effectively placed 2-nitro-p-phenylenediamine in the non-carcinogenic branch of the dendrogram, a conclusion that is consistent with observations made in a 2-year bioassay of this chemical in Fisher F344 rats. Selected genes that were up-regulated by carcinogen treatment included *Abcb1b*, *Aldh1a1*, *Ccng1*, and *Mgmt*, whereas *Fhit*, *Toag1*, and *Nupr1* were down-regulated. Pathway enrichment analysis of the two gene sets suggests the gene expression in response to carcinogen treatment is in large part associated with the p53 signaling pathway. In a follow-up study using 28-day liver gene-expression data from 47 carcinogen and 26 non-carcinogen treatments, a 37-gene signature was identified [106]. Using a hierarchical clustering approach, it was observed that the chemical treatments split into three groups. Group 1 was composed of predominantly hepatic GTCs. Group 2 contained three subgroups that were roughly divided into three compound classes: liver NGTCs, NGTCs that target organ systems other than the liver, and GTCs that target organ systems other than the liver. Group 3 contained predominantly NCs but also included a number of estrogenic carcinogens. The compounds in group 1, which were mainly hepatic GTCs, elicited alterations in expression of a number of xenobiotic metabolism (e.g. *Abcb1b*) and p53-responsive genes (e.g. *Btg2*, *Mgmt*, *Bcl2*, and *Bax*). Genes exhibiting strong differential expression with the group 2 carcinogens did not seem to exhibit a discernable ontological relationship. Genes whose expression was altered by group 3 carcinogens seemed to be linked together at network level by NF-κB and also contained an enrichment of genes associated with steroid-related metabolism and signaling (e.g. *Pbsn*, *Dhrs7*, *Srd5a1*, and *Prlr*).

Fielden and colleagues from Iconix Pharmaceuticals used DrugMatrix hepatic gene-expression data from rats treated for 3, 5, or 7 days with 100 structurally and mechanistically diverse hepatic NGTCs, hepatic GTCs, and non-hepatic carcinogens [96]. Using a modified version of a support vector machine, they derived a 37-gene signature. By split sample cross-validation the signature had an estimated sensitivity of 56% and a specificity of 94%. Independent validation of the signature on 47 test chemicals (21 carcinogens and 26 NCs) administered for 1, 3, or 5 days indicated the genomic signature had a sensitivity and specificity of 86 and 81%, respectively. Short-term in vivo pathological and genomic biomarkers were evaluated in parallel for comparison, including liver weight, hepatocellular hypertrophy, hepatic necrosis, serum alanine aminotransferase activity, induction of cytochrome P450 genes, and repression of Tsc-22 or alpha2-macroglobulin messenger RNA. A comparison with a variety of other endpoints such as pathology, clinical chemistry, and previously identified genomic markers of hepatocarcinogenic activity indicated the 37-gene signature was more accurate than those endpoints in predicting liver cancer outcomes. A hierarchical cluster analysis that included prototypic chemicals representing different hepatocarcinogenic modes of action indicated the signature genes could differentiate chemicals that acted by regenerative proliferation, xenobiotic receptor activation (AhR/CAR/PXR activation), peroxisome proliferation (PPARA activation), or steroid-hormone-mediated mechanisms. A follow-up independent validation study of the 37-gene signature performance using gene-expression data derived from the above-discussed J&J study [99] and additional data contributed by GlaxoSmithKline indicated the accuracy of the 37-gene signature was between 63 and 69% [97]. The latter analysis suggested the signature was relatively robust in relation to its predictive capability when applied across genomic platforms. In a second independent validation on a reduced form of the 37-gene signature, it was noted that there was poor performance in the identification of non-genotoxic hepatocarcinogens from samples derived from rats exposed to a single dose of chemical for 24 hours, with prediction accuracy improved notably if samples were taken from a 5-day repeat dose study.

A study published by a group at Bayer Pharmaceuticals in 2008 described a three-class hepatocarcinogenicity model, which classifies toxicogenomic profiles as GTC, NGTC, and NC [91]. The worked set out in this manuscript is in part borrowed from previous studies by the same group that described signatures that identify genotoxic carcinogens [92] and signatures that differentiate non-genotoxic from genotoxic carcinogens [93]. Initially, gene-expression profiles were generated from the livers of rats treated for 1, 3, 7, or 14 days with chemicals previously studied in a 2-year rat carcinogenicity bioassay. Using this data, multiple feature-selection methods (analysis of variance (ANOVA), support vector machine (SVM), recursive feature (RFE)) were coupled to multiple machine learning algorithms (SVM, sparse linear discriminant analysis (SLDA), and K nearest neighbors (KNN)) in order to identify which combination produced the lowest cross-validation error. The SVM algorithm outperformed KNN and SLDA in this assessment, and was subsequently used in combination with four feature sets identified by the methods described above along with a mechanistic set identified in a previous manuscript [93]. When independently validated, the top performing feature-set–classifier combinations (SVM-SVM and RFE-SVM) achieved 88% accuracy. Genes found in these signatures were primarily induced in the GTC treatments and included genes involved in cell cycle checkpoint, DNA repair, and ATM/p53 signaling pathways (e.g. *Cdkn1a*, *Mdm2*, *Ccng1*, and *Ccnd1*). The non-genotoxic carcinogens elicited a diversity of responses depending on the chemical and its mode of action (e.g. *Acot1* (PPARA activators), *Cyp1a1* (AhR activators), *Abcc3* (microsomal enzyme induction and oxidative stress)).

The Japanese Toxicogenomics Project (http://toxico.nibio.go.jp/) is a public–private collaboration. Since 2002, a large-scale toxicogenomics database has been constructed that can be mined extensively for signatures and used for diagnostic and predicative toxicogenomics [115]. The data has been mined extensively by the consortium members and numbers of studies have been published using the data. In one of the more recent publications, the group explored the identification of a genomic signature of chemical-induced hepatic NGTC activity. In order to identify the signature, the PAM (prediction analysis of microarray) algorithm was used in combination with gene-expression data from two hepatic NGTCs and six NCs. The final signature contained 112 probe sets and gave an overall accuracy of 95% by cross-validation. Independent validation of the signature indicated that it had it exhibited nearly 100% specificity but notably poor sensitivity (five of the eight hepatocarcinogens were not detected using gene expression from any study duration). Despite poor sensitivity, the researchers did note that extending the duration of exposure increased the confidence by which the genomic signature classified the chemicals as hepatocarcinogenic.

An effort based at the National Toxicology Program has used toxicogenomics to prioritize chemicals for carcinogenicity testing [108]. Alkenylbenzenes are a large class of naturally occurring flavoring agents that include the well-studied suspected human carcinogen safrole. In order to identify additional members of this class that may pose a cancer hazard, an ensemble of support-vector-machine-derived genomic signatures were obtained from rat liver gene expression following 2, 14, or 90 days of exposure to a collection of extensively characterized hepatocarcinogens (aflatoxin B1, 1-amino-2,4-dibromoanthraquinone, N-nitrosodimethylamine, methyleugenol) and non-hepatocarcinogens (acetaminophen, ascorbic acid, tryptophan). Seven predictive models were generated based on different combinations of the data. These models were then used to classify gene-expression data from the livers of rats exposed at two dose levels to a collection of alkenylbenzene flavoring agents, which included documented hepatocarcinogens and non-carcinogens. Classification error rates for the alkenylbenzenes that had been previously studied in a cancer bioassay ranged from 47 to 10%. It was noted that the variable with the most notable effect on independent validation accuracy was exposure duration of the alkenylbenzene test data. As noted in some of the above studies, the genomic signatures generally exhibited improved performance as the exposure duration of the alkenylbenzene data increased. The classification of two dose levels (one hepatocarcinogenic and one not) of safrole suggested that a signature score may be capable of differentiating between carcinogenic and non-carcinogenic dose levels as measured by tumor incidence in the bioassay. The genomic signatures predicted that two alkenylbenzenes not previously evaluated in a carcinogenicity bioassay, myristicin and isosafrole, would be weakly hepatocarcinogenic if studied at the high dose level used to elicit the gene expression observed in this study. A number of genes were consistently present across a number of the seven models, including two previously shown to be down-regulated in response to genotoxic stress (*Wwox* and *Fhit*) and a handful that are consistently observed to be induced by certain hepatocarcinogens, including *Adam8*, *Mybl2*, and *RGD1561849*. A pathway analysis of the genes differentially expressed between the training-set hepatocarcinogens and non-carcinogens indicated activation of pathways related to cell matrix remodeling and cell cycle control, both of which have a documented relationship to hepatocarcinogenesis.

A recent study from a group at the National Center for Toxicological Research compared the predictive accuracy of quantitative structure–activity (QSAR) with genomic-based prediction signatures using a set of 62 non-genotoxic carcinogens and non-carcinogens [109]. Gene-expression data for this analysis were derived from a published study that was discussed earlier [96]. Genomic signatures and QSAR models were developed based on the same set of chemicals using the same modeling approach. Signatures containing 90 genes were identified using gene-expression data from 1-, 3-, and 5-day exposures. The authors noted that in all cases the genomics-based signatures outperformed QSAR, with the top genomic signature achieving 82% accuracy as measured by cross-validation. A pathway-based analysis of the genes from the 5-day genomic signature indicated an enrichment of AhR (*Tgm2*, *Mgst1*, and *Nr0b2*), cytoskeletal (*Fn1*, *Arhgef1*, and *Msn*), and p53 (*Ccng1* and *Pik3r1*) signaling. What was particularly striking in the signatures was the presence of many genes typically associated with genotoxicity, including *Ccng1* and *Fhit*. The presence of these genes suggests that either the high doses of non-genotoxic chemicals that were used to generate the genomic data led to significant increases in oxidative-stress-associated genotoxicity or, alternatively, some of the training chemicals are not as free of genotoxic properties as indicated by standard assessments used for classification.

In vivo studies in mice

A study from a group affiliated with the US National Institute of Environmental Health Sciences (NIEHS) evaluated gene expression elicited from hepatocarcinogens and non-carcinogens in B6C3F1 mice [94]. The hepatocarcinogens used in this study represent both modes of action discussed above. Analysis of gene expression after 2 weeks of exposure indicated the genotoxic chemicals caused the down-regulation of *Fhit* and *Wwox* and up-regulation of *Cdkn1a* and *Ccng1*. In the case of the non-genotoxic chemical, there was noted induction of *Gadd45b* and *Cyp2b20* and down-regulation of *Tsc22d1*.

In a series of publications, the Thomas group at the Hamner Institute have explored the application of predictive toxicogenomics at both a qualitative and quantitative level in a number of tissues [110,111,116–118]. In a study published in 2007, this group used both metabonomic and transcriptomic technology to derive biomarkers of lung and liver carcinogenesis hazard. Gene expression and metabolites were derived from lung and liver tissue from female mice exposed for 90 days to two carcinogens (one genotoxic and one non-genotoxic, both lung and liver carcinogens) and two non-carcinogens, and two control groups. Using an iterative SVM approach, a set of genes and metabolites were identified that differentiated carcinogens from non-carcinogens with 100% and 94% accuracy, respectively. The authors noted that the discriminating biomarkers shared a common expression profile, suggesting the 90-day exposure likely reflected universal cellular changes in the transition toward neoplasia. The genes most informative to the liver carcinogenesis signature were *Snhg11*, *Ttc39a*, *Itih1*, *Pls1*, and *Ugdh*. In a second study, this group used sub-chronic mouse liver gene expression to derive transcriptional benchmark dose (BMD) values for five chemicals, and compared the BMDs derived from gene-expression analyses with those derived from traditional toxicological endpoints [111]. Transcriptional BMDs were derived using a previously described software package [119] for all genes exhibiting significant differential expression. A gene-ontology-based assessment was then performed on all genes exhibiting a BMD. Ontologies showing enrichment were assigned an ontology BMD (the median gene BMD for the differentially expressed genes contained in that ontology). A correlation analysis indicated that the non-cancer BMDs and cancer BMDs most strongly correlated with gene ontologies typically associated with inflammation and stress signaling, respectively.

Jonker and colleagues performed a multidimensional assessment of predictive genomic signatures using wild-type and DNA-repair-deficient Xpa(−/−)/p53(+/−) mice [120]. There were six treatments evaluated in this study (two GTCs, two NGTCs, and two NC treatments). Samples from liver, spleen, bladder, blood, and lymph node were taken at 3, 7, and 14 days, and 8205 genes were measured in each sample by microarray. Genomic signatures that differentiated the three classes of compounds were then derived from the different genotype/tissue sets in combination with variety of machine learning approaches. In the end, 16 signatures were generated. The final number of genes that contributed to each signature varied from 4 to 539. When evaluated using a 10-fold cross-validation, all signatures with the exception of one had a $< 5\%$ error rate. The authors noted that signatures derived from the transgenic mice did not perform better than those derived from wild-type mice. In one final component of the study, the authors evaluated expression in blood following exposure to a genotoxic carcinogen. It was noted there was differential expression in the blood from these animals, but none of the signatures from other tissues effectively identified the pattern of gene expression in blood as coming from animals treated with a GTC. Retraining the tissue signatures with genotoxic-carcinogen-responsive genes from blood greatly improved the ability of the non-blood signatures to differentiate the three classes of chemicals. Genes that populated the liver prediction signature were specific to GTC (*Hic2* and *Tlr9*) and NGTC (*S100a10*, *1700034H14Rik*, *Tmem4*, *Pebp1*, *Spcs1*, *Sod*, *Nme6*, and *Smarcal1*) exposure.

2.1.3 Conclusion and future perspective

Overall current studies suggest that toxicogenomics-based prediction of cancer outcomes has promise, but is not at a state that such approaches can replace the rodent cancer bioassay. A couple of specific observations emerge from these studies. With respect to direct-acting genotoxic carcinogens, there seems to be a relatively robust signature that emerges from a number of the studies (Table 2.1.2); however, in the case of non-genotoxic carcinogenic chemicals, a consensus signature is less clear. Owing to the variety of mechanisms that drive non-genotoxic hepatocarcinogenesis, it may be necessary to derive signatures for the different mechanisms of action that can more effectively capture specific subtypes of hepatic NGTC hazard. In addition, hepatic NGTC signature performance is likely to increase substantially if similar study protocol parameters (e.g. dose selection criteria, rodent strain, study duration) are employed in both the training and test data. It was noted in a number of studies that the genomic predictions became more stable and robust as the data used to generate and evaluate the genomic signatures were derived from times that increased with the duration of the study. The researchers who made this observation suggested that the reason for this phenomenon is that the

Table 2.1.2 GHC Signature Genes Identified Across Multiple Studies[a]

Up-Regulated		Down-Regulated	
Ccng1	*Musk*	*Fhit*	*Wwox*
Adam8	*Mybl2*		
Atp6v1d	*Nhej1*		
Bax	*Nrcam*		
Btg2	*Pias1*		
Cdkn1a	*Pln*		
Fat1	*Rab13*		
Lama5	*Zmat3*		
Mdm2	*Znf688*		
Mgmt			

[a]*Survey of genes is limited to in vivo rat studies.*

gene expression at later time points is indicative of secondary changes that are more reflective of pathological changes that better approximate cancer-related processes. Currently, considering the challenges that need to be addressed related to toxicogenomics-based hepatic NGTC signatures, the best use of these in cancer hazard characterization will likely be to couple them with sub-chronic toxicity profiles in weight-of-evidence-based assessments [26]. In the near term, resources should be focused on further developing some nascent applications of toxicogenomics that are highlighted below.

Dose–response is a critical component of hazard characterization and translation of hazard characterization into risk assessment [121]. Dose–response analysis of toxicogenomic data has been done quite extensively in the context of phenotypic anchoring ([122–127]. However, it has been used to a very limited extent in predictive toxicogenomics, particularly in the context of cancer prediction [111]. More needs to be done to develop and evaluate methods for quantifying and relating genomic signatures to disease incidence. Establishment of such quantitative relationships will allow for these new metrics to be used in risk assessment.

In a normal rodent 2-year cancer study performed by the National Toxicology Program (NTP) there are 40 tissues evaluated histologically, in addition to gross observations. To date, most efforts have focused on the liver in relation to prediction of carcinogenic outcomes, as might be expected as a starting point. Hence, a systematic assessment of how genomics can be used to evaluate cancer hazard in other tissues is needed. The question of how many tissues need to be assessed at a genomic level to adequately capture the majority of the critical cancer space needs to be addressed. It is clear that some in the field are already considering this issue in their research [111], but a considerable amount of data remains to be generated that can address this issue. Along this line of thought we must consider evaluating accessible tissues, such as blood, that have the potential to offer a systemic picture of tissue toxicity and potentially carcinogenic hazard, in addition to providing a means of performing in-life assessments of toxicity in animals and potentially humans [128–131].

When chronically exposed to toxic chemicals, humans, and rodents, develop a variety of pathologies other than cancer. For this reason, a major long-term goal of toxicogenomics should be a systems toxicology framework that integrates chemical genomics with the causal molecular biology of disease, at both a quantitative and a qualitative level. Such a framework would allow for toxicogenomics-based assessments to cover a wide swath of disease space and therefore allow for a more comprehensive assessment of potential hazards posed by a chemical. Furthermore, through the use of orthologous gene relationships it will be possible to obtain a better understanding of the concordant and discordant relationships between chemical exposure and disease across different species. These sorts of all-encompassing systems toxicology approaches are beginning to appear in the literature [132–134]. However, these continue to be a minority of studies under the toxicogenomics umbrella. Along with enthusiasm for this line of research, there is a cautionary note. Due to nature's redundant and parsimonious ways, systems toxicology approaches have the potential to create an abundance of associations that are false at a biological level and, in turn, lead to misdirection of valuable resources. Therefore, in the initial phase of systems toxicology, it may be necessary to reconsider the mantra of traditional toxicity assessment, "sensitivity first, specificity second."

References

[1] Chhabra RS, Huff JE, Schwetz BS, Selkirk J. An overview of prechronic and chronic toxicity/carcinogenicity experimental study designs and criteria used by the National Toxicology Program. Environ Health Perspect 1990;86:313–21.

[2] Greaves P, Williams A, Eve M. First dose of potential new medicines to humans: how animals help. Nat Rev Drug Discov 2004;3:226–36.

[3] Olson H, Betton G, Robinson D, Thomas K, Monro A, Kolaja G, et al. Concordance of the toxicity of pharmaceuticals in humans and in animals. Regul Toxicol Pharmacol 2000;32:56–67.

[4] Huff JE. Chemicals causally associated with cancers in humans and in laboratory animals: a perfect concordance. Waalkes MP, Ward JM, editors. Carcinogenesis. New York: Raven Press; 1994. p. xi. 478 p.

[5] Alden CL, Lynn A, Bourdeau A, Morton D, Sistare FD, Kadambi VJ, et al. A critical review of the effectiveness of rodent pharmaceutical carcinogenesis testing in predicting for human risk. Vet Pathol 2011;48:772–84.

[6] Krasowski MD, Yasuda K, Hagey LR, Schuetz EG. Evolutionary selection across the nuclear hormone receptor superfamily with a focus on the NR1I subfamily (vitamin D, pregnane X, and constitutive androstane receptors). Nucl Recept 2005;3:2.

[7] Logan DW, Marton TF, Stowers L. Species specificity in major urinary proteins by parallel evolution. PLoS ONE 2008;3:e3280.

[8] Thomas JH. Rapid birth-death evolution specific to xenobiotic cytochrome P450 genes in vertebrates. PLoS Genet 2007;3:e67.

[9] Holsapple MP, Pitot HC, Cohen SH, Boobis AR, Klaunig JE, Pastoor T, et al. Mode of action in relevance of rodent liver tumors to human cancer risk. Toxicol Sci 2006;89:51–6.

[10] Bucher JR. Doses in rodent cancer studies: sorting fact from fiction. Drug Metab Rev 2000;32:153–63.

[11] Bergeson LL. Chemical management, North American style. Environ Qual Manag 2008;17:89–94.

[12] Lahl U, Hawxwell KA. REACH: the new European chemicals law. Environ Sci Technol 2006;40:7115–21.

[13] Judson R, Richard A, Dix DJ, Houck K, Martin M, Kavlock R, et al. The toxicity data landscape for environmental chemicals. Environ Health Perspect 2009;117:685–95.

[14] Benigni R, Bossa C. Predictivity and reliability of QSAR models: the case of mutagens and carcinogens. Toxicol Mech Methods 2008;18:137–47.

[15] Venkatapathy R, Wang CY, Bruce RM, Moudgal C. Development of quantitative structure-activity relationship (QSAR) models to predict the carcinogenic potency of chemicals I: alternative toxicity measures as an estimator of carcinogenic potency. Toxicol Appl Pharmacol 2009;234:209–21.

[16] Tennant RW, Margolin BH, Shelby MD, Zeiger E, Haseman JK, Spalding J, et al. Prediction of chemical carcinogenicity in rodents from in vitro genetic toxicity assays. Science 1987;236:933–41.
[17] Kirkland D, Aardema M, Henderson L, Muller L. Evaluation of the ability of a battery of three in vitro genotoxicity tests to discriminate rodent carcinogens and non-carcinogens I: sensitivity, specificity and relative predictivity. Mutat Res 2005;584:1–256.
[18] Witt KL, Knapton A, Wehr CM, Hook GJ, Mirsalis J, Shelby MD, et al. Micronucleated erythrocyte frequency in peripheral blood of B6C3F(1) mice from short-term, prechronic, and chronic studies of the NTP carcinogenesis bioassay program. Environ Mol Mutagen 2000;36:163–94.
[19] Sasaki YF, Sekihashi K, Izumiyama F, Nishidate E, Saga A, Ishida K, et al. The comet assay with multiple mouse organs: comparison of comet assay results and carcinogenicity with 208 chemicals selected from the IARC monographs and U.S. NTP Carcinogenicity Database. Crit Rev Toxicol 2000;30:629–799.
[20] Isfort RJ, Kerckaert GA, LeBoeuf RA. Comparison of the standard and reduced pH Syrian hamster embryo (SHE) cell in vitro transformation assays in predicting the carcinogenic potential of chemicals. Mutat Res 1996;356:11–63.
[21] Matthews EJ, Spalding JW, Tennant RW. Transformation of BALB/c-3T3 cells, V: transformation responses of 168 chemicals compared with mutagenicity in *Salmonella* and carcinogenicity in rodent bioassays. Environ Health Perspect 1993;101:347–482.
[22] Allen DG, Pearse G, Haseman JK, Maronpot RR. Prediction of rodent carcinogenesis: an evaluation of prechronic liver lesions as forecasters of liver tumors in NTP carcinogenicity studies. Toxicol Pathol 2004;32:393–401.
[23] Gaylor DW, Gold LS. Quick estimate of the regulatory virtually safe dose based on the maximum tolerated dose for rodent bioassays. Regul Toxicol Pharmacol 1995;22:57–63.
[24] Jacobs A. Prediction of 2-year carcinogenicity study results for pharmaceutical products: how are we doing? Toxicol Sci 2005;88:18–23.
[25] Melnick RL, Kohn MC, Portier CJ. Implications for risk assessment of suggested nongenotoxic mechanisms of chemical carcinogenesis. Environ Health Perspect 1996;104(Suppl. 1):123–34.
[26] Sistare FD, Morton D, Alden C, Christensen J, Keller D, Jonghe SD, et al. An analysis of pharmaceutical experience with decades of rat carcinogenicity testing: support for a proposal to modify current regulatory guidelines. Toxicol Pathol 2011;39:716–44.
[27] Tsuda H, Futakuchi M, Fukamachi K, Shirai T, Imaida K, Fukushima S, et al. A medium-term, rapid rat bioassay model for the detection of carcinogenic potential of chemicals. Toxicol Pathol 2010;38:182–7.
[28] Cohen SM. Alternative models for carcinogenicity testing: weight of evidence evaluations across models. Toxicol Pathol 2001;29(Suppl):183–90.
[29] Flammang TJ, Tungeln LS, Kadlubar FF, Fu PP. Neonatal mouse assay for tumorigenicity: alternative to the chronic rodent bioassay. Regul Toxicol Pharmacol 1997;26:230–40.
[30] Pritchard JB, French JE, Davis BJ, Haseman JK. The role of transgenic mouse models in carcinogen identification. Environ Health Perspect 2003;111:444–54.
[31] Ames BN, Gold LS. Chemical carcinogenesis: too many rodent carcinogens. Proc Natl Acad Sci USA 1990;87:7772–6.
[32] Ames BN, Gold LS. The rodent high-dose cancer test is limited at best: when cell division is ignored, then risk assessment will be flawed. Comments Toxicol 1998;6:271–6.
[33] Beyer LA, Beck BD, Lewandowski TA. Historical perspective on the use of animal bioassays to predict carcinogenicity: evolution in design and recognition of utility. Crit Rev Toxicol 2011;41:321–38.
[34] Cohen SM. Human carcinogenic risk evaluation: an alternative approach to the two-year rodent bioassay. Toxicol Sci 2004;80:225–9.
[35] MacDonald JS. Human carcinogenic risk evaluation, part IV: assessment of human risk of cancer from chemical exposure using a global weight-of-evidence approach. Toxicol Sci 2004;82:3–8.

[36] Borner FU, Schutz H, Wiedemann P. The fragility of omics risk and benefit perceptions. Toxicol Lett 2011;201:249–57.
[37] Pettit S, des Etages SA, Mylecraine L, Snyder R, Fostel J, Dunn Ii RT, et al. Current and future applications of toxicogenomics: results summary of a survey from the HESI genomics state of science subcommittee. Environ Health Perspect 2010;118:992–7.
[38] National Research Council Applications of toxicogenomic technologies to predictive toxicology and risk assessment. Washington DC: National Academies Press (US); 2007.
[39] Afshari CA, Hamadeh HK, Bushel PR. The evolution of bioinformatics in toxicology: advancing toxicogenomics. Toxicol Sci 2011;120(Suppl. 1):S225–37.
[40] Shi L, Campbell G, Jones WD, Campagne F, Wen Z, Walker SJ, et al. The Microarray Quality Control (MAQC)-II study of common practices for the development and validation of microarray-based predictive models. Nat Biotechnol 2010;28:827–38.
[41] Mattes WB. Cross-species comparative toxicogenomics as an aid to safety assessment. Expert Opin Drug Metab Toxicol 2006;2:859–74.
[42] Paules R. Phenotypic anchoring: linking cause and effect. Environ Health Perspect 2003;111:A338–9.
[43] Daston GP, Naciff JM. Predicting developmental toxicity through toxicogenomics. Birth Defects Research Part C—Embryo Today: Reviews 2010;90:110–7.
[44] Robinson JF, van Beelen VA, Verhoef A, Renkens MFJ, Luijten M, van Herwijnen MHM, et al. Embryotoxicant-specific transcriptomic responses in rat postimplantation whole-embryo culture. Toxicol Sci 2010;118:675–85.
[45] van Dartel DA, Pennings JL, Robinson JF, Kleinjans JC, Piersma AH. Discriminating classes of developmental toxicants using gene expression profiling in the embryonic stem cell test. Toxicol Lett 2010;201:143–51.
[46] van Dartel DA, Pennings JL, de la Fonteyne LJ, Brauers KJ, Claessen S, van Delft JH, et al. Evaluation of developmental toxicant identification using gene expression profiling in embryonic stem cell differentiation cultures. Toxicol Sci 2011;119:126–34.
[47] dos Santos CC, Okutani D, Hu P, Han B, Crimi E, He X, et al. Differential gene profiling in acute lung injury identifies injury-specific gene expression. Crit Care Med 2008;36:855–65.
[48] Moreira EG, Yu X, Robinson JF, Griffith W, Hong SW, Beyer RP, et al. Toxicogenomic profiling in maternal and fetal rodent brains following gestational exposure to chlorpyrifos. Toxicol Appl Pharmacol 2010;245:310–25.
[49] Rokushima M, Omi K, Araki A, Kyokawa Y, Furukawa N, Itoh F, et al. A toxicogenomic approach revealed hepatic gene expression changes mechanistically linked to drug-induced hemolytic anemia. Toxicol Sci 2007;95:474–84.
[50] Rokushima M, Omi K, Imura K, Araki A, Furukawa N, Itoh F, et al. Toxicogenomics of drug-induced hemolytic anemia by analyzing gene expression profiles in the spleen. Toxicol Sci 2007;100:290–302.
[51] Uehara T, Kondo C, Yamate J, Torii M, Maruyama T. A toxicogenomic approach for identifying biomarkers for myelosuppressive anemia in rats. Toxicology 2011;282:139–45.
[52] Fielden MR, Eynon BP, Natsoulis G, Jarnagin K, Banas D, Kolaja KL. A gene expression signature that predicts the future onset of drug-induced renal tubular toxicity. Toxicol Pathol 2005;33:675–83.
[53] Jiang Y, Gerhold DL, Holder DJ, Figueroa DJ, Bailey WJ, Guan P, et al. Diagnosis of drug-induced renal tubular toxicity using global gene expression profiles. J Transl Med 2007;5:47.
[54] Kondo C, Minowa Y, Uehara T, Okuno Y, Nakatsu N, Ono A, et al. Identification of genomic biomarkers for concurrent diagnosis of drug-induced renal tubular injury using a large-scale toxicogenomics database. Toxicology 2009;265:15–26.
[55] Thukral SK, Nordone PJ, Hu R, Sullivan L, Galambos E, Fitzpatrick VD, et al. Prediction of nephrotoxicant action and identification of candidate toxicity-related biomarkers. Toxicol Pathol 2005;33:343–55.

[56] Frawley R, White Jr K, Brown R, Musgrove D, Walker N, Germolec D. Gene expression alterations in immune system pathways in the thymus after exposure to immunosuppressive chemicals. Environ Health Perspect 2011;119:371–6.
[57] Mikaelian I, Coluccio D, Morgan KT, Johnson T, Ryan AL, Rasmussen E, et al. Temporal gene expression profiling indicates early up-regulation of interleukin-6 in isoproterenol-induced myocardial necrosis in rat. Toxicol Pathol 2008;36:256–64.
[58] Mori Y, Kondo C, Tonomura Y, Torii M, Uehara T. Identification of potential genomic biomarkers for early detection of chemically induced cardiotoxicity in rats. Toxicology 2010;271:36–44.
[59] Yi X, Bekeredjian R, DeFilippis NJ, Siddiquee Z, Fernandez E, Shohet RV. Transcriptional analysis of doxorubicin-induced cardiotoxicity. Am J Physiol 2006;290:H1098–102.
[60] Rogers JV, Price JA, McDougal JN. A review of transcriptomics in cutaneous chemical exposure. Cutan Ocul Toxicol 2009;28:157–70.
[61] Gold LS, Slone TH, Manley NB, Bernstein L. Target organs in chronic bioassays of 533 chemical carcinogens. Environ Health Perspect 1991;93:233–46.
[62] Gold LS, Manley NB, Slone TH, Ward JM. Compendium of chemical carcinogens by target organ: results of chronic bioassays in rats, mice, hamsters, dogs, and monkeys. Toxicol Pathol 2001;29:639–52.
[63] Guyton KZ, Kyle AD, Aubrecht J, Cogliano VJ, Eastmond DA, Jackson M, et al. Improving prediction of chemical carcinogenicity by considering multiple mechanisms and applying toxicogenomic approaches. Mutat Res 2009;681:230–40.
[64] Butterworth BE. Consideration of both genotoxic and nongenotoxic mechanisms in predicting carcinogenic potential. Mutat Res 1990;239:117–32.
[65] Scott D, Galloway SM, Marshall RR, Ishidate Jr M, Brusick D, Ashby J, et al. International Commission for Protection Against Environmental Mutagens and Carcinogens. Genotoxicity under extreme culture conditions: A report from ICPEMC Task Group 9. Mutation Research - Fundamental and Molecular Mechanisms of Mutagenesis 1991;257:147–205.
[66] Oliver JD, Roberts RA. Receptor-mediated hepatocarcinogenesis: role of hepatocyte proliferation and apoptosis. Pharmacol Toxicol 2002;91:1–7.
[67] Omiecinski CJ, Vanden Heuvel JP, Perdew GH, Peters JM. Xenobiotic metabolism, disposition, and regulation by receptors: from biochemical phenomenon to predictors of major toxicities. Toxicol Sci 2011;120(Suppl. 1): S49–75.
[68] Burczynski ME, McMillian M, Ciervo J, Li L, Parker JB, Dunn RT, et al. Toxicogenomics-based discrimination of toxic mechanism in HepG2 human hepatoma cells. Toxicol Sci 2000;58:399–415.
[69] Hamadeh HK, Bushel PR, Jayadev S, DiSorbo O, Bennett L, Li L, et al. Prediction of compound signature using high density gene expression profiling. Toxicol Sci 2002;67:232–40.
[70] Hamadeh HK, Bushel PR, Jayadev S, Martin K, DiSorbo O, Sieber S, et al. Gene expression analysis reveals chemical-specific profiles. Toxicol Sci 2002;67:219–31.
[71] McMillian M, Nie AY, Parker JB, Leone A, Bryant S, Kemmerer M, et al. A gene expression signature for oxidant stress/reactive metabolites in rat liver. Biochem Pharmacol 2004;68:2249–61.
[72] McMillian M, Nie AY, Parker JB, Leone A, Kemmerer M, Bryant S, et al. Inverse gene expression patterns for macrophage activating hepatotoxicants and peroxisome proliferators in rat liver. Biochem Pharmacol 2004;67:2141–65.
[73] Ruepp S, Boess F, Suter L, De Vera MC, Steiner G, Steele T, et al. Assessment of hepatotoxic liabilities by transcript profiling. Toxicol Appl Pharmacol 2005;207:S161–70.
[74] Steiner G, Suter L, Boess F, Gasser R, de Vera MC, Albertini S, et al. Discriminating different classes of toxicants by transcript profiling. Environ Health Perspect 2004;112:1236–48.
[75] Thomas RS, Rank DR, Penn SG, Zastrow GM, Hayes KR, Pande K, et al. Identification of toxicologically predictive gene sets using cDNA microarrays. Mol Pharmacol 2001;60:1189–94.
[76] Waring JF, Ciurlionis R, Jolly RA, Heindel M, Ulrich RG. Microarray analysis of hepatotoxins in vitro reveals a correlation between gene expression profiles and mechanisms of toxicity. Toxicol Lett 2001;120:359–68.

[77] Waring JF, Jolly RA, Ciurlionis R, Lum PY, Praestgaard JT, Morfitt DC, et al. Clustering of hepatotoxins based on mechanism of toxicity using gene expression profiles. Toxicol Appl Pharmacol 2001;175:28–42.
[78] Werle-Schneider G, Wölfelschneider A, von Brevern MC, Scheel J, Storck T, Müller D, et al. Gene expression profiles in rat liver slices exposed to hepatocarcinogenic enzyme inducers, peroxisome proliferators, and 17α-ethinylestradiol. Int J Toxicol 2006;25:379–95.
[79] Amundson SA, Do KT, Vinikoor L, Koch-Paiz CA, Bittner ML, Trent JM, et al. Stress-specific signatures: expression profiling of p53 wild-type and -null human cells. Oncogene 2005;24:4572–9.
[80] Aubrecht J, Caba E. Gene expression profile analysis: an emerging approach to investigate mechanisms of genotoxicity. Pharmacogenomics 2005;6:419–28.
[81] Boehme K, Dietz Y, Hewitt P, Mueller SO. Genomic profiling uncovers a molecular pattern for toxicological characterization of mutagens and pro-mutagens in vitro. Toxicol Sci 2011;122(1):189–97.
[82] Dickinson DA, Warnes GR, Quievryn G, Messer J, Zhitkovich A, Rubitski E, et al. Differentiation of DNA reactive and non-reactive genotoxic mechanisms using gene expression profile analysis. Mutat Res—Fundam Mol Mech Mutagen 2004;549:29–41.
[83] Ellinger-Ziegelbauer H, Fostel JM, Aruga C, Bauer D, Boitier E, Deng S, et al. Characterization and interlaboratory comparison of a gene expression signature for differentiating genotoxic mechanisms. Toxicol Sci 2009;110:341–52.
[84] Hu T, Gibson DP, Carr GJ, Torontali SM, Tiesman JP, Chaney JG, et al. Identification of a gene expression profile that discriminates indirect-acting genotoxins from direct-acting genotoxins. Mutat Res—Fundam Mol Mech Mutagen 2004;549:5–27.
[85] Islaih M, Li B, Kadura IA, Reid-Hubbard JL, Deahl JT, Altizer JL, et al. Comparison of gene expression changes induced in mouse and human cells treated with direct-acting mutagens. Environ Mol Mutagen 2004;44:401–19.
[86] Le Fevre AC, Boitier E, Marchandeau JP, Sarasin A, Thybaud V. Characterization of DNA reactive and non-DNA reactive anticancer drugs by gene expression profiling. Mutat Res—Fundam Mol Mech Mutagen 2007;619:16–29.
[87] Magkoufopoulou C, Claessen SM, Jennen DG, Kleinjans JC, van Delft JH. Comparison of phenotypic and transcriptomic effects of false-positive genotoxins, true genotoxins and non-genotoxins using HepG2 cells. Mutagenesis 2011;26(5):593–604.
[88] Mathijs K, Brauers KJJ, Jennen DGJ, Lizarraga D, Kleinjans JCS, van Delft JHM. Gene expression profiling in primary mouse hepatocytes discriminates true from false-positive genotoxic compounds. Mutagenesis 2010;25:561–8.
[89] Cunningham ML, Matthews HB. Cell proliferation as a determining factor for the carcinogenicity of chemicals: studies with mutagenic carcinogens and mutagenic noncarcinogens. Toxicol Lett 1995;82–83:9–14.
[90] Kramer JA, Curtiss SW, Kolaja KL, Alden CL, Blomme EAG, Curtiss WC, et al. Acute molecular markers of rodent hepatic carcinogenesis identified by transcription profiling. Chem Res Toxicol 2004;17:463–70.
[91] Ellinger-Ziegelbauer H, Gmuender H, Bandenburg A, Ahr HJ. Prediction of a carcinogenic potential of rat hepatocarcinogens using toxicogenomics analysis of short-term in vivo studies. Mutat Res 2008;637:23–39.
[92] Ellinger-Ziegelbauer H, Stuart B, Wahle B, Bomann W, Ahr HJ. Characteristic expression profiles induced by genotoxic carcinogens in rat liver. Toxicol Sci 2004;77:19–34.
[93] Ellinger-Ziegelbauer H, Stuart B, Wahle B, Bomann W, Ahr HJ. Comparison of the expression profiles induced by genotoxic and nongenotoxic carcinogens in rat liver. Mutat Res—Fundam Mol Mech Mutagen 2005;575:61–84.
[94] Iida M, Anna CH, Holliday WM, Collins JB, Cunningham ML, Sills RC, et al. Unique patterns of gene expression changes in liver after treatment of mice for 2 weeks with different known carcinogens and non-carcinogens. Carcinogenesis 2005;26:689–99.
[95] Iida M, Anna CH, Hartis J, Bruno M, Wetmore B, Dubin JR, et al. Changes in global gene and protein expression during early mouse liver carcinogenesis induced by non-genotoxic model carcinogens oxazepam and Wyeth-14,643. Carcinogenesis 2003;24:757–70.

[96] Fielden MR, Brennan R, Gollub J. A gene expression biomarker provides early prediction and mechanistic assessment of hepatic tumor induction by nongenotoxic chemicals. Toxicol Sci 2007;99:90–100.
[97] Fielden MR, Nie A, McMillian M, Elangbam CS, Trela BA, Yang Y, et al. Forum: interlaboratory evaluation of genomic signatures for predicting carcinogenicity in the rat. Toxicol Sci 2008;103:28–34.
[98] Natsoulis G, Pearson CI, Gollub J, Eynon P, Ferng B, Nair J, et al. The liver pharmacological and xenobiotic gene response repertoire. Mol Syst Biol 2008;4:175.
[99] Nie AY, McMillian M, Parker JB, Leone A, Bryant S, Yieh L, et al. Predictive toxicogenomics approaches reveal underlying molecular mechanisms of nongenotoxic carcinogenicity. Mol Carcinog 2006;45:914–33.
[100] Uehara T, Hirode M, Ono A, Kiyosawa N, Omura K, Shimizu T, et al. A toxicogenomics approach for early assessment of potential non-genotoxic hepatocarcinogenicity of chemicals in rats. Toxicology 2008;250(1):15–26.
[101] Hirode M, Ono A, Miyagishima T, Nagao T, Ohno Y, Urushidani T. Gene expression profiling in rat liver treated with compounds inducing phospholipidosis. Toxicol Appl Pharmacol 2008;229:290–9.
[102] Hirode M, Horinouchi A, Uehara T, Ono A, Miyagishima T, Yamada H, et al. Gene expression profiling in rat liver treated with compounds inducing elevation of bilirubin. Hum Exp Toxicol 2009;28:231–44.
[103] Hirode M, Omura K, Kiyosawa N, Uehara T, Shimuzu T, Ono A, et al. Gene expression profiling in rat liver treated with various hepatotoxic compounds inducing coagulopathy. J Toxicol Sci 2009;34:281–93.
[104] Kiyosawa N, Ito K, Watanabe K, Kanbori M, Niino N, Manabe S, et al. Utilization of a toxicogenomic biomarker for evaluation of chemical-induced glutathione deficiency in rat livers across the GeneChip data of different generations. Toxicol Lett 2006;163:161–9.
[105] Low Y, Uehara T, Minowa Y, Yamada H, Ohno Y, Urushidani T, et al. Predicting drug-induced hepatotoxicity using QSAR and toxicogenomics approaches. Chem Res Toxicol 2011.
[106] Matsumoto H, Yakabe Y, Saito K, Sumida K, Sekijima M, Nakayama K, et al. Discrimination of carcinogens by hepatic transcript profiling in rats following 28-day administration. Cancer Inform 2009;7:253–69.
[107] Nakayama K, Kawano Y, Kawakami Y, Moriwaki N, Sekijima M, Otsuka M, et al. Differences in gene expression profiles in the liver between carcinogenic and non-carcinogenic isomers of compounds given to rats in a 28-day repeat-dose toxicity study. Toxicol Appl Pharmacol 2006;217:299–307.
[108] Auerbach SS, Shah RR, Mav D, Smith CS, Walker NJ, Vallant MK, et al. Predicting the hepatocarcinogenic potential of alkenylbenzene flavoring agents using toxicogenomics and machine learning. Toxicol Appl Pharmacol 2010;243:300–14.
[109] Liu Z, Kelly R, Fang H, Ding D, Tong W. Comparative analysis of predictive models for non-genotoxic hepatocarcinogenicity using both toxicogenomics and quantitative structure-activity relationships. Chem Res Toxicol 2012;13:325.
[110] Thomas RS, O'Connell TM, Pluta L, Wolfinger RD, Yang L, Page TJ. A comparison of transcriptomic and metabonomic technologies for identifying biomarkers predictive of two-year rodent cancer bioassays. Toxicol Sci 2007;96:40–6.
[111] Thomas RS, Clewell III HJ, Allen BC, Wesselkamper SC, Wang NCY, Lambert JC, et al. Application of transcriptional benchmark dose values in quantitative cancer and noncancer risk assessment. Toxicol Sci 2011;120:194–205.
[112] Kojiro M. Pathological evolution of early hepatocellular carcinoma. Oncology 2002;62(Suppl. 1):43–7.
[113] Albihn A, Johnsen JI, Henriksson MA. MYC in oncogenesis and as a target for cancer therapies. Adv Cancer Res 2011;107:163–224.
[114] Nioi P, Pardo ID, Sherratt PJ, Snyder RD. Prediction of non-genotoxic carcinogenesis in rats using changes in gene expression following acute dosing. Chem Biol Interact 2008;172:206–15.
[115] Uehara T, Ono A, Maruyama T, Kato I, Yamada H, Ohno Y, et al. The Japanese toxicogenomics project: application of toxicogenomics. Mol Nutr Food Res 2010;54:218–27.

[116] Thomas RS, Allen BC, Nong A, Yang L, Bermudez E, Clewell Iii HJ, et al. A method to integrate benchmark dose estimates with genomic data to assess the functional effects of chemical exposure. Toxicol Sci 2007;98:240–8.

[117] Thomas RS, Bao W, Chu TM, Bessarabova M, Nikolskaya T, Nikolsky Y, et al. Use of short-term transcriptional profiles to assess the long-term cancer-related safety of environmental and industrial chemicals. Toxicol Sci 2009;112:311–21.

[118] Thomas RS, Pluta L, Yang L, Halsey TA. Application of genomic biomarkers to predict increased lung tumor incidence in 2-year rodent cancer bioassays. Toxicol Sci 2007;97:55–64.

[119] Yang L, Allen BC, Thomas RS. BMDExpress: a software tool for the benchmark dose analyses of genomic data. BMC Genomics 2007;8:387.

[120] Jonker MJ, Bruning O, van Iterson M, Schaap MM, van der Hoeven TV, Vrieling H, et al. Finding transcriptomics biomarkers for in vivo identification of (non-)genotoxic carcinogens using wild-type and Xpa/p53 mutant mouse models. Carcinogenesis 2009;30:1805–12.

[121] Lorentzen RJ. Carcinogenesis, bioassays, and the regulatory modus operandi. Toxicol Pathol 2010;38:506–7.

[122] Boverhof DR, Burgoon LD, Tashiro C, Chittim B, Harkema JR, Jump DB, et al. Temporal and dose-dependent hepatic gene expression patterns in mice provide new insights into TCDD-mediated hepatotoxicity. Toxicol Sci 2005;85:1048–63.

[123] Burgoon LD, Ding Q, N'Jai A, Dere E, Burg AR, Rowlands JC, et al. Automated dose-response analysis of the relative hepatic gene expression potency of TCDF in C57BL/6 mice. Toxicol Sci 2009;112:221–8.

[124] Hamadeh HK, Knight BL, Haugen AC, Sieber S, Amin RP, Bushel PR, et al. Methapyrilene toxicity: anchorage of pathologic observations to gene expression alterations. Toxicol Pathol 2002;30:470–82.

[125] Lobenhofer EK, Cui X, Bennett L, Cable PL, Merrick BA, Churchill GA, et al. Exploration of low-dose estrogen effects: identification of no observed transcriptional effect level (NOTEL). Toxicol Pathol 2004;32:482–92.

[126] Naciff JM, Daston GP. Toxicogenomic approach to endocrine disrupters: identification of a transcript profile characteristic of chemicals with estrogenic activity. Toxicol Pathol 2004;32:59–70.

[127] Naciff JM, Khambatta ZS, Reichling TD, Carr GJ, Tiesman JP, Singleton DW, et al. The genomic response of Ishikawa cells to bisphenol A exposure is dose- and time-dependent. Toxicology 2010;270:137–49.

[128] Dadarkar SS, Fonseca LC, Thakkar AD, Mishra PB, Rangasamy AK, Padigaru M. Effect of nephrotoxicants and hepatotoxicants on gene expression profile in human peripheral blood mononuclear cells. Biochem Biophys Res Commun 2011;401:245–50.

[129] Lobenhofer EK, Auman JT, Blackshear PE, Boorman GA, Bushel PR, Cunningham ML, et al. Gene expression response in target organ and whole blood varies as a function of target organ injury phenotype. Genome Biol 2008;9:R100.

[130] Sukata T, Sumida K, Kushida M, Ogata K, Miyata K, Yabushita S, et al. Circulating microRNAs, possible indicators of progress of rat hepatocarcinogenesis from early stages. Toxicol Lett 2011;200:46–52.

[131] Wetmore BA, Brees DJ, Singh R, Watkins PB, Andersen ME, Loy J, et al. Quantitative analyses and transcriptomic profiling of circulating messenger RNAs as biomarkers of rat liver injury. Hepatology 2010;51:2127–39.

[132] Gohlke JM, Thomas R, Zhang Y, Rosenstein MC, Davis AP, Murphy C, et al. Genetic and environmental pathways to complex diseases. BMC Syst Biol 2009;3:46.

[133] Hu G, Agarwal P. Human disease-drug network based on genomic expression profiles. PLoS ONE 2009;4(8):e6536.

[134] Patel CJ, Butte AJ. Predicting environmental chemical factors associated with disease-related gene expression data. BMC Med Genomics 2010;3:17.

CHAPTER

Unraveling the DNA Damage Response Signaling Network Through RNA Interference Screening

2.2

Louise von Stechow, Bob van de Water and Erik H.J. Danen
Division of Toxicology, LACDR, Leiden University, Leiden, The Netherlands

2.2.1 The DNA-damage-induced signaling response

DNA damage sources and damage sensing

The integrity of our genomes is constantly threatened both by endogenous insults—including spontaneous deamination events and consequences of metabolism, such as reactive oxygen species (ROS) leading to single-strand breaks and subsequent replication stress—and by exogenous sources of DNA damage, such as radiation and genotoxic compounds. Together, these result in up to 1000 lesions per cell per day [1,2]. DNA is the only biomolecule that cannot be recycled and instead is repaired. Failure to repair damaged DNA results in mutations and chromosomal instability. As a safety measure, excessive DNA damage triggers removal of cells from tissues by apoptosis or senescence.

The capacity to recognize and accurately respond to damaged DNA occurs through an evolutionarily conserved and a highly orchestrated signaling response, the DNA damage response (DDR). The DDR comprises signaling components that sense and distribute the damage signal to induce the onset of specific cellular programs, determining DNA repair, cell cycle arrest, apoptosis, cellular senescence, or differentiation [3,4]. Often, the sensing of DNA lesions is associated with transcription and replication, since these processes continuously scan large parts of the genome. Other damage-sensing mechanisms operate transcription- and replication-independently and are frequently linked directly with damage-specific repair programs [5]. The response to DNA damage strongly depends on the cellular and organismal background, as well as on the type and intensity of the damage signal (Figure 2.2.1). The formerly predicted, simple view of a signaling cascade, in which linear interaction flows determine the cellular response to DNA damage, has been replaced by a highly integrated network of intermediates whose expression, interactions, posttranslational modifications, and metabolites change in response to DNA damage, thereby driving the different response pathways [6].

Signal transduction in the DDR depends on a phosphorylation cascade

The PI3-K-related protein kinases (PIKKs) ataxia-telangiectasia mutated (ATM) and ATM and RAD3 related (ATR), as well as DNA-dependent protein kinase (DNA-PK), are crucial for the initial

Toxicogenomics-Based Cellular Models.

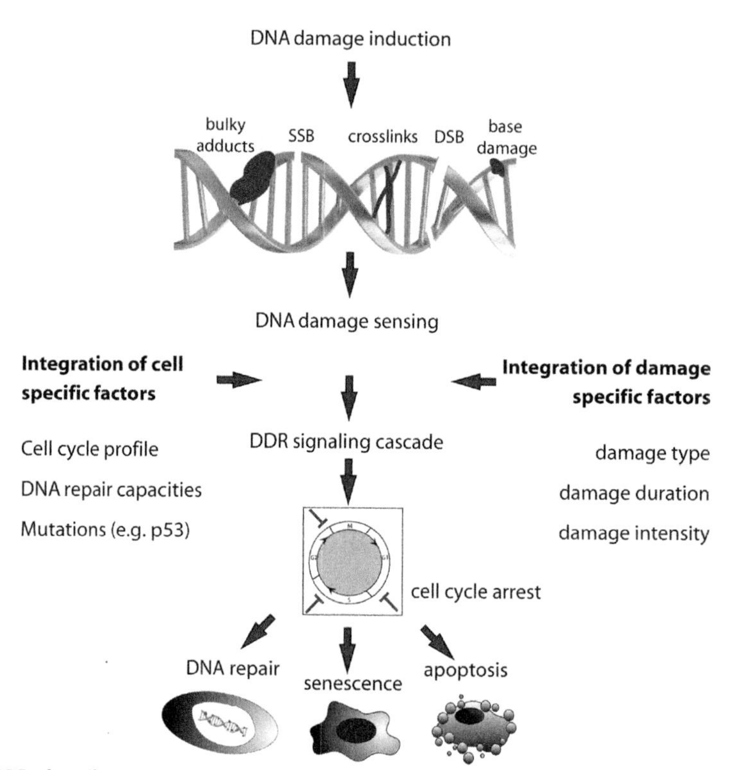

FIGURE 2.2.1 DDR signaling.

Specific types of DNA damage lead to the induction of distinct DNA lesions, including base damage, bulky DNA adducts, single-strand breaks (SSBs), double-strand breaks (DSBs), and intrastrand crosslinks. The presence of such lesions elicits a signaling cascade that is integrated with ongoing cellular activities (e.g. cell cycle, transcription, translation), cell-specific factors (e.g. DNA repair capacity), and damage-specific factors (e.g. type, duration, and intensity of the damage). Altogether, this determines the cellular outcome, which typically starts with a cell cycle arrest and subsequent effective DNA damage repair or, if repair fails, removal of cells from the tissue through senescence or cell death (for example by apoptosis).

recognition and distribution of DNA double-strand breaks (DSBs) and single-strand breaks (SSBs) [7]. After DNA damage, ATM and ATR alone lead to phosphorylation of more than 700 proteins, with a total number of 900 phosphorylation events, demonstrating the magnitude of the response mediated by these kinases [8]. The substrates of ATM and ATR reflect the whole spectrum of the DDR and are involved in cell cycle arrest, DNA repair, and cell survival [7,9]. Whereas ATR is vital for the viability of replicating mammalian cells, ATM-deficient cells are viable. However, mutations in ATM result in the cancer predisposition genetic disorder ataxia–telangiectasia, which is characterized by cerebellar ataxia, immunodeficiency, genomic instability, and cancer predisposition [7].

The signal for the recruitment of ATM is the presence of DSBs, which are recognized by the Mre11, Rad50, and Nbs1 (MRN) complex [2]. The crucial lesion for the recruitment of ATR is replication protein A (RPA)-coated single-strand DNA, which is sensed by ATR interacting protein ATRIP and further requires the Rad9-Rad1-Hus1 (9-1-1) clamp loading complex and the ATR activator topoisomerase binding protein 1 (TOPBP1) [9]. Single-stranded DNA often arises during replication stress, or as a result of resected DSBs. DSBs are a common secondary DNA lesion, resulting, for example, from single-strand breaks that encounter replication forks. Thus, although initially recruited by different kinds of lesions, ATM and ATR frequently cooperate in the DDR signaling response [2]. In the DSB response, ATM is rapidly activated at any stage of the cell cycle, while the activation of ATR occurs more slowly and is mostly limited to S and G2 [9].

ATM and ATR share many common targets and only a few substrates are exclusive for one of the kinases; for example, the checkpoint kinases Chk2 and Chk1 that diffuse through the cell to further convey the damage signal. Similarly to ATM, Chk2 is dispensable for viability, but mutations cause Li–Fraumeni syndrome, a hereditary cancer-susceptibility syndrome, also associated with mutations in p53 [7]. Conversely, Chk1, like ATR, is crucial for viability, with cells of Chk1-knockout mice displaying a phenotype that resembles mitotic catastrophe [10]. ATM and ATR, as well as Chk1 and Chk2, can activate p53 by leading to the disruption of the regulatory loop that exists between p53 and its negative regulator MDM2 through phosphorylation of both proteins [10–12]. Despite the undoubtedly crucial roles of ATM and ATR as key players in the DDR, recent reports have implicated other kinases as important signal transducers later in the genotoxic stress response [13,14]. For instance, p53-deficient cells have been shown to rely on a p38MAPK/MK2 signaling module for checkpoint activation [15,16].

The transcription factor p53 serves as a central hub in the cellular stress response

The transcription factor p53, known as the *guardian of the genome*, regulates a variety of cellular processes ranging from apoptosis, cell cycle arrest, autophagy, DNA repair, and senescence to metabolic processes, angiogenesis, differentiation, and immune responses. It can execute these functions in both in transactivation-dependent and -independent manner [17,18]. Through its action as a transcription factor, p53 regulates a great number of target genes, including not only protein-coding genes but also microRNAs (miRNAs), such as the proapoptotic miR-34 family [19]. Amongst the protein-coding genes, the key proapoptotic target of p53 is the BH3-only protein PUMA. However, unlike the case with p53, PUMA deficiency by itself does not lead to tumor formation in mouse models [20]. Next to this transcriptional effect on apoptosis, p53 can also physically translocate to the mitochondria to directly cause apoptosis via induction of BAK oligomerization and cytochrome c release [21,22].

Besides apoptosis, p53 also regulates G1 arrest via the up-regulation of p21 or GADD45, as well as senescence, which is mediated by induction of plasminogen activator inhibitor, but can also be activated by p21 [17]. Similarly to the effect of PUMA deficiency, lack of p21 is not by itself tumorigenic, indicating that more than one p53 target gene is important for the mediated effect [23]. Moreover, p53 can regulate target genes that are involved in the induction of autophagy [24] and regulate cellular metabolism by acting on mTOR [25]. Interestingly, next to those anti-survival functions, p53 also induces a set of pro-survival pathways, such as DNA repair and antioxidant regulation [26]. The plethora of p53-mediated responses points to a crucial role of p53 as a central hub protein that functions to monitor and integrate a vast amount of information to finally determine the cellular outcome based on the context of cell, tissue, and organism, as well as the type, duration, and amount of stress. Indeed, a lack of p53 or deficiency in p53 function is common for many cancer types, mostly rendering cells resistant to genotoxic-stress-induced killing. However, in certain cellular contexts, lack of p53 can also enhance sensitivity, mostly as a result of deregulated cell cycle checkpoints and resultant mitotic catastrophe [27].

p53 is regulated by posttranslational modifications

The stability, localization, and transactivation capacity of the tumor suppressor p53 is regulated by a complex network of posttranslational modifications, including phosphorylation, ubiquitination, sumoylation, neddylation, methylation, acetylation, and glycosylation. Over 36 different amino acids of p53 have been shown to be modified, sometimes in a stress-induced manner [28]. Regulation by ubiquitination is prominent: in unstressed conditions p53 levels are held low by MDM2-mediated ubiquitination and subsequent proteasomal degradation. The E3 ubiquitin ligase MDM2, in turn, has the capability of inducing self-ubiquitination, but is targeted for proteasomal degradation by other ubiquitin ligases [29]. Moreover, *MDM2* itself is a target gene of p53, providing an autoregulatory feedback loop [30]. The *MDM2* gene is amplified in 7% of human cancers and the critical role of MDM2 in regulation of p53 is underscored by the fact that p53 deletion can rescue embryonic lethality in MDM2-deficient mice [31]. Another player in the ubiquitin-mediated regulation of p53 is MDM4 (MDMX in mice), which, despite being structurally related to MDM2, has no reported ability to ubiquitinate p53, but acts as a structural interactor for p53 and MDM2, and can both activate and repress p53 [32,33].

In addition to the MDM2/MDMX module, other ubiquitin ligases are involved in regulating p53 stability and localization, such as COP1, Pirh2, ARF-BP1, MSL2, and Parc [34]. Ubiquitin removal by deubiquitinases (DUBs) provides a further mode of p53 regulation. The deubiquitinase HAUSP, also known as USP7, can deubiquitinate p53 as well as MDM2 [35]. The main function of USP7 seems to lie in the regulation of MDM2, since its ablation stabilizes p53 levels [35,36]. USP10 removes MDM2-mediated ubiquitin chains from p53, thereby preventing p53 nuclear export and degradation. USP10 itself is stabilized after DNA damage and a fraction of this protein localizes to the nucleus, a process regulated by ATM-mediated phosphorylation [37]. USP4 indirectly regulates p53 by deubiquitinating and thereby stabilizing the negative regulator ARF-BP1 [38].

Other posttranslational modifications can lead to p53 stabilization by hindering the addition of ubiquitin, such as phosphorylation of p53 and MDM2 after stress by kinases such as ATM, ATR, and DNA-PK, as well as the checkpoint kinases [28]. In vivo studies using knock-in mice in which single or double phosphorylation sites in p53 are mutated have shown that no single phosphorylation event is exclusively responsible for stabilization of p53 after stress [28]. Recently, it has been established that DNA-damage-induced stabilization of p53 protein levels occurs as an oscillatory response, pulses depending on activation of p53 by ATM, which leads to induction of negative regulators MDM2 and Wip1, subsequently decreasing p53 levels [39,40]. Finally, posttranslational modifications can change the transactivation features of p53 and affect the interaction with different transcriptional co-activators or repressors. Acetylation of p53 by the histone acetylase CBP/p300 enhances protein stability, sequence-specific DNA binding, and interaction with cofactors and further leads to acetylation-dependent chromatin relaxation in p53 target genes [41].

2.2.2 DNA-damage-induced cellular responses

DNA repair

The variety of DNA repair pathways matches the diversity of different lesions that can be induced in DNA. Generally, one can distinguish repair pathways that act on damage affecting only one strand of the DNA, such as single-strand breaks, mismatches, and smaller base modifications, from pathways that act on damage affecting both strands, such as DSBs and crosslinks [42].

Repair of small DNA lesions

Pathways dealing with smaller kinds of DNA damage include mismatch repair (MMR), a strand-specific repair mechanism that corrects base mismatches occurring during DNA replication but also participates in a variety of other DNA transactions [43], as well as base excision repair (BER). The BER pathway serves to remove small DNA lesions such as base alterations, including oxidative modifications, methylations, or alkylations, as well as single-strand breaks (SSBs) [42,44]. Lesions can be either mutagenic or cytotoxic/cytostatic and, therefore, the failure of different DNA repair pathways will result in different cellular outcomes [1]. Small lesions that are being repaired by MMR and BER do not interfere with the helix, but might, if not repaired, result in mutations.

Repair of bulky and helix-interfering DNA lesions

Helix-distorting, bulky DNA lesions (benzopyrene-induced adducts; UV-induced lesions, e.g. thymine dimers and 6-4-photoproducts; bulky lesions caused by DNA crosslinking agents) are repaired by the nucleotide excision repair (NER) pathway. NER is subclassified into two different types that share a common core pathway, but differ in the damage-sensing mechanism. Global genome repair (GGR) functions both in the transcribed and the untranscribed strand and does not require the gene in which the damage occurs to be active. The GGR pathway makes use of DNA damage-sensor proteins, such as DDB and XPC-Rad23B, in order to recognize helix distortions. Conversely, the transcription-coupled repair (TCR) pathway relies on the sensing of DNA damage by RNA polymerase II, which becomes stalled at bulky lesions [45]. Defects in NER can result in both cancer susceptibility syndromes such as xeroderma pigmentosum (XP) and progeria syndromes such as Cockayne syndrome and trichothiodystrophy [1,46].

The repair of intrastrand crosslinks is achieved with the aid of the Fanconi anemia repair pathway. Fanconi anemia is a rare autosomal recessive genetic disease, caused by mutations in at least 1 out of the 14 complementation group genes. Patients show cancer predisposition; neural, developmental, and skeletal abnormalities; aplastic anemia; and a high sensitivity to crosslinking agents such as mitomycin C or cisplatin [47,48].

DSBs are the most deleterious, however also the most rarely occurring, DNA lesion, resulting either from blockage of replication forks caused by other types of lesions or directly by ionizing radiation. Two pathways are used for the repair of double-strand breaks: homologous recombination (HR), employing a homologous template for strand replacement and thus being restricted to late S and G2-phase of the cell cycle; and the error-prone mechanism non-homologous-end joining (NHEJ), which does not require the presence of a homologous template and can therefore function throughout the cell cycle. The decision for one or the other pathway is dependent on the organism, cell type, cell cycle status, the mode of DSB induction, and the chromatin structure surrounding the break [49]. In the NHEJ pathway the two ends of a DSB are ligated together; crucial for this are the DNA-PK catalytic subunit (DNA-PKcs), as well as the Ku70-Ku80 protein heterodimer that forms complexes on both sides of the DNA ends, which interact and bridge the gap [42]. HR is critically dependent on Rad51, which mediates homology search and strand invasion between the damaged DNA strand and the homologous template, as well as the breast cancer susceptibility genes *BRCA1* and *BRCA2* [42,50]. Many factors are common to more than one repair pathway, such as those involved in HR and Fanconi anemia (FA) pathways. Furthermore, some lesions, such as DNA crosslinks, require more than one repair mechanism for their removal, hinting towards the close interconnections between different repair pathways.

In addition to repair, another way cells can cope with DNA damage is the bypass of the damage during DNA replication, a process known as translesion synthesis (TLS). In this pathway, a switch from a high-fidelity DNA polymerase to the Y-family of DNA polymerases occurs. These polymerases can carry out replication over damaged DNA, but have reduced fidelity on undamaged substrates, making the process of TLS potentially mutagenic. Different types of polymerases are capable of bypassing different kinds of lesions [51].

DNA-damage-induced cell death

In the case of failed DNA repair, cells can induce different forms of cell death, including apoptosis and autophagy, as well as necrosis, senescence, and mitotic catastrophe [52]. The apoptotic process is characterized by chromatin condensation and cellular shrinking, and concludes in the formation of apoptotic bodies, membrane-surrounded cellular particles, which can be phagocytized by neighboring cells. This prevents unwanted immune responses by the release of intracellular factors into the extracellular space, as induced by necrosis, which is characterized by a rapid loss of membrane integrity [53]. Apoptosis features two major pathways. The cell extrinsic or death receptor pathway is regulated by extracellular molecules such as tumor necrosis factor alpha (TNFα) binding to the death receptor family membrane receptors, and shuts on signaling via the Fas-associated death domain protein (FADD). The cell intrinsic or mitochondrial pathway is regulated by intracellular stress signals, acting via the activation of proteins of the Bcl-2 family. This family includes positive regulators of apoptosis such as BAK and BAX and the BH3-only proteins PUMA, NOXA, and Bad, as well as negative regulators such as BCL2 or Mcl1, which modulate the release of cytochrome c from the mitochondria and apoptosome formation [53–55]. Crucial to apoptosis is the activation of the caspase family of cellular proteases, which, together with nucleases, carry out cellular breakdown by degradation of proteins and nucleic acids. Despite differences in the earlier steps of the signaling cascade, both pathways will eventually culminate in the induction of effector caspases 3, 6, and 7 [54].

Dysregulation of many proteins involved in induction of apoptosis, such as p53 or Bcl2 family members, as well as inhibitor of apoptosis proteins (IAPs), has been implicated in cancer formation and therapy response [56]. DNA damage induces apoptosis in normal and cancer cells; however, the extent to which cell death by apoptosis really contributes to the success of chemotherapy is not yet clear. Besides apoptosis, another form of programmed cell death, senescence, and the process of mitotic catastrophe (MC) are important consequences of DNA damage induction. Whereas senescence describes a terminally arrested state in which cells, although not dividing, are still metabolically active and able to affect neighboring cells by secreting factors, in MC abrogation of G1/S or G2/M checkpoints leads to mitotic entry in presence of DNA damage, which results in mitotic abnormalities and subsequent death [57–59]. MC might be especially important in cancer cells, which often carry checkpoint defects. Cells that lack p53 signaling will undergo mitotic catastrophe if the backup checkpoint axis, which is maintained by p38/Mk2, is silenced as well [15].

DNA-damage-induced cell cycle arrest

Different cell cycle checkpoints have evolved that prevent replication of damaged DNA and premature entry to or exit from mitosis, and allow time for DNA repair after encountering DNA damage. The main cell cycle checkpoints are the G1/S checkpoint, the intra-S checkpoint, and the G2/M checkpoint [60]. The transition through stages of the cell cycle is regulated by the action of cyclin-dependent kinases, which are key targets for modulations induced by different cellular stimuli, including DNA damage. G1 arrest, which is the main DNA-damage-induced checkpoint in non-cancerous cells, can be activated via an ATM-Chk2-p53-p21-mediated signaling cascade, leading to silenced cyclinE/Cdk2 kinase and causing G1 arrest [2]. Further, a faster, transcription-independent route for cell cycle regulation exists, via the CDC25 family phosphatases (comprising CDC25A, CDC25B, and CDC25C), which can remove the inhibitory phosphorylation from cyclin-dependent kinases (CDKs). While CDC25A is mainly thought to regulate the G1/S checkpoint by activating cyclinE (A)/Cdk2 kinase, CDC25C acts on cyclinB/Cdk1, thereby mediating the entry into mitosis and regulating the G2/M checkpoint. However, CDC25 phosphatases are not functionally restricted and in many cases have redundant roles [61]. The checkpoint kinases Chk1 and Chk2 phosphorylate the CDC25 phosphatases, which attenuates CDC25 protein stability through priming for proteasomal degradation and, furthermore, induces their interaction with 14-3-3, sequestering them from Cdk1. Both mechanisms result in an induction of cell cycle arrest [61,62].

Next to DNA-damage-induced cell cycle arrest, regulation of checkpoint maintenance and checkpoint recovery is important for cellular survival after genotoxic stress. Polo-like-kinase 1 (Plk1), aurora kinase A, and the phosphatase Wip1 are crucial players in checkpoint recovery after G2 arrest. Plk1 phosphorylates and activates negative regulators of p53 and stimulates nuclear translocation of CDC25B/C, leading to a reinitiating of the cell cycle. Wip1 has been shown to be crucial for cell cycle recovery in p53-proficient, but not -deficient, cells, being itself a target gene of p53 and at the same time capable of regulating p53 on multiple levels. Wip1 assures the low-level expression of cell cycle regulators to enable eventual re-entry into the cell cycle [60].

2.2.3 DNA damage in the context of cancer formation and treatment

DNA damage and cancer formation

Mutations caused by exogenous sources of DNA damage have been closely linked to the incidence of certain cancers, such as cigarette-smoke-related lung cancer development or skin cancer related to an excess of UV exposure. Furthermore, genomic instability is an inherent feature of tumors. The importance of faithful genome maintenance is reflected in inherited cancer susceptibility disorders, which are linked to DNA repair or DNA-damage-signaling genes. These include the MMR-associated syndrome hereditary non-polyposis colorectal cancer (HNPCC or Lynch syndrome), hereditary breast and ovarian cancer incidence linked to the deficiency in the HR genes *BRCA1* and *BRCA2*, skin cancer susceptibility syndromes linked to defects in NER such as XP, and syndromes caused by mutations in DDR signaling molecules such as ataxia telangiectasia (caused by mutations in ATM) and Li–Fraumeni (caused by mutations in p53 or Chk2) [1,7,42,63]. Also, in spontaneously arising tumors there is a high rate of genomic instability, which is described as one of the hallmarks of cancer [64,65]. The consequence of the continuous DNA damage in tumors is the accumulation of mutations. We can distinguish *driver mutations* in crucial genes, that determine the malignant changes

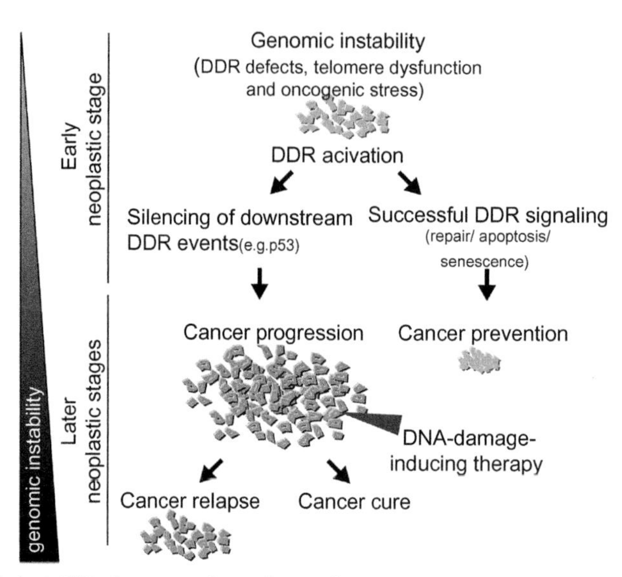

FIGURE 2.2.2 Genomic instability in cancer formation and treatment.

At early neoplastic stages, tumors show genomic instability due to DDR defects, telomere dysfunction, or oncogenic stress. Activation of the DDR can act as a barrier against cancer formation. However, if the tumor progresses, downstream signaling events are often silenced. This leads to an increase in genomic instability and mutation rate in the tumor. DNA-damage-inducing therapy can lead to cure of the cancer, but if relapse occurs, the tumor will likely show even more genomic instability.

within the tumor, and so-called *passenger mutations* that are accumulated as a result of the constant DNA damage but have little impact on the physiology of the tumor [63].

The elevated occurrence of genomic instability in cancers is a result of shortening of telomeres, which provides a site for chromosomal fusion events, as well as of oncogenic stress, resulting in subsequent replication stress. In the early neoplastic stage, there is often a continuous activation of the DDR, serving as a barrier against further transformation. At later neoplastic stages, the early signs of DSB response, such as γH2AX and 53BP1 foci, persist, whereas the signaling response, which in many cases would lead to p53-mediated induction of apoptosis, is lost [66]. Furthermore, at later stages, tumor cells that have been subjected to DNA-damage-inducing cancer therapy show enhanced genomic instability as a result of the treatment itself (Figure 2.2.2) [42]. Genomic instability in cancers is not only seen as a hallmark of cancer initiation and progression, but can also serve as a predictive or prognostic biomarker. HR-deficient tumors, such as BRCA1- and BRCA2-deficient breast and ovarian cancers, have been shown to be highly responsive to platinum agents [42].

Genetic screens might prove useful for identification of the genes that are crucial drivers in cancer formation and novel genes that might predict treatment responses [67]. However, not only gene deletion but also other mechanisms of silencing, such as promoter methylation and posttranscriptional mechanisms of regulation, can lead to tumor phenotypes, such as the one observed for BRCA1 and BRCA2 inactivation. An interesting alternative to studying gene expression might lay in the determination of DNA repair capacity (e.g. by studying formation of Rad51 DSB repair foci) in patient cells, to predict the treatment response [42].

Exploiting the DDR for improved cancer therapy

Treatment of cancers often exploits DNA damage pathways. Radiation, but also chemotherapeutic drugs, such as alkylating agents, antimetabolites, and topoisomerase poisons, can be used to induce DNA damage, eventually leading to the induction of apoptosis in cancer cells [68]. Tumor cell inherent features such as a high proliferation rate and internal genomic instability make them more susceptible to genotoxic treatments than untransformed cells of the human body. However, toxicity of chemotherapeutics or radiation for healthy tissue and primary or acquired resistance constitute rate-limiting factors for the success of DNA-damage-inducing cancer therapy. Furthermore, genomic instability in cancers is a double-edged sword, since on the one hand it makes tumor cells more sensitive to DNA-damaging cancer therapy, but on the other hand the high mutational rate allows for acquiring resistance by selecting for mutations that favor survival in the presence of DNA damage. This phenomenon is reflected in the behavior of BRCA1-deficient ovarian cancers, which initially respond well to platinum compounds, but over the course of treatment acquire resistance, often by genetically or epigenetically reintroducing the mutated DNA repair gene [67].

Clarifying the mechanisms that determine the DDR is therefore of utmost importance to improve the efficacy of therapy. The goal thereby is to specifically kill cancer cells, while sparing cells of healthy tissue from the deleterious effects of DNA damage, preferably by taking into account inter-patient and potentially inter-tumor genetic variability, as well as differences in the microenvironment of tumors and healthy tissue, such as hypoxia, and protective effects of stromal cells within the tumor [10,42,63,69,70].

To find novel genetic interactions that meet the requirements for targeted therapy, research has focused on synthetic lethality approaches; a concept adopted from yeast genetics, where the deletion of one gene alone has no effect whereas the combination of two gene deletions can lead to decreased cell survival. The hope is to specifically kill cancer cells, by exploiting their deficiencies in DNA repair pathways or G1/S checkpoint activation (which makes them depend more heavily on the G2/M checkpoint) [70]. So far, the most potent application of this concept was found in the relation between the inhibition of the ribosylase PARP and players in the DSB repair pathways, i.e. tumors bearing mutations in the repair factors BRCA1 and BRCA2, which are involved in hereditary breast cancer [71,72]. Inhibition of PARP causes an excess of single-strand breaks, leading to secondary DSBs. While those can be repaired by HR in normal cells, HR-deficient cells will be killed specifically. Inhibitors of other DDR factors such as Chk1 and DNA-PK are currently in clinical studies [63]. Although the targeting of key players of the DDR has been proved to be beneficial in some cases, it also bears risks, since those factors often evoke plethoric responses activating not only apoptotic pathways but also DNA repair, cell cycle arrest, or other pro-survival responses and may therefore lead to undesired effects, depending on the genetic context of a cell, tumor, or patient. This has been demonstrated by ambiguous clinical responses to ATM and Chk2 inhibitors [10,27].

2.2.4 RNAi screens to study the DDR signaling network

Mechanism of RNA interference

Small interfering RNA (siRNA)-based screens exploit the endogenous pathway of RNA interference (RNAi). RNAi is characterized by binding of a small RNA molecule to a target mRNA, which induces a reduction in translation of the target transcript, by either the degradation or sequestering of the target mRNA [73]. A plethora of cellular functions are regulated by endogenous RNAi, including many DNA-damage-related processes [74]. In recent years, a growing pool of small RNAs with RNAi function has been identified, including siRNAs, miRNAs, short hairpin RNAs (shRNAs), piwi-interacting RNAs (piRNAs), and a class of small RNAs interacting with the Argonaute protein QDE-2 (qiRNAs), comprising both endogenous and exogenous (such as viral) transcripts. Moreover, RNAi has been used for targeted, gene-family-wide, or "whole-genome-wide" depletion of genes, aimed at understanding gene function in a wide array of cellular processes or aimed at drug target discovery in a variety of disease areas [75–78].

siRNA screening

High-throughput compound screens are a common method in both drug discovery and toxicity screening [79]. Furthermore, genetic screens carried out in model organisms have provided insight into the function of many genes. However, this screening method is very laborious and the function of genes whose depletion is lethal cannot be easily studied [80]. Since its discovery in 2001, the process of RNAi has been established as a research tool, to specifically target genes of interest. Commercial or customized siRNA or shRNA libraries can be used to study the effects of knockdown of a specific gene group (such as all cellular kinases or transcription factors) or the whole genome [81]. Great advantages of siRNAs are the relative easy transfectability and short experimental duration, allowing the study of transient effects. However, the latter must be seen as a double-edged sword, since it limits the assay window to around 72 hours, after which the effect of the knockdown will decrease and, furthermore, downstream effects of the knockdown might dilute the original phenotype [81,82]. Conversely, screens using lentivirus-based stable incorporation of shRNA sequences into the genome allow for the study of gene silencing over a longer period of time [76].

An enormous benefit of RNAi screens lies in the possibility to combine gene knockdown with various endpoints, assessing changes in cell survival or cell cycle profile, cellular morphology or induction of microscopically visibly structures, such as DNA damage foci (Figure 2.2.3) [83–86].

Data analysis of RNAi high-throughput screens can be more laborious than the screening process itself, which especially holds true for high-content screening methods, where the gene knockdown is combined with—often multidimensional—microscopical measurements [79]. For analysis of these large data sets, the use of automated analysis pipelines, such as the one described by Boutros et al., can be of great use [124]. In order to test the robustness of RNAi screening data, signal-to-background ratio or Z'-factors should be determined. Typically the signal-to-noise ratio of RNAi screens is lower than that of compound screens [82].

Many siRNAs have been shown to have off-target effects, targeting another mRNA, next to the mRNA of interest. This effect seems to be mostly dependent on the 2-7/2-8 seed region of the antisense strand of the siRNA [87]. In a study published in 2012, Adamson and coworkers indicated the possible severity of this phenomenon. In an siRNA screen for factors involved in homologous recombination repair (HR), the authors indicated a rate of around 17% false-positive targeting of the DNA repair factor RAD51, due to matches between the siRNA's antisense seed region and the 3' UTR of RAD51 [88].

To exclude such off-target effects, it is crucially important to validate results obtained by siRNA screening. Different methods of validation have been established, including the use of siRNAs with modified seed sequences, which are thought to be less likely to show off-target effects; the use of different individual sequences targeting one mRNA (usually employed with a cut-off of at least two or three sequences showing the same effect); combined use of RNAi and microarray technology (as employed in shRNA barcode-screens); and gene depletion and rescue by overexpression of a non-siRNA-targetable mRNA version of the gene of interest. Although the last method offers the most certain validation of genotype–phenotype correlations observed in siRNA screens, it is laborious and therefore not feasible for high-throughput validation [81].

siRNA screens to study DDR signaling responses

Various RNAi-based screens have aimed at shedding light on the DDR, induced by ionizing radiation, genotoxic drugs, DNA-repair-targeting drugs such as PARP inhibitors, or endogenous DNA damage induction [86,88–90]. Most of these screens used either cell viability or the induction of DNA damage foci as readout, and identified factors implicated in various levels of the DNA damage signaling response (Figure 2.2.3). Interestingly, next to finding crucial new DDR factors, these screens have implicated pathways involved in cellular housekeeping functions, such as translation, transcription, and RNA processing, as well as developmental pathways, as crucial regulators of survival after genotoxic treatment [91].

Different screens have helped to clarify the early signaling dynamics elicited by DNA damage. These include the identification of the 9-1-1 and TOPBP1-interacting protein RHINO, which is required for ATR activation [92]. Furthermore, siRNA-based screens have helped to clarify the dynamics of the DSB-induced signaling cascade. Detection of DSBs evokes a signaling response that recruits repair factors to the site of the damage. This response comprises a hierarchical cascade of additions of posttranslational modifications to chromatin components, as well as mediator and transducer proteins, which appear in discrete, damage-induced foci and are required for the recruitment of repair and checkpoint proteins [93]. These include different ATM target proteins, e.g. the phosphorylated form of histone H2AX, γH2AX, as well as DDR mediators and repair proteins such as MDC1, 53BP1, and BRCA1 [125]. However, ubiquitination-mediated responses have also been implicated as crucial for the proper assembly of DNA-damage-induced foci. Using whole-genome siRNA screens for factors that affect the formation of 53BP1 foci, Kolas et al. identified the E3 RING-type ubiquitinase RNF8 as a key player in the formation of DSB-induced foci [94]. In subsequent studies, another ubiquitinase, RNF168, was identified; both E3 ligases have been shown to ubiquitinate histones in response to DNA damage, an event that precedes the recruitment of the crucial DSB repair factors 53BP1 and BRCA1 into DNA damage foci [95].

Next to these early signaling events, downstream processes in the DDR, such as activation of p53, have also been studied using siRNA screens. Although the p53 gene itself is silenced in many cancers, various other mechanisms have been described for inactivation of the p53 pathway. These interactions could be interesting therapeutic targets, since their inactivation might stabilize p53 and reintroduce sensitivity in p53-proficient tumor cells, as has been shown for small molecule inhibitors of MDM2 such as nutlin [96].

Using barcode screening, Mullenders et al. identified novel p53 regulators, including the ubiquitin ligase Rbck1 and the circadian clock factor ARNTL. ARNTL-deficient cells showed defects in p53 cell cycle arrest, due to an inability to activate p21, linking the circadian clock to the regulation of the cell cycle [97]. Moreover, different screens have aimed at finding novel downstream functions regulated by p53 signaling. Using isogenic HCT116, p53 wild-type, and knockout cells, Krastev et al. uncovered a synthetic lethal interaction of p53 with UNRIP, which is involved in small nuclear ribonucleoprotein assembly [98].

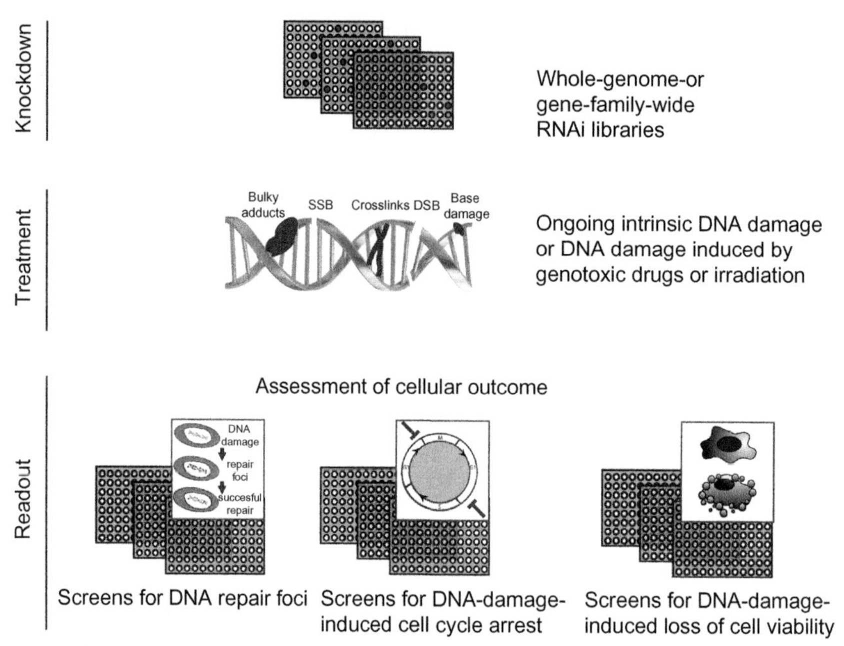

FIGURE 2.2.3 RNAi-Based screening to identify novel factors in the DDR.

RNAi screens to identify novel DDR factors have combined gene silencing with various sources of DNA damage, including, for example, the genotoxic drugs cisplatin and camptothecin, or irradiation, or focused on endogenous DNA damage. Screening endpoints have mostly been cell cycle profiles, the appearance of visible DNA repair foci (such as 53BP1, γH2AX, or RAD51 foci), or loss of cell viability.

siRNA screens for identifying DNA-damage-induced cellular responses

A number of screens have studied DNA-damage-induced cellular responses, including DNA damage repair pathways, as well as cell cycle arrest and apoptosis (Figure 2.2.3). siRNA screens identified factors involved in DSB repair, including the MMS22L-NFKBIIL2 complex as a factor that is involved in the ATR-dependent replication stress response and the subsequent HR-mediated repair, and was found independently by two different groups in 2010 [99,100], as well as the nuclear import factor NUP153, which was shown to be a crucial determinant of the nuclear recruitment of the DSB response factor 53BP1. NUP153-deficient cells showed increased sensitivity to ionizing radiation, which was accompanied by delayed DNA repair [101]. Furthermore, shRNA screens targeting all cellular DUBs led to the identification of the DUB USP1, which is crucial for removing ubiquitin from FANCD2 in the Fanconi anemia pathway [102].

Different RNAi screens have aimed at describing the apoptotic response. Using a shRNA screen to find sensitizers to ligand-induced apoptosis, Dompe et al. identified the two genes *EDD1* and *GRHL2*, which are found 50 kb apart on chromosome 8q22 [103]. Furthermore, in an siRNA screen targeting the cellular phosphorylation machinery, the phosphatase MK-STYX was found to be a regulator of mitochondrion-dependent apoptosis. Cells carrying a knockdown for this catalytically inactive phosphatase were unable to initiate release of cytochrome c [104]. Interestingly, a recently published siRNA screen uncovered a novel interaction between the Wnt signaling pathway, which is a crucial mediator of development as well as of cancer formation and progression, and the X-linked inhibitor of apoptosis XIAP, which was shown to act on canonical Wnt signaling independently of its function in apoptosis [105].

In 2010, a high-content phenotypic siRNA screen was carried out by Neumann et al., intended to define the basics of mitosis in human cells. Whole-genome RNAi was combined with high-throughput microscopy of fluorescently labeled chromosomes. By describing factors that are crucially important for normal progression of the cell cycle, this study provides an important basis for studying mitotic deregulation after genotoxic stress and, in addition, presents a great example for the analysis of high-throughput microscopy data by using vector-machine-based classification of images into morphological classes [85].

Furthermore, siRNA screens for the discovery of modulators of DNA-damage-mediated cell cycle checkpoints have been provided by different groups. Hurov et al. identified factors of the triple T complex as critical modulators of survival in irradiated U2OS cells, using a whole-genome shRNA-based screen. Follow-up validation of the factors TTI1 and TTI2 indicated that they are crucially important for checkpoint signaling by regulating the abundance of PIK-kinases such as ATM [89]. To globally study the DNA-damage-induced G2/M checkpoint network, Kondo and Perrimon performed a genome-wide RNAi study in *Drosophila* S2R cells, implicating signaling hubs that influence G2/M arrest after treatment with doxorubicin, involved in DNA repair, DNA replication, and chromatin, and interestingly also one signaling hub centered on RNA processing [107].

siRNA screens for identification of tumor-development-driving genes

Many RNAi-based screens have been aimed at identifying novel tumor suppressor genes, with the dedicated goal of identifying true driver mutations, which will be able to determine a patient's phenotype and genes that might be suitable for pharmacological inhibition in cancer therapy. In an early study in 2005, McKeighan et al. examined the effect of kinome-wide siRNA screens on apoptosis in unchallenged HeLa cells and, furthermore, determined the role of different phosphatases in chemosensitization to the genotoxic drugs cisplatin and etoposide and the microtubule poison taxol. They could show that the knockdown of survival kinases, such as SGK, mTOR, or FER, sensitized to low doses of chemotherapeutics [108].

Despite the discovery of numerous genes that might function as tumor suppressors, however, in vitro-identified genetic relationships often fail to show the same behavior in an in vivo context. Cell culture systems offer a highly artificial situation. In real tumors, heterogeneity between tumor cells, especially the existence of so-called cancer stem cells, and interaction with non-tumor stromal cells will affect cellular behavior. In recent years, methods for in vivo screening have been developed. These are generally based on shRNA libraries, often combined as so-called barcode screens, in which each shRNA sequence that becomes integrated into the genome carries a small barcode that can later be identified by polymerase chain reaction (PCR), microarray technology, or next-generation sequencing [76,109]. In 2008, Zender et al. illustrated a novel approach for shRNA screening in a mosaic mouse model for the induction of hepatocarcinomagenesis. Using a customized shRNA library based on regions with DNA copy number alterations, they identified a number of tumor suppressor genes whose knockdown led to induction of liver cancer, including the nuclear export protein XPO4, which mediates the shuttling of EIF5A2 [110]. Similar approaches have been developed for other cancer types, such as breast cancer, with the promise of identifying genes whose loss crucially drives cancer formation [111].

siRNA screening to identify novel cancer drug targets and synthetic lethal interactions

Various RNAi-based screening methods have been employed to identify potential new drug targets and synthetic lethal interactions. Given that synthetic lethal interactions do not necessarily result from genes involved in similar pathways or cellular processes, an unbiased approach for gene discovery will be highly relevant [77]. In 2006, Bartz et al. used siRNA screens to identify genes whose knockdown would enhance the cytotoxic effects of cisplatin, gemcitabine, and paclitaxel. Hits were linked to the mechanism of action of the drugs, and knockdown of BRCA1 and BRCA2 was found to specifically sensitize p53-deficient HeLa cells to cisplatin [106]. Furthermore, in an siRNA screen published in 2009, Zhang et al. identified RRM1 and RRM2, factors of the human ribonucleotide reductase complex, which is required for the formation of dNTPs from NTPs, as targets of the checkpoint kinases and sensitizers to topoisomerase inhibitor camptothecin [112].

Since the initial description of synthetic lethality between PARP inhibitors and the BRCA1 and -2 genes, different screens have aimed to discover novel synthetic lethal interactions with PARP, discovering, for example, the cyclin-dependent kinase CDK5, which is required for proper execution of intra-S and G2/M checkpoints, as a potential synthetic lethal interactor of PARP [113]. Next to this HR-dependent synthetic lethal interaction, siRNA screens using mismatch-repair-deficient cells have laid open a synthetic lethal interaction with the PTEN-induced putative kinase1 (PINK1). Knockdown of PINK1 induced oxidative stress in cells deficient for the mismatch repair proteins MSH2, MLH1, and MSH6, leading to DNA damage and loss of cell viability [114]. Interestingly, this kinase had already been identified as a survival kinase in early siRNAs screens performed by MacKeigan et al. [108].

siRNA screens to classify toxic compounds

Genotoxic and, therefore, potentially mutagenic compounds are present in various sources in our environment and their classification and categorization remains challenging [115].

The current state-of-the-art method for classifying genotoxic compounds is a standard 3-battery test (consisting of a bacterial gene mutation assay, a mammalian mutation or chromosome damage assay, and an in vivo chromosome damage assay) that is often accompanied by tests for different DNA adducts and DNA repair efficiency [115,116]. To classify a compound as carcinogenic, it has to be shown to induce tumor formation within a 2-year exposure period in rats [117]. Despite a clear link between genotoxicity and carcinogenicity, not all compounds testing positive for genotoxic potential also induce cancer in animal models and vice versa, numerous established carcinogens are non-genotoxic [116,118]. Described non-genotoxic mechanisms of action for carcinogenic compounds include hormonal changes or epigenetic modifications, such as altered methylation or miRNA profiles [115]. Thus far, large-scale RNAi screening has not been dedicated to clarification of mechanisms of toxicity of compounds for which the mechanism of action is unresolved. siRNA screening data might provide a basis to increase the accuracy of toxicity screening, by identifying, in an unbiased fashion, genes that are functionally and critically involved in the molecular pathways towards toxicity. This may lead to the identification of mechanism-based biomarkers for distinct compound groups or specific target organ toxicities. For example, genes that are activated upon DNA damage and determine the outcome of the DDR: survival or apoptosis. Or genes that define the response to mitochondrial injury and thereby the bioenergetic collapse followed by necrotic cell death.

Outlook of siRNA screens

One limitation in DDR screens performed so far is that many are carried out either in model organisms [107] or in standardized and immortalized cell culture systems (such as HeLa cells) [85,100,119], whose DDR likely differs strongly from that of normal cells in the body or of cancer cells. Since various genes show synthetic lethality or synthetic viability depending on the cellular context, it is crucial to confirm identified hits in different cellular contexts. The use of in vivo screens as discussed might provide an interesting alternative, to study not only tumor formation but also therapy responses and synthetic lethal interactions.

Another way to enhance the knowledge gained from siRNA screens lies in the integration of the data obtained from functional genomics with high-throughput data obtained from other 'omics, such as transcriptomics, proteomics. In such a "systems biology" approach, the goal is to understand a biological system as a whole, be it a specific pathway, cell, or organism, usually by unbiased investigation of high-throughput data sets using sophisticated statistical analyses. Common methods in systems biology analysis are the identification of significantly enriched pathways within a specific data set, or the formation of networks based on the interactions within a data set [120,121]. Statistical and explanatory power might be strongly enhanced through integration of different 'omics techniques [122]. We have recently used this approach for the unraveling of the DDR in embryonic stem (ES) cells treated with the genotoxic drug cisplatin [123]. Integration of siRNA screening data with phosphoproteomics and transcriptomics led to identification of canonical pathways that were significantly enriched in all individual data sets. This allowed the formation of integrated signaling networks based on the "hits" derived from each data set within those pathways. Besides confirmation of DNA-damage- and cell-cycle-related processes, this has led to the identification of a novel mode of DNA-damage-induced Wnt/β-catenin signaling that attenuates p53-mediated apoptosis in ES cells [123].

References

[1] Hoeijmakers JH. DNA damage, aging, and cancer. N Engl J Med 2009;361:1475–85.
[2] Kastan MB, Bartek J. Cell-cycle checkpoints and cancer. Nature 2004;432:316–23.
[3] Jackson SP, Bartek J. The DNA-damage response in human biology and disease. Nature 2009;461:1071–8.
[4] Sherman MH, Bassing CH, Teitell MA. Regulation of cell differentiation by the DNA damage response. Trends Cell Biol 2011;21:312–9.
[5] Rouse J, Jackson SP. Interfaces between the detection, signaling, and repair of DNA damage. Science 2002;297:547–51.
[6] Harper JW, Elledge SJ. The DNA damage response: Ten years after. Mol Cell 2007;28:739–45.
[7] Shiloh Y. ATM and related protein kinases: Safeguarding genome integrity. Nat Rev Cancer 2003;3:155–68.
[8] Matsuoka S, Ballif BA, Smogorzewska A, McDonald III ER, Hurov KE, Luo J, et al. ATM and ATR substrate analysis reveals extensive protein networks responsive to DNA damage. Science 2007;316:1160–6.

[9] Cimprich KA, Cortez D. ATR: An essential regulator of genome integrity. Nat Rev Mol Cell Biol 2008;9:616–27.
[10] Zhou BB, Bartek J. Targeting the checkpoint kinases: Chemosensitization versus chemoprotection. Nat Rev Cancer 2004;4:216–25.
[11] Banin S, Moyal L, Shieh S, Taya Y, Anderson CW, Chessa L, et al. Enhanced phosphorylation of p53 by ATM in response to DNA damage. Science 1998;281:1674–7.
[12] Khosravi R, Maya R, Gottlieb T, Oren M, Shiloh Y, Shkedy D. Rapid ATM-dependent phosphorylation of MDM2 precedes p53 accumulation in response to DNA damage. Proc Natl Acad Sci USA 1999;96:14973–14977.
[13] Bensimon A, Aebersold R, Shiloh Y. Beyond ATM: The protein kinase landscape of the DNA damage response. FEBS Lett 2011;585:1625–39.
[14] Pines A, Kelstrup CD, Vrouwe MG, Puigvert JC, Typas D, Misovic B, et al. Global phosphoproteome profiling reveals unanticipated networks responsive to cisplatin treatment of embryonic stem cells. Mol Cell Biol 2011;31:4964–77.
[15] Reinhardt HC, Aslanian AS, Lees JA, Yaffe MB. p53-deficient cells rely on ATM- and ATR-mediated checkpoint signaling through the p38MAPK/MK2 pathway for survival after DNA damage. Cancer Cell 2007;11:175–89.
[16] Reinhardt HC, Yaffe MB. Kinases that control the cell cycle in response to DNA damage: Chk1, Chk2, and MK2. Curr Opin Cell Biol 2009;21:245–55.
[17] Vousden KH, Prives C. Blinded by the light: The growing complexity of p53. Cell 2009;137:413–31.
[18] Li M, He Y, Dubois W, Wu X, Shi J, Huang J. Distinct regulatory mechanisms and functions for p53-activated and p53-repressed DNA damage response genes in embryonic stem cells. Mol Cell 2012;46:30–42.
[19] Hermeking H. The miR-34 family in cancer and apoptosis. Cell Death Differ 2010;17:193–9.
[20] Michalak EM, Villunger A, Adams JM, Strasser A. In several cell types tumour suppressor p53 induces apoptosis largely via Puma but Noxa can contribute. Cell Death Differ 2008;15:1019–29.
[21] Moll UM, Wolff S, Speidel D, Deppert W. Transcription-independent pro-apoptotic functions of p53. Curr Opin Cell Biol 2005;17:631–6.
[22] Talos F, Petrenko O, Mena P, Moll UM. Mitochondrially targeted p53 has tumor suppressor activities in vivo. Cancer Res 2005;65:9971–81.
[23] Choudhury AR, Ju Z, Djojosubroto MW, Schienke A, Lechel A, Schaetzlein S, et al. Cdkn1a deletion improves stem cell function and lifespan of mice with dysfunctional telomeres without accelerating cancer formation. Nat Genet 2007;39:99–105.
[24] Crighton D, Wilkinson S, O'Prey J, Syed N, Smith P, Harrison PR, et al. DRAM, a p53-induced modulator of autophagy, is critical for apoptosis. Cell 2006;126:121–34.
[25] Budanov AV, Karin M. p53 target genes sestrin1 and sestrin2 connect genotoxic stress and mTOR signaling. Cell 2008;134:451–60.
[26] Bensaad K, Tsuruta A, Selak MA, Vidal MN, Nakano K, Bartrons R, et al. TIGAR, a p53-inducible regulator of glycolysis and apoptosis. Cell 2006;126:107–20.
[27] Jiang H, Reinhardt HC, Bartkova J, Tommiska J, Blomqvist C, Nevanlinna H, et al. The combined status of ATM and p53 link tumor development with therapeutic response. Genes Dev 2009;23:1895–909.
[28] Kruse JP, Gu W. Modes of p53 regulation. Cell 2009;137:609–22.
[29] de Bie P, Ciechanover A. Ubiquitination of E3 ligases: Self-regulation of the ubiquitin system via proteolytic and non-proteolytic mechanisms. Cell Death Differ 2011;18:1393–402.
[30] Wu X, Bayle JH, Olson D, Levine AJ. The p53-mdm-2 autoregulatory feedback loop. Genes Dev 1993;7:1126–32.
[31] Jones SN, Roe AE, Donehower LA, Bradley A. Rescue of embryonic lethality in Mdm2-deficient mice by absence of p53. Nature 1995;378:206–8.

[32] Wang X. p53 regulation: Teamwork between RING domains of Mdm2 and MdmX. Cell Cycle 2011;10:4225–9.
[33] Di Conza G, Mancini F, Buttarelli M, Pontecorvi A, Trimarchi F, Moretti F. MDM4 enhances p53 stability by promoting an active conformation of the protein upon DNA damage. Cell Cycle 2012;11:749–60.
[34] Brooks CL, Gu W. p53 ubiquitination: Mdm2 and beyond. Mol Cell 2006;21:307–15.
[35] Coutts AS, Adams CJ, La Thangue NB. p53 ubiquitination by Mdm2: A never ending tail? DNA Repair (Amst) 2009;8:483–90.
[36] Meulmeester E, Pereg Y, Shiloh Y, Jochemsen AG. ATM-mediated phosphorylations inhibit Mdmx/Mdm2 stabilization by HAUSP in favor of p53 activation. Cell Cycle 2005;4:1166–70.
[37] Yuan J, Luo K, Zhang L, Cheville JC, Lou Z. USP10 regulates p53 localization and stability by deubiquitinating p53. Cell 2010;140:384–96.
[38] Zhang X, Berger FG, Yang J, Lu X. USP4 inhibits p53 through deubiquitinating and stabilizing ARF-BP1. EMBO J 2011;30:2177–89.
[39] Batchelor E, Mock CS, Bhan I, Loewer A, Lahav G. Recurrent initiation: A mechanism for triggering p53 pulses in response to DNA damage. Mol Cell 2008;30:277–89.
[40] Batchelor E, Loewer A, Lahav G. The ups and downs of p53: Understanding protein dynamics in single cells. Nat Rev Cancer 2009;9:371–7.
[41] Brooks CL, Gu W. The impact of acetylation and deacetylation on the p53 pathway. Protein Cell 2011;2:456–62.
[42] Vollebergh MA, Jonkers J, Linn SC. Genomic instability in breast and ovarian cancers: Translation into clinical predictive biomarkers. Cell Mol Life Sci 2012;69:223–45.
[43] Kunkel TA, Erie DA. DNA mismatch repair. Annu Rev Biochem 2005;74:681–710.
[44] Fortini P, Dogliotti E. Base damage and single-strand break repair: Mechanisms and functional significance of short- and long-patch repair subpathways. DNA Repair (Amst) 2007;6:398–409.
[45] Fousteri M, Mullenders LH. Transcription-coupled nucleotide excision repair in mammalian cells: Molecular mechanisms and biological effects. Cell Res 2008;18:73–84.
[46] Diderich K, Alanazi M, Hoeijmakers JH. Premature aging and cancer in nucleotide-excision-repair disorders. DNA Repair (Amst) 2011;10:772–80.
[47] Alpi AF, Patel KJ. Monoubiquitylation in the Fanconi anemia DNA damage response pathway. DNA Repair (Amst) 2009;8:430–5.
[48] Jacquemont C, Taniguchi T. The Fanconi anemia pathway and ubiquitin. BMC Biochem 2007;8(Suppl 1):S10.
[49] Shibata A, Conrad S, Birraux J, Geuting V, Barton O, Ismail A, et al. Factors determining DNA double-strand break repair pathway choice in G2 phase. EMBO J 2011;30:1079–92.
[50] Li X, Heyer WD. Homologous recombination in DNA repair and DNA damage tolerance. Cell Res 2008;18:99–113.
[51] Sale JE, Lehmann AR, Woodgate R. Y-family DNA polymerases and their role in tolerance of cellular DNA damage. Nat Rev Mol Cell Biol 2012;13:141–52.
[52] Fuchs Y, Steller H. Programmed cell death in animal development and disease. Cell 2011;147:742–58.
[53] Cotter TG. Apoptosis and cancer: The genesis of a research field. Nat Rev Cancer 2009;9:501–7.
[54] Taylor RC, Cullen SP, Martin SJ. Apoptosis: Controlled demolition at the cellular level. Nat Rev Mol Cell Biol 2008;9:231–41.
[55] Green DR, Evan GI. A matter of life and death. Cancer Cell 2002;1:19–30.
[56] Vermeulen K, van Bockstaele DR, Berneman ZN. Apoptosis: Mechanisms and relevance in cancer. Ann Hematol 2005;84:627–39.
[57] d'Adda di Fagagna F. Living on a break: Cellular senescence as a DNA-damage response. Nat Rev Cancer 2008;8:512–22.

[58] Singh R, George J, Shukla Y. Role of senescence and mitotic catastrophe in cancer therapy. Cell Div 2010;5:4.
[59] Vakifahmetoglu H, Olsson M, Zhivotovsky B. Death through a tragedy: Mitotic catastrophe. Cell Death Differ 2008;15:1153–62.
[60] Medema RH, Macurek L. Checkpoint control and cancer. Oncogene 2012;31:2601–13.
[61] Kiyokawa H, Ray D. In vivo roles of CDC25 phosphatases: Biological insight into the anti-cancer therapeutic targets. Anticancer Agents Med Chem 2008;8:832–6.
[62] Bartek J, Lukas J. Chk1 and Chk2 kinases in checkpoint control and cancer. Cancer Cell 2003;3:421–9.
[63] Lord CJ, Ashworth A. The DNA damage response and cancer therapy. Nature 2012;481:287–94.
[64] Hanahan D, Weinberg RA. Hallmarks of cancer: The next generation. Cell 2011;144:646–74.
[65] Negrini S, Gorgoulis VG, Halazonetis TD. Genomic instability: An evolving hallmark of cancer. Nat Rev Mol Cell Biol 2010;11:220–8.
[66] Halazonetis TD, Gorgoulis VG, Bartek J. An oncogene-induced DNA damage model for cancer development. Science 2008;319:1352–5.
[67] Bouwman P, Jonkers J. The effects of deregulated DNA damage signalling on cancer chemotherapy response and resistance. Nat Rev Cancer 2012;12:587–98.
[68] Helleday T, Petermann E, Lundin C, Hodgson B, Sharma RA. DNA repair pathways as targets for cancer therapy. Nat Rev Cancer 2008;8:193–204.
[69] Al-Ejeh F, Kumar R, Wiegmans A, Lakhani SR, Brown MP, Khanna KK. Harnessing the complexity of DNA-damage response pathways to improve cancer treatment outcomes. Oncogene 2010;29:6085–98.
[70] Begg AC, Stewart FA, Vens C. Strategies to improve radiotherapy with targeted drugs. Nat Rev Cancer 2011;11:239–53.
[71] Bryant HE, Schultz N, Thomas HD, Parker KM, Flower D, Lopez E, et al. Specific killing of BRCA2-deficient tumours with inhibitors of poly(ADP-ribose) polymerase. Nature 2005;434:913–7.
[72] Farmer H, McCabe N, Lord CJ, Tutt AN, Johnson DA, Richardson TB, et al. Targeting the DNA repair defect in BRCA mutant cells as a therapeutic strategy. Nature 2005;434:917–21.
[73] McManus MT, Sharp PA. Gene silencing in mammals by small interfering RNAs. Nat Rev Genet 2002;3:737–47.
[74] Pothof J, Verkaik NS, van Ijcken W, Wiemer EA, Ta VT, van der Horst GT, et al. MicroRNA-mediated gene silencing modulates the UV-induced DNA-damage response. EMBO J 2009;28:2090–9.
[75] Alvarez-Garcia I, Miska EA. MicroRNA functions in animal development and human disease. Development 2005;132:4653–62.
[76] Bernards R, Brummelkamp TR, Beijersbergen RL. shRNA libraries and their use in cancer genetics. Nat Methods 2006;3:701–6.
[77] Iorns E, Lord CJ, Turner N, Ashworth A. Utilizing RNA interference to enhance cancer drug discovery. Nat Rev Drug Discov 2007;6:556–68.
[78] Lee HC, Chang SS, Choudhary S, Aalto AP, Maiti M, Bamford DH, et al. qiRNA is a new type of small interfering RNA induced by DNA damage. Nature 2009;459:274–7.
[79] Rausch O. High content cellular screening. Curr Opin Chem Biol 2006;10:316–20.
[80] Nagy A, Perrimon N, Sandmeyer S, Plasterk R. Tailoring the genome: The power of genetic approaches. Nat Genet 2003;33(Suppl):276–84.
[81] Lord CJ, Martin SA, Ashworth A. RNA interference screening demystified. J Clin Pathol 2009;62:195–200.
[82] Birmingham A, Selfors LM, Forster T, Wrobel D, Kennedy CJ, Shanks E, et al. Statistical methods for analysis of high-throughput RNA interference screens. Nat Methods 2009;6:569–75.
[83] Bettencourt-Dias M, Giet R, Sinka R, Mazumdar A, Lock WG, Balloux F, et al. Genome-wide survey of protein kinases required for cell cycle progression. Nature 2004;432:980–7.
[84] Kittler R, Pelletier L, Heninger AK, Slabicki M, Theis M, Miroslaw L, et al. Genome-scale RNAi profiling of cell division in human tissue culture cells. Nat Cell Biol 2007;9:1401–12.

[85] Neumann B, Walter T, Heriche JK, Bulkescher J, Erfle H, Conrad C, et al. Phenotypic profiling of the human genome by time-lapse microscopy reveals cell division genes. Nature 2010;464:721–7.

[86] Paulsen RD, Soni DV, Wollman R, Hahn AT, Yee MC, Guan A, et al. A genome-wide siRNA screen reveals diverse cellular processes and pathways that mediate genome stability. Mol Cell 2009;35:228–39.

[87] Birmingham A, Anderson EM, Reynolds A, Ilsley-Tyree D, Leake D, Fedorov Y, et al. 3' UTR seed matches, but not overall identity, are associated with RNAi off-targets. Nat Methods 2006;3:199–204.

[88] Adamson B, Smogorzewska A, Sigoillot FD, King RW, Elledge SJ. A genome-wide homologous recombination screen identifies the RNA-binding protein RBMX as a component of the DNA-damage response. Nat Cell Biol 2012;14:318–28.

[89] Hurov KE, Cotta-Ramusino C, Elledge SJ. A genetic screen identifies the triple T complex required for DNA damage signaling and ATM and ATR stability. Genes Dev 2010;24:1939–50.

[90] Lord CJ, McDonald S, Swift S, Turner NC, Ashworth A. A high-throughput RNA interference screen for DNA repair determinants of PARP inhibitor sensitivity. DNA Repair (Amst) 2008;7:2010–9.

[91] Ravi D, Wiles AM, Bhavani S, Ruan J, Leder P, Bishop AJ. A network of conserved damage survival pathways revealed by a genomic RNAi screen. PLoS Genet 2009;5:e1000527.

[92] Cotta-Ramusino C, McDonald III ER, Hurov K, Sowa ME, Harper JW, Elledge SJ. A DNA damage response screen identifies RHINO, a 9-1-1 and TopBP1 interacting protein required for ATR signaling. Science 2011;332:1313–7.

[93] Huen MS, Chen J. Assembly of checkpoint and repair machineries at DNA damage sites. Trends Biochem Sci 2010;35:101–8.

[94] Kolas NK, Chapman JR, Nakada S, Ylanko J, Chahwan R, Sweeney FD, et al. Orchestration of the DNA-damage response by the RNF8 ubiquitin ligase. Science 2007;318:1637–40.

[95] Stewart GS, Stankovic T, Byrd PJ, Wechsler T, Miller ES, Huissoon A, et al. RIDDLE immunodeficiency syndrome is linked to defects in 53BP1-mediated DNA damage signaling. Proc Natl Acad Sci USA 2007;104:16910–15.

[96] Brown CJ, Lain S, Verma CS, Fersht AR, Lane DP. Awakening guardian angels: Drugging the p53 pathway. Nat Rev Cancer 2009;9:862–73.

[97] Mullenders J, Fabius AW, Madiredjo M, Bernards R, Beijersbergen RL. A large-scale shRNA barcode screen identifies the circadian clock component ARNTL as putative regulator of the p53 tumor suppressor pathway. PLoS ONE 2009;4:e4798.

[98] Krastev DB, Slabicki M, Paszkowski-Rogacz M, Hubner NC, Junqueira M, Shevchenko A, et al. A systematic RNAi synthetic interaction screen reveals a link between p53 and snoRNP assembly. Nat Cell Biol 2011;13:809–18.

[99] O'Connell BC, Adamson B, Lydeard JR, Sowa ME, Ciccia A, Bredemeyer AL, et al. A genome-wide camptothecin sensitivity screen identifies a mammalian MMS22L-NFKBIL2 complex required for genomic stability. Mol Cell 2010;40:645–57.

[100] Piwko W, Olma MH, Held M, Bianco JN, Pedrioli PG, Hofmann K, et al. RNAi-based screening identifies the Mms22L-Nfkbil2 complex as a novel regulator of DNA replication in human cells. EMBO J 2010;29:4210–22.

[101] Moudry P, Lukas C, Macurek L, Neumann B, Heriche JK, Pepperkok R, et al. Nucleoporin NUP153 guards genome integrity by promoting nuclear import of 53BP1. Cell Death Differ 2012;19:798–807.

[102] Nijman SM, Huang TT, Dirac AM, Brummelkamp TR, Kerkhoven RM, D'Andrea AD, et al. The deubiquitinating enzyme USP1 regulates the Fanconi anemia pathway. Mol Cell 2005;17:331–9.

[103] Dompe N, Rivers CS, Li L, Cordes S, Schwickart M, Punnoose EA, et al. A whole-genome RNAi screen identifies an 8q22 gene cluster that inhibits death-receptor-mediated apoptosis. Proc Natl Acad Sci USA 2011;108:E943–51.

[104] Niemi NM, Lanning NJ, Klomp JA, Tait SW, Xu Y, Dykema KJ, et al. MK-STYX, a catalytically inactive phosphatase regulating mitochondrially dependent apoptosis. Mol Cell Biol 2011;31:1357–68.

[105] Hanson AJ, Wallace HA, Freeman TJ, Beauchamp RD, Lee LA, Lee E. XIAP monoubiquitylates Groucho/TLE to promote canonical Wnt signaling. Mol Cell 2012;45:619–28.
[106] Bartz J, Imakura M, Martin M, Palmieri A, Needham R, Guo J, et al. Small interfering RNA screens reveal enhanced cisplatin cytotoxicity in tumor cells having both BRCA network and TP53 disruptions. Mol Cell Biol 2006;26:9377–86.
[107] Kondo S, Perrimon N. A genome-wide RNAi screen identifies core components of the G(2)-M DNA damage checkpoint. Sci Signal 2011;4:rs1.
[108] MacKeigan JP, Murphy LO, Blenis J. Sensitized RNAi screen of human kinases and phosphatases identifies new regulators of apoptosis and chemoresistance. Nat Cell Biol 2005;7:591–600.
[109] Sims D, Mendes-Pereira AM, Frankum J, Burgess D, Cerone MA, Lombardelli C, et al. High-throughput RNA interference screening using pooled shRNA libraries and next-generation sequencing. Genome Biol 2011;12:R104.
[110] Zender L, Xue W, Zuber J, Semighini CP, Krasnitz A, Ma B, et al. An oncogenomics-based in vivo RNAi screen identifies tumor suppressors in liver cancer. Cell 2008;135:852–64.
[111] Iorns E, Ward TM, Dean S, Jegg A, Thomas D, Murugaesu N, et al. Whole-genome in vivo RNAi screening identifies the leukemia inhibitory factor receptor as a novel breast tumor suppressor. Breast Cancer Res Treat 2012;135:79–91.
[112] Zhang YW, Jones TL, Martin SE, Caplen NJ, Pommier Y. Implication of checkpoint kinase-dependent up-regulation of ribonucleotide reductase R2 in DNA damage response. J Biol Chem 2009;284:18085–18095.
[113] Turner NC, Lord CJ, Iorns E, Brough R, Swift S, Elliott R, et al. A synthetic lethal siRNA screen identifying genes mediating sensitivity to a PARP inhibitor. EMBO J 2008;27:1368–77.
[114] Martin SA, Hewish M, Sims D, Lord CJ, Ashworth A. Parallel high-throughput RNA interference screens identify PINK1 as a potential therapeutic target for the treatment of DNA mismatch-repair-deficient cancers. Cancer Res 2011;71:1836–48.
[115] Ellinger-Ziegelbauer H, Aubrecht J, Kleinjans JC, Ahr HJ. Application of toxicogenomics to study mechanisms of genotoxicity and carcinogenicity. Toxicol Lett 2009;186:36–44.
[116] Brambilla G, Mattioli F, Robbiano L, Martelli A. Genotoxicity and carcinogenicity testing of pharmaceuticals: Correlations between induction of DNA lesions and carcinogenic activity. Mutat Res 2010;705:20–39.
[117] Chhabra RS, Huff JE, Schwetz BS, Selkirk J. An overview of prechronic and chronic toxicity/carcinogenicity experimental study designs and criteria used by the National Toxicology Program. Environ Health Perspect 1990;86:313–21.
[118] Snyder RD, Green JW. A review of the genotoxicity of marketed pharmaceuticals. Mutat Res 2001;488:151–69.
[119] Raman M, Havens CG, Walter JC, Harper JW. A genome-wide screen identifies p97 as an essential regulator of DNA-damage-dependent CDT1 destruction. Mol Cell 2011;44:72–84.
[120] Chuang HY, Hofree M, Ideker T. A decade of systems biology. Annu Rev Cell Dev Biol 2010;26:721–44.
[121] Vidal M, Cusick ME, Barabasi AL. Interactome networks and human disease. Cell 2011;144:986–98.
[122] Cavill R, Kamburov A, Ellis JK, Athersuch TJ, Blagrove MS, Herwig R, et al. Consensus-phenotype integration of transcriptomic and metabolomic data implies a role for metabolism in the chemosensitivity of tumour cells. PLoS Comput Biol 2011;7:e1001113.
[123] Puigvert J, von Stechow L, Siddappa R, Pines A, Bahjat M, Haazen L, et al. Systems biology approach identifies the kinase Csnk1a1 as a regulator of the DNA damage response in embryonic stem cells. Sci Signal 2013;6.
[124] Boutros M, Bras LP, Huber W. Analysis of cell-based RNAi screens. Genome Biol 2006;7:R66.
[125] Shilo Y. The ATM-mediated DNA-damage response: taking shape. Trends Biochem Sci 2006;31:402–10.

SECTION

Immunotoxicity

3

CHAPTER

3.1 Immunotoxicity Testing: Implementation of Mechanistic Understanding, Key Pathways of Toxicological Concern, and Components of These Pathways

Erwin L. Roggen*, Emanuela Corsini†, Henk van Loveren and Robert Luebke‡**

**3Rs Management and Consultancy, Kongens Lyngby, Denmark, †Department of Pharmacological Sciences, University of Milan, Milano, Italy, **Center for Health Protection Research, National Institute of Public Health and the Environment, Bilthoven, The Netherlands, ‡Cardiopulmonary and Immunotoxicology Branch, US Environmental Protection Agency, Research Triangle Park, North Carolina, USA*

3.1.1 Introduction

Globally, growing societal and ethical concerns [1], European legislation (7th Amendment of the Cosmetics Directive, REACH, European Directive 2010/63/EU), and current research demands by industry have accelerated the development of new visions for toxicity testing (e.g. the Tox21 program; see *Toxicity Testing in the 21st Century*, [2]). The strategy of the Tox21 vision builds upon an in-depth understanding of the in vivo physiological and toxicological processes in humans as they relate to toxicological endpoints, identification of relevant key pathways and components of these pathways involved in the response to toxicant exposure, and establishing for each key event representative in vitro models.[1]

A forum of scientists, industry representatives, key opinion leaders, and developers and users of animal-free tests and testing strategies have called for immediate implementation of new technologies and paradigms, which are considered essential for transforming the vision into a toxicity-testing system to assess biologically significant perturbations in key pathways that are relevant for human biology. In this context, the implementation of so-called 'omics technologies was considered a

[1] This report has been reviewed by the National Health and Environmental Effects Research Laboratory, US Environmental Protection Agency, and approved for publication. Approval does not signify that the contents necessarily reflect the views and policies of the Agency nor does mention of trade names or commercial products constitute endorsement or recommendation for use.

Toxicogenomics-Based Cellular Models.

prerequisite for the acquisition of data that provide insight into the pathways and key events of toxicity within organisms and biological systems, and thus can feed into systems biology techniques [3].

Toxicogenomics provides a library of generic expression profiles for different classes of toxicity that allows the characterization of an unknown compound based upon the profiles with which it fits. Genomics is currently used on a large scale for pathway analysis and marker identification, and its ability to reduce the need for animal studies is already being examined. However, the concept has not yet been fully implemented in toxicity testing strategies or accepted by the risk assessment community as suitable for setting exposure limits.

Carcinogenicity testing is in this respect an interesting case study. The use of toxicogenomics for identifying the mechanisms of action of genotoxic and non-genotoxic carcinogens has resulted in training sets for carcinogens and non-hepatotoxic non-carcinogens [15,23]. In addition, the US Food and Drug Administration (FDA) invites the submission of genomics data for control compounds for interpretation by toxicogenomics experts. The knowledge and experience gained with carcinogen training sets and control compounds can be applied to other toxicological endpoints of concern, including immunotoxicity.

Immunotoxic compounds are capable of initiating, facilitating, or exacerbating adverse immune processes. Toxicity to the immune system includes two main adverse effects. *Immunosuppression* is defined as a decreased capacity to neutralize external organisms, which may result in repeated, more severe, or prolonged infections, as well as an increased susceptibility to cancer development. *Immunostimulation* is defined by an exaggerated expression of the immune response, which, as an adverse effect, may lead to immune-mediated diseases such as hypersensitivity reactions or autoimmune diseases. Being an important aspect of the safety evaluation of drugs and chemicals, the assessment of immunotoxic potency has become an integral part of chemical and drug development. The currently applied International Conference on Harmonization (ICH) safety guidance number 8 (S8) advocates a weight-of-evidence approach, meaning that specific immunotoxicity testing should be conducted on compounds of concern that were identified in routine toxicology studies, rather than by routine screening. The preclinical assessment of immunotoxicity is, at present, restricted to animal models and assays that predict immunosuppression and contact hypersensitivity [4]. On the other hand, the US Environmental Protection Agency currently requires testing of pesticides for the potential to cause immunosuppression (OPPTS 870.7800 Immunotoxicity) and allergic contact hypersensitivity (OPPTS 870.2600 Skin Sensitization).

3.1.2 Animal-free assays to detect immunotoxicological endpoints

A review of the state of the art in the field of in vitro immunotoxicity has revealed a limited availability of mechanism-based cellular assays for predicting the toxicity of xenobiotics towards the immune system. There are indications that genomics techniques can be used to detect immunosuppression. Significant progress in the development of physiologically relevant in vitro alternatives has been made in the area of skin and respiratory sensitization, where novel mechanistic understanding, key pathways, and components of these pathways involved in responses to toxin exposure were implemented in test development and hazard prediction of xenobiotics and proteins [4,5].

Non-animal test methods for the identification of immunosuppressive chemicals

Activation, oxidative stress, and apoptosis were detected in a human a T-cell line (Jurkat) exposed to the immunosuppressive chemical tributyltin oxide using transcriptomics. The observed effects are similar to those observed in vivo [6]. Furthermore, changes in gene expression in cord blood samples from newborns were correlated with maternal exposure to dioxin and PCBs, and with reduced responses to measles vaccine at 3 years of age.

Non-animal test methods for the identification of chemicals with the potential to induce skin sensitization

An understanding of the biological processes involved in the induction of skin sensitization has led to the development of test methods aligned to the key mechanisms of skin sensitization including bioavailability, haptenation, epidermal inflammation, dendritic cell activation and migration, and T-cell proliferation [7]. With the exception of bioavailability, non-animal tests covering each of these key mechanisms have been developed. Some of these test methods are rather advanced in terms of standardization and/or number of chemicals tested and some of them are currently being evaluated in a formal pre-validation exercise to determine their robustness and reproducibility. An overview of the performance of such test methods, as reported in the literature, is illustrated in Table 3.1.1, together with some indications of whether a particular method may contribute to generate sensitizer potency information.

Cell-based tests addressing epidermal inflammation

An understanding of the mechanisms behind chemical-induced epidermal inflammation resulted in the development of two different approaches [8]. One approach assesses the impact of chemicals on the oxidative stress response pathways represented by the Nrf2-Keap1-ARE pathway [9,10]. The other approach measures the increase in intracellular interleukin 18 (IL-18) levels in primary keratinocytes or keratinocyte cell lines exposed to chemicals [11]. When tested on the same set of chemicals (n = 35), Nrf2 induction correctly identified 87% of the compounds. IL-18 detection identified 97% of the compounds correctly, including pre- and pro-haptens. The higher accuracy may be explained by the fact that IL-18 induction occurs downstream from the Nrf2-Keap1 pathway and better reflects other mechanisms involved in epidermal inflammation, e.g. Toll-like-receptor- and inflammasome-mediated processes [8].

The reconstituted human epidermis (RHE) assay evaluates the irritancy strength of a compound by determining the concentration that reduces the viability of a human epidermal equivalent culture by 50% as well as by IL-1α secretion. This test does not distinguish between irritants and sensitizers. However, when applied to compounds identified as sensitizers, the test allows classification of skin sensitizers with 92% accuracy according to their sensitizing potency as determined by the local lymph node assay (LLNA). It can be performed with different commercially available RHEs [26].

Dendritic cell activation

Dendritic cell (DC) activation was originally assessed on the basis of up-regulation of well-known markers, such as CD86, CD80, and CD54, on the surface of primary dendritic cells of various origins as well as on the surface of dendritic cell lines [12–14]. An alternative approach employed differential

Table 3.1.1 Non-Animal Test Methods for the Identification of Chemicals with the Potential to Induce Skin Sensitization

Test Method	Mechanism	Compounds Tested (n)	Predictive Performance (%) (Against In Vivo Data)	Comments	References
Direct peptide reactivity assay (DPRA)	Peptide reactivity	81 underwent ring-trials, undergoing ECVAM pre-validation with 24 coded chemicals	Sensitivity: 88 Specificity: 90 Accuracy: 89	Yes/no answer Classifies chemicals into reactivity categories	[10]
KeratinoSens	Epidermal inflammatory responses	67 ring-trial with 28 substances (21 coded substances)	Sensitivity: 86.4 Specificity: 82.6 Accuracy: 85.1	Yes/no answer	Emter et al., 2010; [10]
SenCeeTox + glutathione depletion	Epidermal inflammatory responses + peptidereactivity	58	Sensitivity: 81 Specificity: 92 Accuracy: 84	Yes/no answer Classifies chemicals into potency categories	[9]
Human cell line activation test (h-CLAT)	DC activation	100 ring-trials, undergoing ECVAM pre-validation with 24 coded chemicals	Sensitivity: 88 Specificity: 75 Accuracy: 84		[12]
MUSST	DC activation	Undergoing ECVAM pre-validation with 24 coded chemicals	Information available but not yet published.	Yes/no answer	
THP-1 cell-surface thiols chances	DC activation	52	Not calculated but high correlation with reported in vivo data	Yes/no answer	Suzuki et al., 2009
VitoSens	DC activation	21	Sensitivity: 82 Specificity: 97 Accuracy: 89	Yes/no answer 15 chemicals used to assess the test method capability for potency characterization	Hooyberghs et al., 2008; Lambrechts et al., 2010

DC, dendritic cell; ECVAM, European Center for Validation of Alternative Methods.

genomic analysis to identify 200 genes found to be affected by skin sensitizers in the dendritic MUTZ-3 cell line. This gene signature was subsequently reduced to 10 genes without significant loss of accuracy. When tested on the same set of chemicals (n = 80), tests based on surface markers correctly identified 80% of the compounds, while the gene signature was correct in 95% of the cases. Overall, the gene signature was found to be less sensitive to (1) irritants in general, and strong irritants in particular, and (2) the potency of the compounds to be tested [15].

Dendritic cell migration

The DC migration assay was developed based on acquired knowledge and understanding of the impact of fibroblasts on the fate of Langerhans cells after exposure to chemicals. The resulting test uses the Langerhans cell (LC)-like maturation form of MUTZ-3 cells in a two-chamber culture system [16]. The LCs are fluorescently labeled and deposited in the upper chamber. Upon treatment with sensitizers (both skin and respiratory) or irritants, the cells migrate towards the recombinant trophic chemokines CXCL12 or CCL5, respectively, and are quantified by fluorescence in the lower chamber. To date, no misclassifications have been observed with this test (n = 12 chemicals).

T-cell priming

The T-cell priming assay [17,18] addresses the final and specificity-determining steps of allergic sensitization, i.e. the hapten-specific activation of T cells. Development of standard operation procedures (SOP) for the preparation of effector, memory, and regulatory-depleted human T cells; of shortcuts for the generation of autologous dendritic cells; and the application of haptens either free or coupled to human serum albumin are quite advanced. As a readout, the determination of interferon gamma (IFN-γ) and tumor necrosis factor alpha (TNF-α) appears superior to T-cell proliferation.

Non-animal test methods for the identification of chemicals with the potential to induce respiratory sensitization

Human precision-cut lung slices (PCLS) represent the material of choice to correlate ex vivo chemical toxicity with in vitro data. This technology revealed an interdependence of irritation and inflammation for sensitizing processes very similar to what has been demonstrated for the skin [8,19].

Cell-based tests addressing inflammation in the lung mucosa

The PCLS method is able to verify in vitro-detected induction of mediators and biomarkers under ex vivo conditions in human lung tissue. Moreover, the system allows assessment of the significance of markers detected in proteomic or genomic analyses of other exposed cell-based test models [19]. In the context of a German government-approved (Federal Ministry of Education and Research, BMBF) project, the method is presently being pre-validated in three different laboratories.

An alveolar–endothelial cell line model, developed at the University of Mainz, Germany, was implemented by Sens-it-iv. Evaluation of this model with chemical sensitizers and proteins revealed it to be potentially useful for the identification of respiratory sensitizers [20].

An in-house model employing primary human bronchial epidermal cells was developed by Epithelix (MucilAir™). The characteristics of the bronchial cells in the MucilAir model are similar to those found in primary cells, but they exhibit a shelf life of up to 1 year, and allow studying both immediate (up to 72 hrs) and delayed (several weeks after exposure) responses. The impact of a compound on the transepithelial electrical resistance (TEER), and cilia-beating frequency and recovery were found to be good indicators for the sensitization potency of the tested compounds [21].

Dendritic cell activation

As for the skin, up-regulation of CD86, CD80, or CD54 can be used for assessing the sensitizing capacity of respiratory sensitizers. The genomic approach has revealed that respiratory sensitizers trigger different genes (n = 300) and pathways as compared to skin sensitizers. On a set of well-characterized chemical respiratory sensitizers (n = 10), 8 were correctly identified, while none of the irritants and control compounds (n = 9) were positive (authors' unpublished data).

3.1.3 Toxicogenomics approaches to predicting chemical safety

With respect to sensitization, proteomic and genomic studies, designed to identify molecular initiators, adverse outcome pathways, and new opportunities for test development, that differentiate sensitizers from irritants and non-sensitizers have become highly successful during recent years. Proteome analyses and gene-chip-based analyses of primary human keratinocytes as well as of MUTZ-3 dendritic cells revealed large numbers of clearly sensitizer-specific markers. Markers identified by proteomics techniques were analyzed in both systems according to their assignment to defined cellular signaling pathways. The most prominent of these pathways for keratinocytes and MUTZ-3 cells revealed several interesting overlaps. In particular, markers relating to the Nrf2-mediated oxidative response, and oxidative stress in general, were identified repeatedly, very much in line with the prominent position that oxidative stress holds in chemical sensitization. Gene profiling also showed a sensitizer-specific response of the Toll-like receptor signaling pathways. Preliminary data suggest a higher specificity for prediction on the basis of Toll-like receptor reactivity as compared to Nrf2 induction [15,22,23].

Genomics approaches have identified 200 genes that in MUTZ-3 cells are affected by skin sensitizers, but not by irritants or control chemicals. From these 200 genes, a signature profile was identified allowing for the identification of skin sensitizers. This profile contains genes representing six pathways known to be involved in cellular responses to xenobiotic exposure. It revealed a 95% accuracy (one false negative) when blind tested on 69 chemicals [15].

A gene expression signature was recently identified for the detection of chemical respiratory sensitizers. Preliminary studies indicate a good concordance (8/10) with in vivo data when using a gene profile containing 300 genes (n = 10 chemicals). With respect to proteins, a preliminary gene profile (300 genes) was obtained with the potential to discriminate environmental and food allergens (n = 19) from proteins without any human evidence for causing allergy (n = 3) (authors' unpublished data).

3.1.4 Gaps and hurdles on the way to risk assessment and human safety

Based on a major analysis of the status of alternative methods [7] and its independent review [24], a roadmap was proposed for overcoming the acknowledged scientific gaps for the full replacement of systemic toxicity testing using animals [25]. For skin sensitization, seven recommendations were formulated. One recommendation addressed the importance of an expanded list of biomarkers or chemical signatures that are quantitatively associated with the acquisition of skin sensitization. The importance of better understanding the mechanisms defining sensitizing potency was considered of high priority. Finally, investment in developing a more detailed understanding of the cellular and molecular events that initiate, orchestrate, and control immune responses to skin-sensitizing chemicals should be encouraged, to foster development of definitive assays and improve our understanding of the biology that leads to skin sensitization.

3.1.5 An applied systems toxicology approach to predicting chemical safety

One initiative that addresses some of these recommendations is a project funded by The Netherlands Genomics Initiative (NGI) aimed at designing mechanism-based alternatives to animal tests for predicting skin sensitization. The human keratinocyte cell line HaCat is exposed to a set of known chemical sensitizers; or alternatively irritants; or specific agents that may affect certain pathways in the keratinocytes, yet are not sensitizers. Chemicals known to be false positives or false negatives in the animal test currently used for prediction of sensitizers, i.e. the LLNA, are also being tested. Gene expression profiling has been performed, and based on this a gene set that appears predictive has been selected, and is being validated using quantitative polymerase chain reaction (qPCR). Validation of the gene set will also be performed by applying gene silencing. An additional approach is the investigation of proteins, including phosphoproteins, that may be differentially expressed in keratinocytes exposed to sensitizers or control agents.

Whereas such approaches may be useful for identification of sensitizers and discrimination of sensitizers from irritants, they may be less relevant for assessing the potency of sensitizing activity, as in vitro exposure of keratinocytes may reflect skin exposure less well. Relative potency may be better determined by assessing dose–response relationships of the chemical-induced expression of relevant genes in a three-dimensional model of skin.

The Netherlands Toxicogenomics Center (NTC) project also aims to develop a tiered in vitro approach to detect direct toxicity of chemicals, pharmaceuticals, and biologicals to the immune system The approach taken is primarily based on human cell systems to avoid interspecies extrapolation. A battery of in vitro assays is employed to predict different aspects of the immune system in vivo. The first tier will generate toxicokinetics data for each test compound. These toxicokinetics data will be used as input for the immunotoxicity high-throughput screens (HTS) in a second tier. This second tier consists of HTS for the in vitro identification of direct immunotoxicity. When the compound is known to undergo biotransformation, it would be necessary to perform co-cultures with either microsomal S9 fractions, or with metabolically competent cells, such as cytochrome P450 transgenic HepG2 cells. The second tier addresses the basic endpoints, including cytotoxicity, viability, and proliferation, as well as specific effect markers for direct immunotoxicity, i.e. myelotoxic and lymphotoxic potential, for which primary cells, such as peripheral blood mononuclear cell (PBMC)-derived lymphocytes, could be used, but for which Jurkat T cells, RPMI-1788 B-cells, or KG-1 myeloblasts, respectively, may also be useful. Chemicals expressing cytotoxic and/or antiproliferative potentials will be further assessed by a set of 10–40 functional biomarkers of direct immunotoxicity that was developed by NTC as a predictive in vitro assay. The assay, known as the Messenger RNA Signature for Direct Immunotoxicity in High-Throughput Screening format (MSDI-HTS; patent pending; Volger et al., 2012) is undergoing pre-validation studies to determine whether it is sufficiently accurate to replace animal-based immunotoxicity testing. This proposed tiered testing approach requires further research in order to assess its validity and practicability, and may need further refinement before it can adequately be put in place for animal-free immunotoxicity testing.

References

[1] Zurlo J, Rudacille D, Golderg AM. The three Rs: the way forward. Environ Health Perspect 1996; 104:878–80.

[2] National Research Council. Toxicity testing in the 21st century: a vision and a strategy. Washington DC: National Academies Press; 2007.

[3] Berg N, De Wever B, Fuchs HW, Gaca M, Krul C, Roggen EL. Toxicology in the 21st century: working our way towards a visionary reality. Toxicol In Vitro 2011;25:874–81.

[4] Corsini E, Roggen EL. Immunotoxicology: opportunities for non-animal test development. ATLA 2009;37:387–97.

[5] Roggen EL. Application of the acquired knowledge and implementation of the Sens-it-iv toolbox for identification and classification of skin and respiratory sensitizers. Toxicol In Vitro 2012 doi:10.1016/j.tiv.2012.09.019.

[6] Katika MR, Hendriksen PJ, van Loveren H, Peijnenburg A. Exposure of Jurkatt cells to bis(tri-n-butyltin) oxide (TBTO) induces transcriptomic changes for ER- and oxidative stress, T cell activation and apoptosis. Toxicol Appl Pharmacol 2011;254:311–22.

[7] Adler S, Basketter D, Creton S, Pelkonen O, van Benthem J, Zuang V, et al. Alternative (non-animal) methods for cosmetics testing: current status and future prospects—2010. Arch Toxicol 2011;85:367–485.

[8] Martin SF, Esser PR. Innate immune mechanisms in contact dermatitis and the resulting T cell responses. ALTEX 2010;27:293–5.

[9] McKim Jr JM, Keller 3rd DJ, Gorski JR. A new in vitro method for identifying chemical sensitizers combining peptide binding with ARE/EpRE-mediated gene expression in human skin cells. Cutan Ocul Toxicol 2010;29:171–92.

[10] Natsch A, Bauch C, Foertsch L, Gerberick F, Norman K, Hilberer A, et al. The intra- and interlaboratory reproducibility and predictivity of the KeratinoSens assay to predict skin sensitizers in vitro: results of a ring-study in five laboratories. Toxicol In Vitro 2010;25:733–44.

[11] Corsini E, Mitjans M, Galbiati V, Lucchi L, Gall,i CL, Marinovich M. Use of IL-18 production in a human keratinocyte cell line to discriminate contact sensitizers from irritants and low molecular weight respiratory allergens. Toxicol In Vitro 2009;23:789–96.

[12] Ashikaga T, Sakaguchi H, Nukada Y, Kosaka N, Sono S, Nishiyama N, et al. Database of h-CLAT (cell-based skin sensitization test) for clarification of applicability domain. Abstracts of the 45th Congress of the European Societies of Toxicology. Toxicol Lett 2008;180(Suppl. 1):S95.

[13] dos Santos GG, Reinders J, Ouwehand K, Rustemeyer T, Scheper RJ, Gibbs S. Progress on the development of human in vitro dendritic-cell-based assays for assessment of the sensitizing potential of a compound. Toxicol Appl Pharmacol 2009;236:372–82.

[14] Sakaguchi H, Ashikaga T, Miyazawa M, Kosaka N, Ito Y, Yoneyama K, et al. The relationship between CD86/CD54 expression and THP-1 cell viability in an in vitro skin sensitization test: human cell line activation test (h-CLAT). Cell Biol Toxicol 2009;25:109–26.

[15] Johansson H, Lindstedt M, Albrekt A-S, Borrebaeck CAK. A genomic biomarker signature can predict skin sensitizers using a cell-based in vitro alternative to animal tests. BMC Genomics 2011;12:399.

[16] Ouwehand K, Scheper RJ, de Gruijl TD, Gibbs S. Epidermis-to-dermis migration of immature Langerhans cells upon topical irritant exposure is dependent on CCL2 and CCL5. Eur J Immunol 2010;40(7):2026–34.

[17] Dietz L, Esser PR, Schmucker SS, Goette I, Richter A, Schnölzer M, et al. Tracking human contact allergens: from mass spectrometric identification of peptide-bound reactive small chemicals to chemical-specific naive human T-cell priming. Toxicol Sci 2010;117:336–47.

[18] Martin SF, Esser PR, Schmucker S, Dietz L, Naisbitt DJ, Park BK, et al. T cell recognition of chemicals, protein allergens and drugs: towards the development of in vitro assays. Cell Mol Life Sci 2010;67:4171–84.

[19] Switalla S, Lauenstein L, Prenzler F, Knothe S, Förster C, Fieguth HG, et al. Natural innate cytokine response to immunomodulators and adjuvants in human precision-cut lung slices. Toxicol Appl Pharmacol 2010;246:107–15.

[20] Kasper J, Hermanns MI, Bantz C, Maskos M, Stauber R, Pohl C, et al. Inflammatory and cytotoxic responses of an alveolar-capillary coculture model to silica nanoparticles: comparison with conventional monocultures. Part Fibre Toxicol 2011;8:6–21.

[21] Huang S, Wisznieski L, Constant S, Roggen. EL. Potential of in vitro reconstituted 3D human airway epithelia (MucilAirTM) to assess respiratory sensitizers. Toxicol In Vitro 2012 doi:10.1016/j.tiv.2012.10.010.

[22] Dietz L, Kinzebach S, Ohnesorge S, Franke B, Goette I, Koenig-Gressel D, et al. Proteomic allergen-peptide/protein interaction assay for the identification of human skin sensitizers. Toxicol In Vitro 2012;27:1157–62.

[23] van der Veen JW, Pronk TE, van Loveren H, Ezendam J. Applicability of a keratinocyte gene signature to predict skin sensitizing potential. Toxicol In Vitro 2013;27:314–22.

[24] Hartung T, Blaauboer BJ, Bosgra S, Carney E, Coenen J, Conolly RB, et al. An expert consortium review of the EC-commissioned report "Alternative (non-animal) methods for cosmetics testing: current status and future prospects—2010". ALTEX 2011;28(3):183–209.

[25] Basketter DA, Clewell H, Kimber I, Rossi A, Blaauboer B, Burrier R, et al. A roadmap for the development of alternative (non-animal) methods for systemic toxicity testing. ALTEX 2012;29:1–91.

[26] dos Santos GG, Spiekstra SW, Sampat-Sardjoepersad SC, Reinders J, Scheper RJ, Gibbs S. A potential in vitro epidermal equivalent assay to determine sensitizer potency. Toxicol In Vitro 2011;25:347–57.

[27] Hochstenbach K, van Leeuwen DM, Gmuender H, Gottschalk RW, Stølevik SB, Nygaard UC, et al. Toxicogenomic profiles in relation to maternal immunotoxic exposure and immune functionality in newborns. Toxicol Sci 2012;129:315–24.

[28] US EPA NCCT (National Center for Computational Toxicology), <http://www.epa.gov/ncct/toxcast/>.

CHAPTER

3.2

Chemical Sensitization

Marjam Alloul-Ramdhani, Cornelis P. Tensen and Abdoelwaheb El Ghalbzouri

Department of Dermatology, Leiden University Medical Center, Leiden, The Netherlands

3.2.1 Introduction

Allergic contact dermatitis (ACD) is one of the most common occupational skin diseases, with a great socioeconomic impact [1,2]. The disease is a T-cell mediated skin inflammation that is caused by repeated skin exposure to allergens. These contact allergens are non-protein molecules called haptens. As the human skin is the outermost barrier of our body, it is the first organ to encounter external factors such as chemicals. These chemicals (e.g. nickel, chrome) are able to activate antigen-specific acquired immunity, resulting in the development of effector T cells that mediate skin inflammation. A patient is diagnosed with contact dermatitis when the person has a rash that typically itches. The rash classically has small blisters containing clear fluid, but can swell, crust, leak, or peel in other cases [3]. Systemic administration of hapten to sensitized patients may possibly result in systemic ACD [4,5].

There are a limited number of strong contact sensitizers and thousands of weak haptens responsible for human ACD [6]. The contact allergens are low molecular weight chemicals (haptens), that are not antigenic by themselves and need to bind to proteins present in the skin. These bound proteins act as carriers to form the so-called hapten-carrier complex that eventually acts as the antigen [6,7]. Depending on how these allergens are "transformed," either via their environmental condition (e.g. oxidation) or via enzymes present in the skin, they are converted into highly reactive metabolites. These molecules are referred to as pre-haptens and pro-haptens, respectively [8,9].

ACD is a typical hypersensitivity response that develops in two phases, (1) the sensitization phase and (2) the elicitation phase. In the sensitization phase, the haptens penetrate through the stratum corneum of the skin, where the hapten will be loaded by dendritic cells (DCs) [10–12]. At the same time, they will induce an inflammatory reaction with the release of inflammatory mediators (e.g. interleukin 18 (IL-18), IL-1β, tumor necrosis factor alpha (TNF-α), reactive oxygen species (ROS)), the DCs are activated and will migrate to the draining lymph nodes. In the draining lymph nodes they can present haptenated peptides on major histocompatibility complex (MHC) class I and II molecules to CD8+ and CD4+ T cells, respectively [13]. These specific T-cell precursors expand clonally in the draining lymph nodes and diffuse to the bloodstream, thereby generating skin-homing CD8+ Tc1/Tc17 and CD4+ Th1/Th17 effector T cells [13,14]. During this process, they acquire skin-specific homing antigens (cutaneous leukocyte-associated receptor (CLA) and C–C chemokine receptor 4 (CCR4)) and become memory

Toxicogenomics-Based Cellular Models.

T cells. The primed T cells diffuse into the skin after migration, where they are ready for the development of a contact hypersensitivity (CS) reaction upon exposure with the relevant hapten. In the elicitation phase, the hapten is marked for a second time and diffuses through the epidermis where it can be loaded by skin cells that express MHC molecules, such as Langerhans cells, keratinocytes, and dermal dendritic cells. Upon activation, CD8+ cytotoxic T cells will initiate the inflammatory process through keratinocyte apoptosis and production of cytokines/chemokines, such as interferon gamma (IFN-γ), IL-12, IL-17, and IL-23, which will eventually lead to the development of skin lesions [15,16].

In order to guarantee the safety of compounds used in the chemical and cosmetic industries, it is essential to identify and predict the sensitizing capacity of these ingredients. Currently, the murine local lymph node assay (LLNA) is the only in vivo validated assay that can predict the skin sensitization potential of chemicals [17,18]. This assay was approved according to the Organisation for Economic Co-operation and Development (OECD) guidelines (OECD TG 429). The LLNA is an alternative approach to the traditional guinea pig method (guinea pig maximization test or Buehler test) [19]. In addition, the LLNA has also proven to be very useful in assessing the skin sensitizing potency of test chemicals, and this has provided invaluable information to risk assessors [20].

Nowadays, there is increasing interest in reducing and ultimately replacing current animal tests, including skin sensitization [21]. Therefore, as of 2013, the European Union (EU) Cosmetics Directive has passed a ban on animal testing of cosmetic ingredients. Within the Registration, Evaluation, Authorisation and Restriction of Chemical substances (REACH) program, which places greater responsibility on industry, thousands of chemical substances will be tested [22]. This has triggered the development of predictive and robust in vitro test alternatives [21]. These in vitro tests must mimic the very complex interactions between the chemical and the different parts of the skin. For this purpose, different human cell lines (e.g. HaCat, MUTZ-3, THP-1, NCTC2544) have been used to identify genome- or transcriptome-level biomarker signatures that can predict skin sensitization [14,23–26]. A number of in vitro and in vivo studies have shown that the most prominent pathway that may be activated by chemical sensitizers is the oxidative-stress-related Keap1 (Kelch-like ECH-associated protein 1)/Nrf2 (nuclear factor E2 p45-related factor 2) pathway [27]. Chemical sensitizers, like 1-chloro 2,4-dinitrobenzene (DNCB), have been shown to up-regulate the antioxidant responsive element (ARE) genes (e.g. *HO-1*, *NQO1*) [28–30]. This signaling pathway plays a significant role in protecting cells from endogenous and exogenous stresses.

Although conventional mono-cell cultures are an excellent in vitro tool to screen for skin sensitizers, they still lack the three-dimensional (3D) microenvironment of human skin. One can imagine that that a chemical will trigger a different mode of action when the skin barrier, which consists of lipid layers, ceramides, and cholesterol, is absent. Therefore, it is important to understand the underlying mechanisms of skin sensitization. For this purpose, reconstructed human skin equivalents (HSEs), may provide a promising tool to search and identify new biomarkers that can not only discriminate between skin sensitizers and skin irritants, but also predict the potency of chemical sensitizers.

3.2.2 Three-dimensional human skin equivalent as a tool for safety testing purposes

In the last few years, the use of 3D cultures of human keratinocytes on a "dermal substitute" has gained increasing interest as a tool to conduct fundamental research. The reason for this is that the quality of these skin cultures has significantly improved, which makes them also suitable for skin toxicity testing. Next to the field of toxicity and safety testing, HSEs are an excellent tool to study different basic aspects of skin biology, such as molecular mechanisms dealing with keratinocyte differentiation, proliferation, and migration; wound healing; pathogenesis of skin diseases; and skin cancer.

Mimicking native human skin

HSEs, also designated "cultured skin substitute" or "organotypic co-cultures," are 3D systems that are engineered by seeding fibroblasts into or onto a 3D dermal matrix (e.g. inert filters, de-epidermized dermis, collagen matrices, lyophilized collagen-GAG membranes, and fibroblast-derived matrices) [31–34]. Such a dermal equivalent is subsequently seeded with human keratinocytes. After cell attachment, the culture is kept first under submerged conditions to allow keratinocyte proliferation. Thereafter, the culture is lifted to the air–liquid interface to expose the epidermal compartment to the air, and to further induce keratinocyte differentiation [35–39]. During the air exposure, nutrients from the medium will diffuse through the underlying dermal substrate towards the epidermal compartment and support keratinocyte proliferation, differentiation, and migration. These HSEs have the advantage that they harbor different cell types (e.g. melanocytes, Langerhans cells, endothelial cells) that are surrounded with a local environment that is highly similar to that of the in vivo tissue, and in which cellular processes may be normalized compared to conventional monolayer cultures. Under these conditions, a skin equivalent is formed that recapitulates most of the in vivo characteristics of the skin (Figure 3.2.1) [36,37]. The generated epidermis is composed of

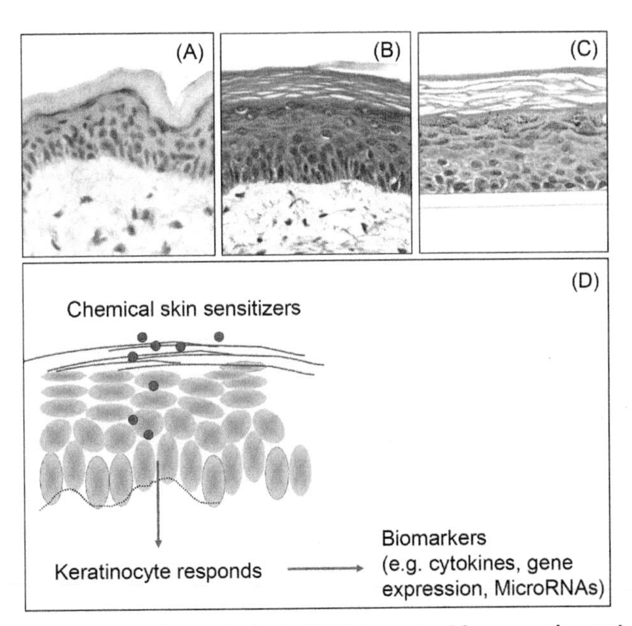

FIGURE 3.2.1 Reconstructed human skin equivalents (HSEs) as a tool for screening and research purposes.

Shown are cross-sections of (A) native skin; (B) HSEs that were generated on a dermal matrix and that are mostly used for research purposes; (C) an epidermal skin equivalent generated on an inert filter membrane, which is used as a tool in safety testing, such as skin irritation and skin corrosion tests. (D) Similar epidermal models are used in toxicogenomics for the identification of biomarkers (e.g. cytokines, gene expression profiles) for chemical skin sensitization.

an organized stratum basale, stratum spinosum, stratum granulosum, and a stratum corneum. This multilayered epithelium displays characteristic epidermal ultrastructures and expresses markers of epidermal differentiation [36,37,39]. In addition, the epidermis shows features of a functional permeability barrier, which is one of the main functions of viable skin [39–41]. Therefore, these HSEs are an attractive tool to study cell–cell, cell–matrix, dermal–epidermal interactions and other processes that are involved in epidermal morphogenesis. Since the HSEs are easy to handle and modulate, they are also an excellent tool to mimic skin disorders in vitro (e.g. recessive epidermolysis bullosa simplex, cutaneous squamous cell carcinoma, melanoma, eczema, bacterial wound infections, and wound healing) in order to test and develop new therapeutics [36,37,42–48].

Skin barrier properties in human skin equivalents

Human skin protects all organs from external factors, thereby being itself a target for many exogenous compounds (including skincare products, household detergents, drugs and chemicals). Depending on the mode of action of these constituents, the active molecules should be delivered to their target site, which may be the stratum corneum, epidermis, dermis, or blood vessels, respectively [49,50]. The ultimate effect of such a molecule is influenced by a number of processes, including time-dependent processes of penetration through the stratum corneum, distribution, binding, metabolism, and eradication [51]. The most frequent skin reactions to barrier disturbance at an early stage is the loss of the water (dryness) and skin irritation manifested by formation of erythema and edema [52]. Adverse responses of sensitization leading to pigmentation and contact allergy are observed as well [53]. Some of these constituents entering the stratum corneum may cause lipid degradation and protein denaturation [54]. In addition, they disrupt membranes of viable cells, resulting in the release of cytoplasm into extracellular spaces [55]. Penetration of these chemicals through the skin barrier is a complex process in which the conditions, such as the organization and composition of the stratum corneum, play a crucial role [56]. The permeability of the skin barrier resides in multiple lipid lamellae that fill the extracellular spaces between the corneocytes of the stratum corneum [57,58]. The stratum corneum lipid matrix is composed of free fatty acids, cholesterol, and ceramides [58,59]. To ensure that the penetration pathway is restricted to the intercorneocyte space, the presence of an impermeable cornified envelope is crucial. The cornified envelope is composed of several cross-linked precursors and is rather impermeable, which reduces the penetration of most substances into the corneocytes [60,61]. Therefore, the pathway for compound penetration is thought to mainly proceed via the stratum corneum lipid domains [62–64].

Changes in the composition and organization of the stratum corneum lipids are known to have a harmful effect on the barrier function of the skin [56,65,66]. Damage to the epidermal barrier leads to enhanced transepidermal water loss and, more seriously, enhanced penetration of toxic compounds to viable keratinocytes.

One of the prerequisites for the use of HSEs for screening purposes is that their barrier function is similar to that of native human skin. Using reconstructed human epidermis generated on various scaffolds, such as de-epidermized dermis, fibroblast-populated collagen matrix, or an inert filter, it has been shown that the formation of the stratum corneum barrier in vitro proceeds similarly as in vivo [34,37,38,67]. The extensive production of lamellar bodies, their complete extrusion at the stratum granulosum/stratum corneum interface, and the formation of multiple broad lamellar structures in the inter-corneocyte space demonstrate great resemblance to the in vivo situation [65]. However, recent data clearly show that differences in barrier properties are still present and that the stratum corneum lipid composition and organization of HSEs can be effected by the composition of, for example, the culture medium [41,68]. Therefore, optimization of the current culture conditions is essential in order to develop a new generation of HSEs that harbor a competent barrier that even more closely mimics the stratum corneum lipids and organization of native human skin than the current HSEs.

Validated safety tests using human skin equivalents

In 2007, the EU launched a program that aims to improve the protection of human health and the environment through the better and earlier identification of the intrinsic properties of 10,000 known chemical substances. Registration, Evaluation, Authorisation and Restriction of Chemicals (REACH) places greater responsibility on industry to manage the risks from chemicals and to provide safety information on chemical substances [22]. Within the REACH program, a large number of animal studies are foreseen, and necessary. Alternatives to these tests are welcomed by REACH and, therefore, a number of in vitro models are under development to reduce or replace the number of animals used in experimentation [69,70].

For the prediction of the potential toxic effects of various chemicals on human skin, preferably easy, reliable, and low-cost test methods are required [71,72]. As mentioned earlier, the human skin barrier plays a key role in the outcome of these toxicity tests. One can imagine that these barrier properties differ between native human skin, animals, and non-animal test systems. In addition, it is also well known that differences in some stages of the metabolic routes of exogenous chemicals do exist between humans and several types of animal models.

The in vivo skin corrosivity test is known to cause severe pain to test animals. This painful test would preferably be conducted in an in vitro or ex vivo model [52,73]. Ex vivo human skin might be regarded as the perfect (and preferred) alternative test system. However, the supply of fresh human skin by hospitals and cosmetic clinics is minimal and irregular. In addition, there are also legal and ethical issues linked to the use of human tissues for commercial activities, such as routine testing for commercial purposes. In recent years, the general public as well as the relevant authorities have demanded the reduction of the use of animals for product testing. In March 2009, this resulted in a European ban on the use of animals in cosmetic testing, and a complete sales ban effective in 2013. Therefore, a considerable interest has focused on three-dimensional HSEs, which became available in the 1980s [35,74,75]. This has triggered the development of in vitro alternatives. Today, several HSEs are commercially available, e.g. EPISKIN™ (L'Oreal, France), EpiDerm™ (MatTek, USA) and the EST-1000™ (Cell Systems GmbH, Germany) [76–79]. Some of these HSEs have been validated according to the European Center for the Validation of Alternative Methods (ECVAM) guidelines, and finally implemented in Europe and the OECD guidelines for testing of dangerous substances, OECD TG 430 (TER assay) and OECD TG 431 (human skin models) [80,81]. Interestingly, several studies have showed that if the generated HSEs show a well-developed epidermal compartment and barrier function, they will perform quite similarly to the validated models for skin corrosivity testing [77,81,82]. Next to the skin corrosion tests, the ECVAM Skin Irritation Task Force (ESITF) has put a lot of effort into replacing the Draize skin irritation test with in vitro skin models. In 2008, the SkinEthic-RHE™ and EpiDerm™ models were validated as an in vitro tool for the screening of irritants [78,79]. Currently, ECVAM is also putting effort in replacing the LLNA by in vitro tests [17]. For this purpose, a number of alternatives are under development that aim to determine the potency of and discriminate between sensitizers and irritants [83–85]. Another test that is supported by ECVAM and Cosmetics Europe is the prediction of phototoxicity. Phototoxicity is defined as an acute toxic response that is elicited after initial exposure of the skin to certain chemicals and subsequent exposure to light (*OECD Guidelines for the Testing of Chemicals/Section 4: Health Effects*, 2004). This test gained regulatory acceptance in the EU in 2000 and was accepted as a new OECD test guideline (OECD TG 432) in 2004 [86–88]. In this test, a mouse 3T3 fibroblast cell line, Balb/c 3T3, was used to predict phototoxicity; as long as the tested chemicals provided a negative result, additional tests were not necessary. However, when chemicals gave a positive result, additional tests were obligatory to obtain information on the phototoxicity and bioavailability of the chemical in the skin (e.g. phototoxic potency) [89]. In 2002, ECVAM and the European Medicine Evaluation Agency (EMEA) approved a study in which an epidermal model was used to assess whether the potencies of phototoxic substances could correctly be predicted when subjected to the pre-validated in vitro EpiDerm™ phototoxicity test [90].

In conclusion, HSEs have been shown to be a very powerful tool for screening purposes, since they mimic human skin to a high degree, including the presence of a competent barrier. The use of HSEs has become indispensable as test systems for many purposes (scientific and commercial), including for the cosmetic, chemical, and pharmaceutical industries. The incorporation of various immunological components (e.g. T and B cells) and the improvement of barrier properties will eventually contribute to the development of a new generation of HSEs that can be deployed for the validation of more sophisticated safety tests.

3.2.3 Skin sensitization in keratinocytes

Most chemical allergens act like haptens, meaning they can only activate the immune system by binding to carrier proteins. The first contact of a chemical allergen (hapten) with the skin of a susceptible individual leads to sensitization (priming of the immune system), the first phase of ACD. When penetrating the skin through the stratum corneum, the first cells to be encountered are the keratinocytes. In the cell, the chemical covalently binds with a carrier protein, which is a crucial step in order to induce the innate immune system and leads to induction of several proinflammatory cytokines (e.g. IL-18, IL-1β, TNF-α, ROS) resulting in the activation of skin-resident DCs. The activated DCs migrate to the skin draining lymph nodes, where they present the antigen (hapten–protein complex) to naïve T cells. This results in the formation of antigen-specific effector and memory T-cell clones, which subsequently enter the blood circulation. Upon a second exposure, memory T cells will be recruited to the site of exposure on the skin and elicit an inflammatory process (elicitation phase).

Activation of the Keap1-Nrf2-ARE pathway by sensitizers

A significant body of evidence supports the activation of the Keap1-Nrf2-ARE pathway by sensitizers and not by irritants, reported in several different cell types such as THP-1 (dendritic cells), PBMc (monocyte-derived dendritic cells), the MUTZ-3 cell line (monocytic dendritic cells), HaCat (keratinocytes), primary keratinocytes, and primary HSEs [28,70,91–95].

The electrophilic quality of a hapten makes it possible for it to react with cysteine or lysine residues of a sensor protein and undergo covalent binding. This sensor protein is called Keap1 and is

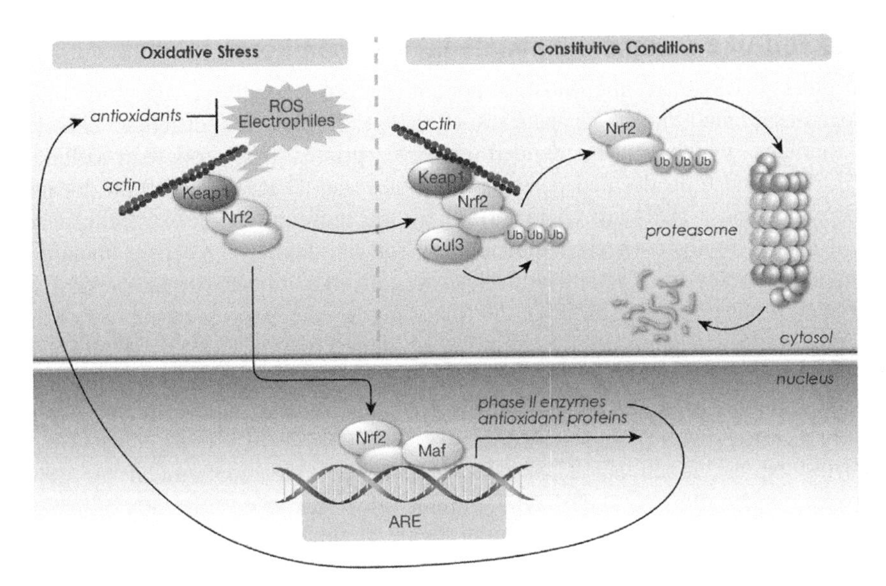

FIGURE 3.2.2 Activation of the Keap1-Nrf2-ARE pathway.

Keap1, a cysteine-rich protein that is anchored to actin in the cytosol, interacts with Nrf2, acting as an adaptor protein for the Cul3-dependent E3 (Cul3) ubiquitin ligase complex. Under normal conditions, Keap1 promotes ubiquitination and eventual degradation of Nrf2. Under oxidative and electrophilic stresses, the Nrf2 signaling pathway is activated. Nrf2 enters the nucleus, where it activates transcription of a host of cytoprotective genes (antioxidant and phase II enzymes), including the components of an antioxidant system that can balance high ROS levels. This figure has been provided by Cayman Chemicals.

a cytosolic inhibitor of the transcription factor Nrf2, promoting its degradation through Cullin-3-mediated ubiquitination. Covalent binding of chemical allergens to Keap1 leads to reactivity of cysteine residues of Keap1 and ROS-mediated oxidation. This will eventually result in the dissociation of the protein complex and stabilization of Nrf2 levels. In addition, this leads to its subsequent translocation to the nucleus and binding to the ARE consensus binding sites, together with a small Maf proteins, and other cofactors. *As a consequence, antioxidant genes and phase II enzymes, such as heme oxygenase 1 (HMOX-1), NADPH quinone oxidoreductase 1 (NQO1), glutamate–cysteine ligase modifier subunit (GCLM), thioredoxin reductase 1 (TXNRD1), and others, will be expressed* (Figure 3.2.2) [27,96,97].

A large number of haptens are unable to form a direct covalent binding with a carrier protein, but need to be transformed by detoxifying enzymes resident in the skin in order to be reactive; these are known as pro-haptens. The enzymes are responsible for the metabolism of pro-haptens into electrophilic reactive haptens. Activation of pro-haptens is induced by large panel of phase I detoxifying enzymes, such as the cytochrome P450 isoenzymes, sulfotransferases, acyltransferases, glutathione *S*-transferases, glucuronosyltransferases, and alcohol and aldehyde dehydrogenases. Another class of haptens is the pre-haptens, which need to be transformed by oxidization in order to be reactive with carrier proteins [98–100].

Activation of Toll-like receptors by haptens in human keratinocytes

Toll-like receptors (TLRs) are the most important pattern-recognition receptors that can recognize pathogen-associated molecular patterns (PAMPs). So far, ten different TLRs have been identified in humans, which can be divided into two groups based on their cellular localization. TLR1, 2, 4, 5, and 6 are located on the cell surface and TLR3, 7, and 9 are located intracellularly on the endosomes. TLRs are characterized by the presence of an extracellular leucine-rich repeat domain and a highly conserved intracellular Toll-interleukin-1 receptor domain [101,102]. The most important events in TLR activation lead to activation of nuclear factor κB (NF-κB) and mitogen-activated protein kinase (MAPK), both leading to the expression of genes involved in immune responses and activation of, for example, cytokines (e.g. TNF-α, IL-1β, IL-6, IL-12) and chemokines (IL-8; growth regulated oncogene-alpha; monocyte chemoattractant protein-1, -2, -3, -4; macrophage inflammatory protein 1 alpha/beta; and regulated upon activation normal T cell expressed and presumably secreted (RANTES)) [103–105].

Keratinocytes express functional TLRs 1–6, 9 and 10, which are crucial for inducing skin immune responses leading to typical Th1-type immune responses [101]. Recently, it has been shown that TLR2 and TLR4 play an important role in ACD. Double knockout mice lacking functional IL-12RB2/TLR4, IL-12RB2/TLR2, or TLR2/TLR4 are resistant to ACD, suggesting a role for TLR2 and TLR4 [106]. In addition, haptens cause ROS production, leading to hyaluronic acid (HA) degradation in the skin and damage-associated molecular patterns (DAMPs), which are recognized by TLR2 and TLR4 (Figure 3.2.3) [107]. This was shown in a study where TLR2 and TLR4 were activated in vitro in DCs by low-molecular-weight derivatives of HA [108,109]. Furthermore, nickel ions are one of the most prevalent contact allergens and are found to be a direct activator of human TLR4, emphasizing the relevance of TLRs in sensitization [110].

Activation of the inflammasome by haptens in keratinocytes

Several studies indicate the importance of the inflammasome in the activation of ACD [111–113]. The NACHT, LRR and PYD domains-containing protein 3 (NALP3) inflammasome is a cytoplasmic protein complex that activates cleavage of proinflammatory cytokines IL-1β and IL-18 by caspase 1 (Figure 3.2.3). Recognition of danger signals, such as ATP and other DAMPs, activates the P2 × 7 receptor (NOD-like receptor) and triggers proinflammatory signals promoting T-cell responses in the skin through active caspase-1 cleavage of pro-IL-1β and IL-18 [107,111,112]. The exact mechanism is not yet fully understood.

In summary, we can conclude that keratinocytes are crucial for the development of ACD. Since topical exposure is a very important event in inducing ACD, we emphasize the development of an in vitro human 3D skin equivalent as a tool to identify sensitizers.

3.2.4 Toxicogenomic analysis of cutaneous responses

Studies using so-called 'omics techniques to describe toxicogenomic responses of the human skin can roughly be divided into two categories: (1) research directed towards the identification of robust and reproducible genomic/protein markers and biological systems that allow classification and risk assessment of (chemical) compounds, and (2) investigations that aim to use genomics and/or proteomics to elucidate (toxicological) mechanisms operational in the human skin.

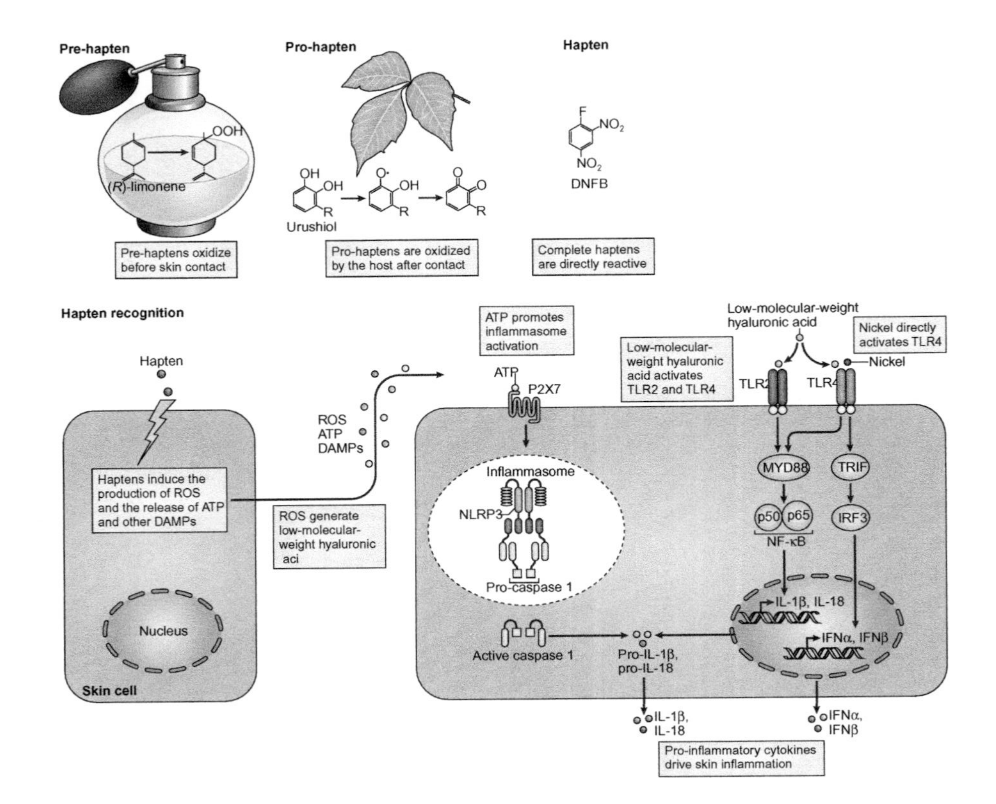

FIGURE 3.2.3 Innate recognition of haptens in keratinocytes.

The earliest event in allergic contact dermatitis is the formation of hapten–self complexes. Haptens induce the production of reactive oxygen species (ROS), which leads to the release of ATP and possibly other damage-associated molecular patterns (DAMPs), as well as to the generation of low-molecular-weight hyaluronic acid. Low-molecular-weight hyaluronic acid is sensed by neighboring cells via TLR2 and TLR4, resulting in increased expression of inactive pro-interleukin-1β (pro-IL-1β) and pro-IL-18. ATP is sensed through the purinergic receptor P2 × 7 and activates the inflammasome, resulting in caspase 1 activity and the generation of active IL-1β and IL-18. Nickel has a unique adjuvant property and can directly bind to histidine residues in the extracellular domain of TLR4, triggering the activation of this receptor. IRF3, IFN-regulatory factor 3; MYD88, myeloid differentiation primary response gene 88; NF-κB, nuclear factor-κB; NLRP3, NOD-, LRR- and pyrin domain-containing 3; TIR-domain-containing adaptor-inducing interferon-β (TRIF) (adapted from Ref. [107]).

Ideally, these two toxicogenomic approaches should be complementary, leading to mechanism-based safety evaluation of (chemical) compounds. To reach this goal the field clearly needs further development, in which the use of epidermal skin models might play a crucial role.

A few studies have focused on toxicogenomic analysis of chemical-induced stimulation of monocyte- or CD34-derived dendritic cells (moDCs; CD34-DC) or myeloid-/DC-like cell lines such as THP-1 and MUTZ-3. A comparative DNA microarray analysis of moDCs and MUTZ-3 cells exposed to the sensitizer cinnamaldehyde revealed that genes downstream of the antioxidant response are up-regulated, which was confirmed in CD34-DC and THP-1 by real time polymerase chain reaction (PCR) for a selection of genes [94,114]. In more detail, Ott et al. demonstrated that selected genes involved in oxidative and metabolic stress responses (*CES1*, *NQO1*, *GCML*), in immunological reactions (*IL8*, *IL1B*), or in transcriptional regulation processes (*PIR*, *TRIM16*), are up-regulated after treatment of moDCs with either of the contact sensitizers 2,4,6-trinitrobenzenesulfonic acid (TNBS) and cinnamaldehyde [115]. From follow-up experiments, it was concluded that most likely the same pathways are triggered in the different dendritic cell systems, but that they differ with regard to their sensitivity [116]. More elaborate toxicogenomic studies have been carried out using epidermal cells, in vitro HSEs, and in vivo/ex vivo skin.

Microarray-based gene expression analysis of human epidermal cells, in HSEs and ex vivo skin models

Gene expression analysis using high-density oligonucleotide microarrays (representing 47,000 transcripts) were used to screen the response of human keratinocytes, simulated by exposure of a cell line (HaCaT) to a series of irritants and sensitizers [70]. In this study, biologically relevant pathways regulated by skin sensitizers were identified. In addition, it was shown that an expression signature of 13 genes, including the oxidative stress gene *HMOX1* and members of the Keap1-Nrf2 signaling pathway, can identify contact sensitizers with 70% accuracy. These findings were confirmed in a follow-up study using an increased number of compounds and time-points, providing more statistical power [95]. The potential for accurate prediction of irritation or sensitization in humans was further tested by inclusion of false-positive and false-negative chemicals. Although the gene signature had improved prediction accuracy compared to the LLNA, the misclassified compounds were comparable between the two assays. The gene expression data also showed a sensitizer-specific response of the Keap1-Nrf2 and Toll-like receptor signaling pathways. It was concluded that keratinocyte (cell-line)-based prediction assays may provide essential information on the properties of compounds [95].

Fletcher et al. were the first to use gene expression analysis in combination with HSEs to analyze irritation responses [117]. The EpiDerm™ skin model was challenged with sodium dodecyl sulfate (SDS), and differences in mRNA expression levels were determined using a complementary DNA (cDNA) nitrocellulose filter array. A number of genes involved in transportation and receptors were up-regulated at early time-points (15–30 min), while DNA repair genes (e.g. UV excision repair protein, DNA-directed RNA polymerase II), transforming growth factor beta 3 (TGF-β3; *TGFB3*), and tumor suppressors were upregulated after 1–3 h; at later time-points of 4–24 h, genes involved in protein translation (e.g. cathepsin D receptor) and metabolism (e.g. cytochrome P450 enzyme, sterol 27-hydroxylase (*CYP27A*)) were upregulated.

Using dedicated low-density DNA microarrays, Borlon et al. identified approximately 50 genes as significantly and differentially expressed between another human reconstituted epidermis model (EPISKIN) exposed to irritants and non-irritants [118]. These genes are involved in different cellular processes such as cell signaling, stress response, cell cycle, protein metabolism, and cell structure. The expression of 16 of these genes was altered in the same direction irrespective of the irritant tested, and this expression was suggested to represent endpoints for assessment of the skin irritation potential of compounds.

A similar approach was taken by Niwa et al., by determining the gene expression profiles altered by a topically applied mild irritant (0.075% SDS) in a HSE (EPI-MODEL, LabCyte™) [119]. By using a DNA microarray consisting of 205 genes related to inflammation, immunity, and stress, and as well as housekeeping genes, the expression of 10 of these genes—including interleukin-1 receptor antagonist (*IL1RN*), FOS-like antigen 1 (*FOSL1*), heat shock 70 kDa protein 1A (*HSPA1A*), myeloid differentiation primary response gene 88 (*MYD88*), and the known marker genes for irritation *IL1B* and *IL8*—was significantly induced. Remarkably, common denominators, i.e. genes identified all three studies using skin models, are not present. This might be owing to either the different types of epidermal models or the different array platforms used.

Commercially available high-density arrays (Affymetrix) were used to analyze the skin of human volunteers exposed to either the irritant SDS or the control nonanoic acid (NON) [120]. It was demonstrated that SDS transiently down-regulated cellular energy metabolism pathways, while NON transiently induced the IL-6 pathway as well as a number of mitogen-activated signaling cascades, including extracellular signal-regulated kinase and growth factor receptor signaling.

Similar high-density gene expression arrays revealed an essentially different pathway in the human skin during elicitation of ACD [121]. Gene expression profiles were generated from normal and nickel-exposed skin from nickel-allergic patients and non-allergic controls at different time-points during elicitation of eczema. Bioinformatics analyses indicated a possible involvement of signal transducers and activators of transcription (STATs) and small/mothers against decapentaplegic (SMAD) transcription factors in the late-stage phase (48–72 h) of ACD elicitation.

Proteome analysis of human epidermal cells and in vivo/*ex vivo* skin

Proteome analysis of ex vivo skin (models) is not only confirming gene expression studies but also has the advantage that modifications of proteins resulting from exposure to chemicals (i.e. posttranslational modifications but also covalent binding) can be detected. Despite these potential advantages there have been very few proteomic studies aimed at characterizing human skin tissue, which is in part due to the difficulties with solubilization and extracting proteins with the skin, but also due to limitations of the current proteomic technologies.

Pioneering work in unraveling the proteome of (primary) keratinocytes was carried out by Celis and coworkers. Much of the data gathered and maintained was initially published in a database [122]. However, access to this database via the internet is, to our best knowledge, at present not possible (http://biobase.dk/cgi-bin/celis). Proteomic analysis of toxicological responses after exposure of a human keratinocyte cell line to heavy metals was carried out by Zhang et al. [123]. This approach identified nine proteins with altered expression in HaCaT cells in response to the treatment with heavy metals. Among these proteins, heat-shock protein 27 (HSP27) was up-regulated most significantly, and the ratio of phosphorylated vs non-phosphorylated HSP27 significantly decreased in treated keratinocytes. It was concluded that the ratio of p-HSP27 and HSP27 may be a sensitive marker or additional endpoint for the hazard assessment of potential skin irritation caused by the chemicals tested.

To discover other endpoints for skin irritation, a proteomics approach was used to analyze the protein expression in human keratinocytes exposed to SDS [124]. Among the 20 proteins identified with altered expression, HSP27 and superoxide dismutase (such as Cu–Zn, SOD1) were down-regulated, while cofilin-1 was up-regulated significantly in response to the chemical challenge. HSP27 displayed the most significant alteration in levels of both mRNA and protein, accompanied by nuclear translocation, thereby confirming earlier results from Boxman et al. [125,126].

The number of studies in the field of what collectively could be referred to as exposed intact human skin is, in contrast to murine studies, surprising small. Boxman et al. determined changes in the proteome of human ex vivo skin after exposure to the irritant SDS [125]. By using a very rigorous solubilization, followed by a combination of two-dimensional (2D) gel electrophoresis and mass spectrometry, seven proteins were identified as potentially new epidermal markers for skin irritation. Among these seven proteins, the heat-shock protein HSP27 was identified and validated as the most prominent marker for skin irritation. A similar approach by the same group was instrumental in the identification of proteins responding to UV radiation of the human skin. In addition to up-regulation of HSP27, also higher expression of proteins involved in oxidative stress response were identified, as well as altered phosphorylation of cofilin-1 and dextrin [127]. This latter observation suggests that the seemingly increased expression of cofilin-1 as a response to skin irritation found by Zhang et al. in fact represented differential phosphorylation of the protein (and hence altered migration characteristics on a 2D gel electrophoresis) [124].

In summary, it can be concluded that, in contrast to the gene expression data, suggesting many different signaling pathways and effectors involved, the proteome data converge to a few candidates. In particular, altered expression and/or phosphorylation/translocation of HSP27 is a common denominator in epidermal cells, HSEs, and in vivo human skin. Such similarities on the proteome level were also observed by van Eijl et al. when investigating the profile of xenobiotic metabolizing enzymes present in human skin and in vitro models thereof [128]. Using a broad-based unbiased proteomics approach, it was demonstrated that the human skin contains a range of defined enzymes capable of metabolizing different classes of chemicals and having a high degree of similarity with the profiles determined in HSEs, thereby indicating their suitability for epidermal toxicity testing. For three phase II enzymes (glutathione *S*-transferase, UDP-glucuronosyltransferase, and *N*-acetyltransferase) Götz et al. confirmed that 3D-epidermal models are far better suited than monolayer cultures in mimicking the human skin xenobiotic metabolism [129].

3.2.5 Alternatives for animal testing of chemical sensitization: an overview

Due to the legislation, the development of new in vitro models for sensitization has gained high priority. Several approaches have been implemented in order to fully replace the LLNA. A wide range of models has emerged, from computer-based to cell-based models, in many different cell types such as keratinocytes, dendritic cells, and T cells. So far, there is no validated in vitro model available for sensitization determination [130]. In Table 3.2.1, an overview is shown of the current in vitro models that are either in the validation process or research phase, according to ECVAM and OECD guidelines.

The majority of assays using reporters as a read-out for sensitization are based on the observation that electrophilic compounds can modify the repressor protein Keap1 by covalent binding. As a

Table 3.2.1 Overview of In Vitro Models for Assessment of Sensitization or Potency

In Vitro Model	Cell Type	Read Out	References
DPRA	n.a.	Reactivity of chemicals with lysine or cysteine peptide	[131]
hCLAT	THP-1 cell line (human monocytic leukemia)	CD86 and CD54 surface expression	[132, 133]
mMMUSST	U937 cell line (histiocytic lymphoma)	CD86 surface expression	[132,133]
Keratinosens	HaCat cell line (keratinocyte)	ARE-mediated gene reporter	[134, 93]
AREc32 assay	MCF7 cell line (human breast adenocarcinoma)	ARE-mediated gene reporter	[93]
NCTC2544 IL-18 assay	Contaminated Hela dendritic cell line	IL-18 cytokine secretion	[23]
Vitosens	Primary naïve dendritic cells	*CREM* and *CCR2* gene expression	[135–137]
QSAR	n.a. (computer based)	In silico sensitizing prediction	[138]
HaCat assay	HaCat cell line (keratinocyte)	Gene expression: *DNMT3B*, *HMOX1*, *STC2*, *ADM*, *SRD1*, *cFos*, *FosL1*, *RBM5*, *CDK12*, *ARD37*	[70, 95]
T-cell proliferation assay	Naïve T cells	Proliferation, IFN-γ secretion	[139]
Dendritic cell migration assay	MUTZ-LC cell line (CD34+ human acute myeloid leukemia, differentiated into Langerhans cells)	Migration ratio of CXCL12: CCL5	[140, 24]
GARD	MUTZ-3 cell line (CD34+ human acute myeloid leukemia)	Gene signature of 200 genes	[25,141]
SensCeeTox	HaCat cell line (keratinocyte	ARE/EpRE-XRE-MRE-mediated gene expression	[91]
SensCeeTox	3D-human epidermal model MatTek Epiderm® SkinEthic RHE (primary keratinocytes)	ARE/EpRE-XRE-MRE-mediated gene expression	[92]
VUMC-EE	3D-human epidermal model (primary keratinocytes)	Potency assessment of chemicals	[84]

n.a., not applicable.

result, the transcription factor Nrf2 can bind to ARE in the promoter of many phase II detoxification genes (Figure 3.2.2). Initially, a reporter construct (a luciferase gene under the control of the ARE from the human *AKR1C2* gene) was stably expressed in murine hepatoma cell lines (Hepa1c1c7) and used to screen skin sensitizers [93]. Subsequently, this assay was further refined by using HaCaT cells (immortalized keratinocytes) stably expressing the reporter construct [134]. The adjusted assay was subjected to a detailed ring study in five laboratories on a total of 28 reference substances and currently known under the name Keratinosens [142]. However, phase III validation by ECVAM is not yet complete.

Although monolayer keratinocyte tests are in pre-validation for skin sensitization assessment, the use of monolayer has major drawbacks. As mentioned previously, epidermal models are superior over monolayer cultures in mimicking human skin xenobiotic metabolism, and epidermal tissue (containing layers of different functional compartments) might as a whole respond in a different way to environmental exposures as compared to monoculture. In addition, many toxic substances cannot be tested in cell cultures because of their chemical properties, including solubility, while topical application on skin tissue reflects better the in vivo situation. An additional drawback would be the use of primary cells, since the biological diversity results in variable data. So far, there are only 3D skin models available constructed from primary keratinocytes. It would be desirable to use 3D skin models from a cell line since this would reduce the variability that is seen in primary models. Furthermore, cell lines can easily be manipulated (e.g. transfection of reporter genes) offering the possibility for high-throughput screening. Such models might also provide a better way of potency assessment of sensitizers, as was shown by previous studies by dos Santos et al. (Table 3.2.1) [84]. Thus, intermediate models containing the best of both worlds (reconstructed skin models expressing reporter genes) are highly warranted.

References

[1] Coenraads PJ, Gonc alo M. Skin diseases with high public health impact: contact dermatitis. Eur J Dermatol 2007;17:564–5.

[2] Uter W, Schnuch A, Geier J, Frosch PJ. Epidemiology of contact dermatitis. The information network of departments of dermatology (IVDK) in Germany. Eur J Dermatol 1998;8:36–40.

[3] Saint-Mezard P, Rosieres A, Krasteva M, Berard F, Dubois B, Kaiserlian D, et al. Allergic contact dermatitis. Eur J Dermatol 2004;14:284–95.

[4] Asano Y, Makino T, Norisugi O, Shimizu T. Occupational cobalt-induced systemic contact dermatitis. Eur J Dermatol 2009;19:166–7.

[5] Nicolas JF, Testud F, Vocanson M. [Sensitisation versus tolerance in contact eczema]. Ann Dermatol Venereol 2008;135:733–6. [in French].

[6] Scott AE, Kashon ML, Yucesoy B, Luster MI, Tinkle SS. Insights into the quantitative relationship between sensitization and challenge for allergic contact dermatitis reactions. Toxicol Appl Pharmacol 2002;183:66–70.

[7] Lepoittevin JP, Cribier B. [Physiopathology of contact eczema]. Ann Dermatol Venereol 1998;125:775–82. [in French].

[8] Griem P, Wulferink M, Sachs B, Gonzalez JB, Gleichmann E. Allergic and autoimmune reactions to xenobiotics: how do they arise? Immunol Today 1998;19:133–41.

[9] Lepoittevin JP. Metabolism versus chemical transformation or pro- versus prehaptens? Contact Dermatitis 2006;54:73–4.

[10] De Benedetto A, Kubo A, Beck LA. Skin barrier disruption: a requirement for allergen sensitization? J Invest Dermatol 2012;132:949–63.

[11] Furio L, Guezennec A, Ducarre B, Guesnet J, Peguet-Navarro J. Differential effects of allergens and irritants on early differentiating monocyte-derived dendritic cells. Eur J Dermatol 2008;18:141–7.

[12] Martin SF, Jakob T. From innate to adaptive immune responses in contact hypersensitivity. Curr Opin Allergy Clin Immunol 2008;8:289–93.

[13] Xu H, DiIulio NA, Fairchild RL. T cell populations primed by hapten sensitization in contact sensitivity are distinguished by polarized patterns of cytokine production: interferon gamma-producing (Tc1) effector

CD8+ T cells and interleukin (Il) 4/Il-10-producing (Th2) negative regulatory CD4+ T cells. J Exp Med 1996;183:1001–12.
[14] Vandebriel RJ, van Loveren H. Non-animal sensitization testing: State-of-the-art. Crit Rev Toxicol 2010;40:389–404.
[15] Akiba H, Kehren J, Ducluzeau MT, Krasteva M, Horand F, Kaiserlian D, et al. Skin inflammation during contact hypersensitivity is mediated by early recruitment of CD8+ T cytotoxic 1 cells inducing keratinocyte apoptosis. J Immunol 2002;168:3079–87.
[16] Kehren J, Desvignes C, Krasteva M, Ducluzeau MT, Assossou O, Horand F, et al. Cytotoxicity is mandatory for CD8(+) T-cell-mediated contact hypersensitivity. J Exp Med 1999;189:779–86.
[17] Basketter DA, Scholes EW, Kimber I. The performance of the local lymph node assay with chemicals identified as contact allergens in the human maximization test. Food Chem Toxicol 1994;32:543–7.
[18] Kimber I, Dearman RJ, Scholes EW, Basketter DA. The local lymph node assay: developments and applications. Toxicology 1994;93:13–31.
[19] Gerberick GF, Vassallo JD, Foertsch LM, Price BB, Chaney JG, Lepoittevin JP. Quantification of chemical peptide reactivity for screening contact allergens: a classification tree model approach. Toxicol Sci 2007;97:417–27.
[20] Kimber I, Basketter DA, Berthold K, Butler M, Garrigue JL, Lea L, et al. Skin sensitization testing in potency and risk assessment. Toxicol Sci 2001;59:198–208.
[21] Mehling A, Eriksson T, Eltze T, Kolle S, Ramirez T, Teubner W, et al. Non-animal test methods for predicting skin sensitization potentials. Arch Toxicol 2012;86:1273–95.
[22] van der Wielen A. REACH: next step to a sound chemicals management. J Expo Sci Environ Epidemiol 2007;17(Suppl. 1):S2–S6.
[23] Corsini E, Mitjans M, Galbiati V, Lucchi L, Galli CL, Marinovich M. Use of IL-18 production in a human keratinocyte cell line to discriminate contact sensitizers from irritants and low molecular weight respiratory allergens. Toxicol In Vitro 2009;23:789–96.
[24] dos Santos GG, Reinders J, Ouwehand K, Rustemeyer T, Scheper RJ, Gibbs S. Progress on the development of human in vitro dendritic cell based assays for assessment of the sensitizing potential of a compound. Toxicol Appl Pharmacol 2009;236:372–82.
[25] Johansson H, Lindstedt M, Albrekt AS, Borrebaeck CA. A genomic biomarker signature can predict skin sensitizers using a cell-based in vitro alternative to animal tests. BMC Genomics 2011;12:399.
[26] Sakaguchi H, Ashikaga T, Miyazawa M, Kosaka N, Ito Y, Yoneyama K, et al. The relationship between CD86/CD54 expression and THP-1 cell viability in an in vitro skin sensitization test: human cell line activation test (h-CLAT). Cell Biol Toxicol 2009;25:109–26.
[27] Natsch A. The Nrf2-Keap1-ARE toxicity pathway as a cellular sensor for skin sensitizers: functional relevance and a hypothesis on innate reactions to skin sensitizers. Toxicol Sci 2010;113:284–92.
[28] Ade N, Leon F, Pallardy M, Peiffer JL, Kerdine-Romer S, Tissier MH, et al. HMOX1 and NQO1 genes are upregulated in response to contact sensitizers in dendritic cells and THP-1 cell line: role of the Keap1/Nrf2 pathway. Toxicol Sci 2009;107:451–60.
[29] Majewska M, Zajac K, Dulak J, Szczepanik M. Heme oxygenase (HO-1) is involved in the negative regulation of contact sensitivity reaction. Pharmacol Rep 2008;60:933–40.
[30] Matsue H, Edelbaum D, Shalhevet D, Mizumoto N, Yang C, Mummert ME, et al. Generation and function of reactive oxygen species in dendritic cells during antigen presentation. J Immunol 2003;171:3010–8.
[31] Auger FA, Lopez Valle CA, Guignard R, Tremblay N, Noel B, Goulet F, et al. Skin equivalent produced with human collagen. In Vitro Cell Dev Biol Anim 1995;31:432–9.
[32] Bell E, Parenteau N, Gay R, Nolte C, Kemp P, Bilbo P, et al. The living skin equivalent: its manufacture, its organotypic properties and its responses to irritants. Toxicol In Vitro 1991;5:591–6.
[33] Bell E, Rosenberg M, Kemp P, Gay R, Green GD, Muthukumaran N, et al. Recipes for reconstituting skin. J Biomech Eng 1991;113:113–9.

[34] Boyce S, Michel S, Reichert U, Shroot B, Schmidt R. Reconstructed skin from cultured human keratinocytes and fibroblasts on a collagen–glycosaminoglycan biopolymer substrate. Skin Pharmacol 1990;3:136–43.
[35] Bell E, Ivarsson B, Merrill C. Production of a tissue-like structure by contraction of collagen lattices by human fibroblasts of different proliferative potential in vitro. Proc Natl Acad Sci USA 1979;76:1274–8.
[36] El Ghalbzouri A, Lamme E, Ponec M. Crucial role of fibroblasts in regulating epidermal morphogenesis. Cell Tissue Res 2002;310:189–99.
[37] El Ghalbzouri A, Gibbs S, Lamme E, van Blitterswijk CA, Ponec M. Effect of fibroblasts on epidermal regeneration. Br J Dermatol 2002;147:230–43.
[38] Ponec M, Boelsma E, Gibbs S, Mommaas M. Characterization of reconstructed skin models. Skin Pharmacol Appl Skin Physiol 2002;15(Suppl. 1):4–17.
[39] Ponec M, Weerheim A, Kempenaar J, Mulder A, Gooris GS, Bouwstra J, et al. The formation of competent barrier lipids in reconstructed human epidermis requires the presence of vitamin C. J Invest Dermatol 1997;109:348–55.
[40] Bouwstra JA, Dubbelaar FE, Gooris GS, Ponec M. The lipid organisation in the skin barrier. Acta Derm Venereol Suppl (Stockh) 2000;208:23–30.
[41] Thakoersing VS, Gooris GS, Mulder A, Rietveld M, El Ghalbzouri A, Bouwstra JA. Unraveling barrier properties of three different in-house human skin equivalents. Tissue Eng Part C Methods 2012;18:1–11.
[42] Commandeur S, de Gruijl FR, Willemze R, Tensen CP, El Ghalbzouri A. An in vitro three-dimensional model of primary human cutaneous squamous cell carcinoma. Exp Dermatol 2009;18:849–56.
[43] de Breij A, Haisma EM, Rietveld M, El Ghalbzouri A, van den Broek PJ, Dijkshoorn L, et al. Three-dimensional human skin equivalent as a tool to study *Acinetobacter baumannii* colonization. Antimicrob Agents Chemother 2012;56:2459–64.
[44] El Ghalbzouri A, Jonkman M, Kempenaar J, Ponec M. Recessive epidermolysis bullosa simplex phenotype reproduced in vitro: ablation of keratin 14 is partially compensated by keratin 17. Am J Pathol 2003;163:1771–9.
[45] Engelhart K, El Hindi T, Biesalski HK, Pfitzner I. In vitro reproduction of clinical hallmarks of eczematous dermatitis in organotypic skin models. Arch Dermatol Res 2005;297:1–9.
[46] Tjabringa G, Bergers M, van Rens D, de Boer R, Lamme E, Schalkwijk J. Development and validation of human psoriatic skin equivalents. Am J Pathol 2008;173:815–23.
[47] Carlson MW, Alt-Holland A, Egles C, Garlick JA. Three-dimensional tissue models of normal and diseased skin. Curr Protoc Cell Biol 2008; Chapter 19, Unit 19.9.
[48] Garlick JA. Engineering skin to study human disease: tissue models for cancer biology and wound repair. Adv Biochem Eng Biotechnol 2007;103:207–39.
[49] Rubio L, Alonso C, Lopez O, Rodriguez G, Coderch L, Notario J, et al. Barrier function of intact and impaired skin: percutaneous penetration of caffeine and salicylic acid. Int J Dermatol 2011;50:881–9.
[50] Gibbs S, van de Sandt JJ, Merk HF, Lockley DJ, Pendlington RU, Pease CK. Xenobiotic metabolism in human skin and 3D human skin reconstructs: a review. Curr Drug Metab 2007;8:758–72.
[51] Roper CS, Howes D, Blain PG, Williams FM. Prediction of the percutaneous penetration and metabolism of dodecyl decaethoxylate in rats using in vitro models. Arch Toxicol 1995;69:649–54.
[52] Basketter D, Reynolds F, Rowson M, Talbot C, Whittle E. Visual assessment of human skin irritation: a sensitive and reproducible tool. Contact Dermatitis 1997;37:218–20.
[53] Hengge UR, Ruzicka T, Schwartz RA, Cork MJ. Adverse effects of topical glucocorticosteroids. J Am Acad Dermatol 2006;54:1–15. quiz 16-18.
[54] Welss T, Basketter DA, Schroder KR. In vitro skin irritation: facts and future. State-of-the-art review of mechanisms and models. Toxicol In Vitro 2004;18:231–43.
[55] Osborne R, Perkins MA. An approach for development of alternative test methods based on mechanisms of skin irritation. Food Chem Toxicol 1994;32:133–42.

[56] Bouwstra JA, Ponec M. The skin barrier in healthy and diseased state. Biochim Biophys Acta 2006;1758:2080–95.
[57] Schurer NY, Elias PM. The biochemistry and function of stratum corneum lipids. Adv Lipid Res 1991;24:27–56.
[58] Wertz PW, Squier CA. Cellular and molecular basis of barrier function in oral epithelium. Crit Rev Ther Drug Carrier Syst 1991;8:237–69.
[59] Feingold KR. The outer frontier: the importance of lipid metabolism in the skin. J Lipid Res 2009;50(Suppl): S417–22.
[60] Kalinin A, Marekov LN, Steinert PM. Assembly of the epidermal cornified cell envelope. J Cell Sci 2001;114:3069–70.
[61] Marshall D, Hardman MJ, Nield KM, Byrne C. Differentially expressed late constituents of the epidermal cornified envelope. Proc Natl Acad Sci USA 2001;98:13031–6.
[62] Johnson ME, Blankschtein D, Langer R. Evaluation of solute permeation through the stratum corneum: lateral bilayer diffusion as the primary transport mechanism. J Pharm Sci 1997;86:1162–72.
[63] Meuwissen ME, Janssen J, Cullander C, Junginger HE, Bouwstra JA. A cross-section device to improve visualization of fluorescent probe penetration into the skin by confocal laser scanning microscopy. Pharm Res 1998;15:352–6.
[64] Elias PM. Epidermal lipids, barrier function, and desquamation. J Invest Dermatol 1983;80:44s–9s.
[65] Bouwstra JA, Honeywell-Nguyen PL, Gooris GS, Ponec M. Structure of the skin barrier and its modulation by vesicular formulations. Prog Lipid Res 2003;42:1–36.
[66] Ponec M. In vitro cultured human skin cells as alternatives to animals for skin irritancy screening. Int J Cosmet Sci 1992;14:245–64.
[67] Ponec M, Weerheim A, Kempenaar J, Mommaas AM, Nugteren DH. Lipid composition of cultured human keratinocytes in relation to their differentiation. J Lipid Res 1988;29:949–61.
[68] Thakoersing VS, van Smeden J, Mulder AA, Vreeken RJ, El Ghalbzouri A, Bouwstra JA. Increased presence of monounsaturated fatty acids in the stratum corneum of human skin equivalents. J Invest Dermatol 2012. doi:10.1038/jid.2012.262.
[69] Benigni R, Bossa C. Alternative strategies for carcinogenicity assessment: an efficient and simplified approach based on in vitro mutagenicity and cell transformation assays. Mutagenesis 2011;26:455–60.
[70] Vandebriel RJ, Pennings JL, Baken KA, Pronk TE, Boorsma A, Gottschalk R, et al. Keratinocyte gene expression profiles discriminate sensitizing and irritating compounds. Toxicol Sci 2010;117:81–9.
[71] Balls M, Combes RD. The OECD health effects test guidelines: a challenge to the sincerity of commitment to the three Rs. Altern Lab Anim 2006;34:105–8.
[72] Combes RD. The current OECD Health Effects Test Guidelines are in urgent need of revision. Altern Lab Anim 2004;32:463–4.
[73] Whittle E, Basketter DA. The in vitro skin corrosivity test: development of method using human skin. Toxicol In Vitro 1993;7:265–8.
[74] Fartasch M, Ponec M. Improved barrier structure formation in air-exposed human keratinocyte culture systems. J Invest Dermatol 1994;102:366–74.
[75] Vicanova J, Mommaas AM, Mulder AA, Koerten HK, Ponec M. Impaired desquamation in the in vitro reconstructed human epidermis. Cell Tissue Res 1996;286:115–22.
[76] Kandarova H, Liebsch M, Genschow E, Gerner I, Traue D, Slawik B, et al. Optimisation of the EpiDerm test protocol for the upcoming ECVAM validation study on in vitro skin irritation tests. ALTEX 2004;21:107–14.
[77] Kandarova H, Liebsch M, Gerner I, Schmidt E, Genschow E, Traue D, et al. The EpiDerm test protocol for the upcoming ECVAM validation study on in vitro skin irritation tests: an assessment of the performance of the optimised test. Altern Lab Anim 2005;33:351–67.

[78] Kandarova H, Liebsch M, Schmidt E, Genschow E, Traue D, Spielmann H, et al. Assessment of the skin irritation potential of chemicals by using the SkinEthic reconstructed human epidermal model and the common skin irritation protocol evaluated in the ECVAM skin irritation validation study. Altern Lab Anim 2006;34:393–406.
[79] Kandarova H, Liebsch M, Spielmann H, Genschow E, Schmidt E, Traue D, et al. Assessment of the human epidermis model SkinEthic RHE for in vitro skin corrosion testing of chemicals according to new OECD TG 431. Toxicol In Vitro 2006;20:547–59.
[80] Liebsch M, Barrabas C, Traue D, Spielmann H. Development of a new in vitro test for dermal phototoxicity using a model of reconstituted human epidermis. ALTEX 1997;14:165–74.
[81] Liebsch M, Spielmann H. Currently available in vitro methods used in the regulatory toxicology. Toxicol Lett 2002;127:127–34.
[82] Cotovio J, Grandidier MH, Portes P, Roguet R, Rubinstenn G. The in vitro skin irritation of chemicals: optimisation of the EPISKIN prediction model within the framework of the ECVAM validation process. Altern Lab Anim 2005;33:329–49.
[83] Gibbs S. In vitro irritation models and immune reactions. Skin Pharmacol Physiol 2009;22:103–13.
[84] dos Santos GG, Spiekstra SW, Sampat-Sardjoepersad SC, Reinders J, Scheper RJ, Gibbs S. A potential in vitro epidermal equivalent assay to determine sensitizer potency. Toxicol In Vitro 2011;25:347–57.
[85] Teunis M, Corsini E, Smits M, Madsen CB, Eltze T, Ezendam J, et al. Transfer of a two-tiered keratinocyte assay: IL-18 production by NCTC2544 to determine the skin sensitizing capacity and epidermal equivalent assay to determine sensitizer potency. Toxicol In Vitro 2012. doi:10.1016/j.tiv.2012.06.004.
[86] Spielmann H, Balls M, Brand M, Doring B, Holzhutter HG, Kalweit S, et al. EEC/COLIPA project on in vitro phototoxicity testing: first results obtained with a Balb/c 3T3 cell phototoxicity assay. Toxicol In Vitro 1994;8:793–6.
[87] Spielmann H, Balls M, Dupuis J, Pape WJ, Pechovitch G, de Silva O, et al. The international EU/COLIPA In Vitro Phototoxicity Validation Study: results of phase II (blind trial). Part 1: the 3T3 NRU phototoxicity test. Toxicol In Vitro 1998;12:305–27.
[88] Spielmann H, Liebsch M, Reinhardt C. [ERGATT/ECVAM workshop on acceptance of validated alternative methods: Amden III]. ALTEX 1998;15:18–22.
[89] Liebsch M, Spielmann H, Pape W, Krul C, Deguercy A, Eskes C. UV-induced effects. Altern Lab Anim 2005;33(Suppl. 1):131–46.
[90] Jirova D, Kejlova K, Bendova H, Ditrichova D, Mezulanikova M. Phototoxicity of bituminous tars—correspondence between results of 3T3 NRU PT, 3D skin model and experimental human data. Toxicol In Vitro 2005;19:931–4.
[91] McKim Jr JM, Keller 3rd DJ, Gorski JR. A new in vitro method for identifying chemical sensitizers combining peptide binding with ARE/EpRE-mediated gene expression in human skin cells. Cutan Ocul Toxicol 2010;29:171–92.
[92] McKim Jr. JM, Keller 3rd DJ, Gorski JR. An in vitro method for detecting chemical sensitization using human reconstructed skin models and its applicability to cosmetic, pharmaceutical, and medical device safety testing. Cutan Ocul Toxicol 2012. doi:10.3109/15569527.2012.667031.
[93] Natsch A, Emter R. Skin sensitizers induce antioxidant response element dependent genes: application to the in vitro testing of the sensitization potential of chemicals. Toxicol Sci 2008;102:110–9.
[94] Python F, Goebel C, Aeby P. Comparative DNA microarray analysis of human monocyte derived dendritic cells and MUTZ-3 cells exposed to the moderate skin sensitizer cinnamaldehyde. Toxicol Appl Pharmacol 2009;239:273–83.
[95] van der Veen JW, Pronk TE, van Loveren H, Ezendam J. Applicability of a keratinocyte gene signature to predict skin sensitizing potential. Toxicol In Vitro 2012. doi:10.1016/j.tiv.2012.08.023.
[96] Kensler TW, Wakabayashi N, Biswal S. Cell survival responses to environmental stresses via the Keap1-Nrf2-ARE pathway. Annu Rev Pharmacol Toxicol 2007;47:89–116.

[97] Martin SF, Esser PR, Weber FC, Jakob T, Freudenberg MA, Schmidt M, et al. Mechanisms of chemical-induced innate immunity in allergic contact dermatitis. Allergy 2011;66:1152–63.

[98] Dinkova-Kostova AT, Talalay P. Direct and indirect antioxidant properties of inducers of cytoprotective proteins. Mol Nutr Food Res 2008;52(Suppl. 1):S128–38.

[99] Divkovic M, Pease CK, Gerberick GF, Basketter DA. Hapten–protein binding: from theory to practical application in the in vitro prediction of skin sensitization. Contact Dermatitis 2005;53:189–200.

[100] Gotz C, Pfeiffer R, Tigges J, Blatz V, Jackh C, Freytag EM, et al. Xenobiotic metabolism capacities of human skin in comparison with a 3D epidermis model and keratinocyte-based cell culture as in vitro alternatives for chemical testing: activating enzymes (phase I). Exp Dermatol 2012;21:358–63.

[101] Nestle FO, Di Meglio P, Qin JZ, Nickoloff BJ. Skin immune sentinels in health and disease. Nat Rev Immunol 2009;9:679–91.

[102] Lebre MC, van der Aar AM, van Baarsen L, van Capel TM, Schuitemaker JH, Kapsenberg ML, et al. Human keratinocytes express functional Toll-like receptor 3, 4, 5, and 9. J Invest Dermatol 2007;127:331–41.

[103] Akira S, Uematsu S, Takeuchi O. Pathogen recognition and innate immunity. Cell 2006;124:783–801.

[104] Albanesi C. Keratinocytes in allergic skin diseases. Curr Opin Allergy Clin Immunol 2010;10:452–6.

[105] Miller LS. Toll-like receptors in skin. Adv Dermatol 2008;24:71–87.

[106] Martin SF, Dudda JC, Bachtanian E, Lembo A, Liller S, Durr C, et al. Toll-like receptor and IL-12 signaling control susceptibility to contact hypersensitivity. J Exp Med 2008;205:2151–62.

[107] Kaplan DH, Igyarto BZ, Gaspari AA. Early immune events in the induction of allergic contact dermatitis. Nat Rev Immunol 2012;12:114–24.

[108] Scheibner KA, Lutz MA, Boodoo S, Fenton MJ, Powell JD, Horton MR. Hyaluronan fragments act as an endogenous danger signal by engaging TLR2. J Immunol 2006;177:1272–81.

[109] Termeer C, Benedix F, Sleeman J, Fieber C, Voith U, Ahrens T, et al. Oligosaccharides of hyaluronan activate dendritic cells via toll-like receptor 4. J Exp Med 2002;195:99–111.

[110] Schmidt M, Raghavan B, Muller V, Vogl T, Fejer G, Tchaptchet S, et al. Crucial role for human Toll-like receptor 4 in the development of contact allergy to nickel. Nat Immunol 2010;11:814–9.

[111] Antonopoulos C, Cumberbatch M, Mee JB, Dearman RJ, Wei XQ, Liew FY, et al. IL-18 is a key proximal mediator of contact hypersensitivity and allergen-induced Langerhans cell migration in murine epidermis. J Leukoc Biol 2008;83:361–7.

[112] Watanabe H, Gaide O, Petrilli V, Martinon F, Contassot E, Roques S, et al. Activation of the IL-1beta-processing inflammasome is involved in contact hypersensitivity. J Invest Dermatol 2007;127:1956–63.

[113] Yazdi AS, Ghoreschi K, Rocken M. Inflammasome activation in delayed-type hypersensitivity reactions. J Invest Dermatol 2007;127:1853–5.

[114] Aaltonen T, Adelman J, Akimoto T, Alvarez Gonzalez B, Amerio S, Amidei D, et al. Search for the production of narrow *tb* resonances in 1.9 fb^{-1} of p{Pbar} collisions at $\sqrt{s} = 1.96$ TeV. Phys Rev Lett 2009;103:041801.

[115] Ott H, Wiederholt T, Bergstrom MA, Heise R, Skazik C, Czaja K, et al. High-resolution transcriptional profiling of chemical-stimulated dendritic cells identifies immunogenic contact allergens, but not prohaptens. Skin Pharmacol Physiol 2010;23:213–24.

[116] Sebastian K, Ott H, Zwadlo-Klarwasser G, Skazik-Voogt C, Marquardt Y, Czaja K, et al. Evaluation of the sensitizing potential of antibiotics in vitro using the human cell lines THP-1 and MUTZ-LC and primary monocyte-derived dendritic cells. Toxicol Appl Pharmacol 2012;262:283–92.

[117] Fletcher ST, Baker VA, Fentem JH, Basketter DA, Kelsell DP. Gene expression analysis of EpiDerm (TM) following exposure to SLS using cDNA microarrays. Toxicology In Vitro 2001;15:393–8.

[118] Borlon C, Godard P, Eskes C, Hartung T, Zuang V, Toussaint O. The usefulness of toxicogenomics for predicting acute skin irritation on in vitro reconstructed human epidermis. Toxicology 2007;241:157–66.

[119] Niwa M, Nagai K, Oike H, Kobori M. Evaluation of the skin irritation using a DNA microarray on a reconstructed human epidermal model. Biol Pharm Bull 2009;32:203–8.

[120] Clemmensen A, Andersen KE, Clemmensen O, Tan Q, Petersen TK, Kruse TA, et al. Genome-wide expression analysis of human in vivo irritated epidermis: differential profiles induced by sodium lauryl sulfate and nonanoic acid. J Invest Dermatol 2010;130:2201–10.

[121] Pedersen MB, Skov L, Menne T, Johansen JD, Olsen J. Gene expression time course in the human skin during elicitation of allergic contact dermatitis. J Invest Dermatol 2007;127:2585–95.

[122] Celis JE, Rasmussen HH, Olsen E, Madsen P, Leffers H, Honore B, et al. The human keratinocyte two-dimensional protein database (update 1994): towards an integrated approach to the study of cell proliferation, differentiation and skin diseases. Electrophoresis 1994;15:1349–458.

[123] Zhang Q, Zhang L, Xiao X, Su Z, Zou P, Hu H, et al. Heavy metals chromium and neodymium reduced phosphorylation level of heat shock protein 27 in human keratinocytes. Toxicol In Vitro 2010;24:1098–104.

[124] Zhang Q, Dai T, Zhang L, Zhang M, Xiao X, Hu H, et al. Identification of potential biomarkers for predicting acute dermal irritation by proteomic analysis. J Appl Toxicol 2011;31:762–72.

[125] Boxman IL, Hensbergen PJ, van der Schors RC, Bruynzeel DP, Tensen CP, Ponec M. Proteomic analysis of skin irritation reveals the induction of HSP27 by sodium lauryl sulphate in human skin. Br J Dermatol 2002;146:777–85.

[126] Boxman IL, Kempenaar J, de Haas E, Ponec M. Induction of HSP27 nuclear immunoreactivity during stress is modulated by vitamin C. Exp Dermatol 2002;11:509–17.

[127] Hensbergen P, Alewijnse A, Kempenaar J, van der Schors RC, Balog CA, Deelder A, et al. Proteomic profiling identifies a UV-induced activation of cofilin-1 and destrin in human epidermis. J Invest Dermatol 2005;124:818–24.

[128] van Eijl S, Zhu Z, Cupitt J, Gierula M, Gotz C, Fritsche E, et al. Elucidation of xenobiotic metabolism pathways in human skin and human skin models by proteomic profiling. PLoS One 2012;7:e41721.

[129] Götz C, Pfeiffer R, Tigges J, Ruwiedel K, Hubenthal U, Merk HF, et al. Xenobiotic metabolism capacities of human skin in comparison with a 3D-epidermis model and keratinocyte-based cell culture as in vitro alternatives for chemical testing: phase II enzymes. Exp Dermatol 2012;21:364–9.

[130] Bauch C, Kolle SN, Ramirez T, Eltze T, Fabian E, Mehling A, et al. Putting the parts together: combining in vitro methods to test for skin sensitizing potentials. Regul Toxicol Pharmacol 2012;63:489–504.

[131] Gerberick GF, Vassallo JD, Bailey RE, Chaney JG, Morrall SW, Lepoittevin JP. Development of a peptide reactivity assay for screening contact allergens. Toxicol Sci 2004;81:332–43.

[132] Ashikaga T, Yoshida Y, Hirota M, Yoneyama K, Itagaki H, Sakaguchi H, et al. Development of an in vitro skin sensitization test using human cell lines: the human Cell Line Activation Test (h-CLAT). I. Optimization of the h-CLAT protocol. Toxicol In Vitro 2006;20:767–73.

[133] Sakaguchi H, Ashikaga T, Miyazawa M, Yoshida Y, Ito Y, Yoneyama K, et al. Development of an in vitro skin sensitization test using human cell lines: the human Cell Line Activation Test (h-CLAT). II. An inter-laboratory study of the h-CLAT. Toxicol In Vitro 2006;20:774–84.

[134] Emter R, Ellis G, Natsch A. Performance of a novel keratinocyte-based reporter cell line to screen skin sensitizers in vitro. Toxicol Appl Pharmacol 2010;245:281–90.

[135] Hooyberghs J, Schoeters E, Lambrechts N, Nelissen I, Witters H, Schoeters G, et al. A cell-based in vitro alternative to identify skin sensitizers by gene expression. Toxicol Appl Pharmacol 2008;231:103–11.

[136] Lambrechts N, Vanheel H, Hooyberghs J, De Boever P, Witters H, van den Heuvel R, et al. Gene markers in dendritic cells unravel pieces of the skin sensitization puzzle. Toxicol Lett 2010;196:95–103.

[137] Lambrechts N, Vanheel H, Nelissen I, Witters H, van den Heuvel R, van Tendeloo V, et al. Assessment of chemical skin-sensitizing potency by an in vitro assay based on human dendritic cells. Toxicol Sci 2010;116:122–9.

[138] Roberts DW, Patlewicz G, Kern PS, Gerberick F, Kimber I, Dearman RJ, et al. Mechanistic applicability domain classification of a local lymph node assay dataset for skin sensitization. Chem Res Toxicol 2007;20:1019–30.

[139] Martin SF, Esser PR, Schmucker S, Dietz L, Naisbitt DJ, Park BK, et al. T-cell recognition of chemicals, protein allergens and drugs: towards the development of in vitro assays. Cell Mol Life Sci 2010;67:4171–84.

[140] Gibbs S, Spiekstra S, Corsini E, McLeod J, Reinders J. Dendritic cell migration assay: a potential prediction model for identification of contact allergens. Toxicol In Vitro 2012.

[141] Johansson H, Albrek AS, Borrebaeck CA, Lindstedt M. The GARD assay for assessment of chemical skin sensitizers. Toxicol In Vitro 2012. doi:10.1016/j.tiv.2012.05.016.

[142] Andreas N, Caroline B, Leslie F, Frank G, Kimberly N, Allison H, et al. The intra- and inter-laboratory reproducibility and predictivity of the KeratinoSens assay to predict skin sensitizers in vitro: results of a ring-study in five laboratories. Toxicol In Vitro 2011;25:733–44.

CHAPTER

3.3 'Omics-Based Testing for Direct Immunotoxicity

Oscar L. Volger

RIKILT Institute of Food Safety, Wageningen University and Research Center, Wageningen, The Netherlands

3.3.1 Introduction to immunotoxicity

Immunotoxicity is defined as deleterious effects of a xenobiotic on the functioning of the immune system. Immunotoxic xenobiotics can be natural toxins, chemicals, or pharmaceuticals. Immunotoxicity can be elicited either by suppression or by activation of the immune system. These separate processes are termed immunosuppression and immune-enhancement, respectively [1–3]. Immunotoxicity may be brought about by a direct effect of the xenobiotic on the cells of the immune system, alternatively by a primary effect on functions of non-immune-related organs or physiological systems, such as the endocrine system and the nervous system. For example, endocrine disruptors can have effects on the immune system by altering the levels of hormones that alter the growth of immune organs such as the thymus or spleen. Immunosuppression can lead to a reduced resistance against infectious diseases or to an increased susceptibility to non-genotoxic carcinogens. For instance, in organ transplant patients, treatment with immunosuppressive drugs is often associated with infectious complications and a more frequent development of malignancies [4,5]. Immunostimulation is defined as the unwanted overactivation of the immune system, upon exposure to either pathogens or to xenobiotics. Prolonged exposure to immunostimulants may predispose a person to the development of certain allergies, or inflammation-related or autoimmune diseases. Indirect immunotoxicity is a topic that will not be dealt with in this chapter, and is defined as an allergic reaction upon exposure to a xenobiotic that causes tissue damage.

3.3.2 Current guidelines for immunotoxicity testing

Immunotoxicants are considered to be a high risk to human health, because immunotoxicity may be associated with high morbidity and mortality [6], and humans are continuously exposed to a range of immunotoxic substances, including dioxins [7], mycotoxins [8], organotin compounds [9], and various pesticides [10]. A considerable proportion of these immunotoxicants are man-made industrial chemicals. In addition, certain pharmaceuticals have the potential to have immunotoxic side effects. In order

Toxicogenomics-Based Cellular Models.

to minimize the risk of public exposure to such immunotoxic chemicals and drugs, the regulatory authorities have developed guidelines for safety testing, before these compounds come to the market.

In 2004, the regulatory authorities of the EU (European Medicines Agency, EMA), US (Food and Drug Administration, FDA; FDA Center for Drug Evaluation and Research, CDER), and Japan agreed to adhere to the guidelines for immunotoxicity evaluation of pharmaceuticals published by the International Conference on Harmonization (ICH S8, CHMP/167235/2004), detailing technical requirements for registration of pharmaceuticals for human use [11]. In line with this, risk assessment guidelines for evaluating the immunotoxic potential of chemicals have been developed by the World Health Organization in cooperation with the International Programme on Chemical Safety (WHO/IPCS; harmonization project document no. 10). In addition, since 1998, the Organisation for Economic Co-operation and Development (OECD) has published and updated guidelines for conducting animal-based in vivo standard toxicity studies (STS).

Of these OECD guidelines, those for extended one-generation reproductive toxicity studies (guideline 443 [12]) and 28-day repeated dose studies (guideline 407 TG [13]) also include guidelines for assessing the immunotoxic potential of chemicals and drugs. These guidelines have a tiered approach, in which the immunotoxic effects are identified in Tier I, and the underlying mode of action (MOA) is characterized in Tier II. Such a tiered approach was first proposed in 1979 [14]. In the guidelines for both chemicals and pharmaceuticals, it is mandatory to base data on the immunotoxic potential in Tier I on at least one animal-based STS. The international guidelines on preclinical immunotoxicity testing of pharmaceuticals also state that multiple dose levels are recommended in order to determine dose–response relationships [11]. STS in rodents usually comprise a repeated-dose 28-day or 90-day study. In these guidelines, which overlap with the guidelines for chemicals, the signs of immunotoxicity that should be taken into consideration are hematological changes; alterations in the weights of the thymus and spleen; changes in serum globulins occurring without a plausible reason, or effects on the liver or kidney that could be an indication of changes in serum globins; an increased incidence of infections; and tumor formation. Formation of tumors can be a sign of immunosuppression in the absence of other plausible causes, such as genotoxicity, hormonal effects, or liver enzyme induction. In practice, this means that if a standard toxicity study detects signs of immunotoxicity of the compound, additional immunotoxicity studies should be performed to verify the immunotoxic potential of the compound. The guidelines also state that such additional studies should aim to identify the cell type affected, the reversibility, and the MOA. However, no additional studies are required when information can be obtained from pre-existing non-clinical studies. If such data are not available, additional immunotoxicity tests are to be considered and could consist of the following tests, depending on the potential weight of evidence from the STS: T-cell-dependent antibody response (TDAR), immunophenotyping, natural killer (NK) cell activity, host resistance, macrophage neutrophil function, or cell-mediated immunity.

The international guidelines of the IPCS on the non-clinical testing of chemicals state that if a compound shows signs of immunotoxicity in Tier I, a more in-depth characterization of the immunotoxic MOA has to be carried out in Tier II by employing tests on functionality of the immune system [15]. In line with this, the US Environmental Protection Agency (EPA) has proposed additional guidelines (EPA-OPPTS 870.7800, EPA-OPPTS 880.3550, and EPA-OPPTS 880.3800) for functional immunotoxicity assays in Tier II tests, including specific host-resistance assays (*Listeria monocytogenes* and *Trichinella spiralis*) and the sheep red blood cell (SRBC) assay.

In addition to traditional in vivo exposure assessment, in vitro approaches can play a role as screening tools [16]. At the moment, in vitro approaches for identifying direct immunotoxicity have not matured enough to fully replace animal testing. In the past, international regulatory authorities have recommended various in vitro assays for the detection of direct immunotoxicity as alternative methods for in vivo STS. In 1994, the Immunotoxicology and In Vitro Possibilities workshop was organized to review the status of in vitro methods for assessing immunotoxicity, and the possibilities for development of in vitro alternatives to reduce, refine, and replace the use of animals were also discussed [17]. In 1997, the US National Institute of Environmental Health Sciences (US NIEHS), in cooperation with the Interagency Coordinating Committee on the Validation of Alternative Methods (ICCVAM), was directed to develop and validate improved alternative toxicological testing methods (US Public Law 103-43) to replace animal studies. In 2003, a workshop hosted by the European Centre for the Validation of Alternative Methods (EURL ECVAM) reported on the state of the art of in vitro systems for evaluating immunotoxicity [18]. Based on the recommendations from this workshop, pre-validation studies for detection of immunosuppressive activity in vitro were initiated and showed promising results [19]. Within Europe, the current development and validation of in vitro immunotoxicity tests primarily focuses on chemical sensitization [20–23], mainly owing to the implementation of the 7th Amendment to the Cosmetics Directive (76/768/EEC), banning the use of in vivo tests for skin-sensitizing properties of chemicals [24]. With respect to the other types of immunotoxicity—immunostimulation, autoimmunity, and developmental immunotoxicity—little or no effort in the area of in vitro testing has been made as yet. The processes involved in immunostimulation, autoimmunity, and developmental immunotoxicity are not only complex, involving delicately balanced interactions encompassing many tissues, but also often badly understood, and cell systems that can mimic some of the processes involved are difficult to devise.

3.3.3 Toxicogenomics

Toxicogenomics is defined as the application of genomics technology in toxicology research. Toxicogenomics gives an overview of the genomic responses of a cell, organ, or organism to chemical compounds, and improves the understanding of the molecular basis of the adverse effects of toxic compounds. In addition, the combination of genomics with other so-called 'omics technologies, such as proteomics and metabolomics, provides deeper insights into the mechanistic action of toxic compounds.

The basic principle of the toxicogenomics approach is that the effect of a toxic compound is determined by comparing gene expression profiles of exposed samples to non-exposed samples. The gene expression changes corresponding to specific classes of compounds with varying modes of action are expected to reveal compound-specific gene expression profiles that can subsequently be used to classify the compounds with respect to their toxicity (e.g. hepatotoxicity, neurotoxicity, carcinogenicity, reproductive toxicity, and immunotoxicity).

For instance, toxicity-specific gene expression profiles could potentially be used to discriminate immunotoxic from non-immunotoxic compounds, and carcinogens (genotoxic or non-genotoxic) from noncarcinogenic compounds. Therefore, in the US, the EPA and the Occupational Safety and Health Administration (OSHA) are developing regulatory guidelines for risk assessment based on genomic data analysis.

First, an introduction will be given on the structure of the human transcriptome. Then we will discuss the tools that are available to quantify alterations of the human transcriptome.

The human transcriptome

A transcriptome is defined as the complete set of RNA molecules of a species. RNA consists of different functional classes, including the precursor and mature forms of messenger RNAs (mRNA), microRNA (miRNA), and transfer RNA (tRNA). The size of the human transcriptome is vast and currently estimated to consist of 300,000 molecules with a total base-length of more than 3 billion, encoding approximately 23,000 protein encoding genes, approximately 2042 mature miRNAs (the University of Manchester hosts a searchable database of published miRNA sequences, miRBase 19, http://www.mirbase.org), and 497 tRNA molecules [25].

Messenger RNA molecules (mRNA) represent copies of genomic DNA that encode for peptide sequences of eventually formed proteins. The formation of mRNA is a multistep process. It starts with the formation of pre-mRNA molecules by transcription from genomic DNA. Pre-mRNA molecules comprise two parts, introns and exons. Introns are noncoding sequences that are not used by subsequent translation to form protein. Exons are the coding sequences eventually used for translation. In brief, the RNA splicing machinery uses enzymes to excise the introns from the pre-mRNA molecules, and then ligates the exons to form mature mRNA molecules.

In contrast with mRNA, miRNAs do not encode peptide sequences. Instead, miRNAs inhibit gene expression at the posttranscriptional level. In recent years, the role of miRNAs as modulators of immune and inflammatory responses has become apparent. Examples of miRNAs that respond to bacterial infection are miR-29, which controls innate and adaptive immune responses by targeting intracellular interferon-γ mRNA [26]; and miR-21, miR-146, and miR-155, that modulate Toll-like receptor (TLR) responses [27]. Certain miRNAs serve as important negative feedback loops in the immune system, such as miR-223 [28], whereas others amplify the response of the immune system by repressing inhibitors of the response, such as miR-301a [29].

Transfer RNA molecules (tRNA), with a typical length between 73 and 93 nucleotides, are shorter than mRNA and play an important role in the translational machinery, as templates between mRNA and newly formed peptide sequences [30]. During protein translation, upon attachment of amino acids by aminoacyl tRNA synthetases (ARS), specific proteins named elongation factors deliver the tRNAs to the ribosome.

3.3.4 Transcriptome quantification tools

Tools for quantifying individual RNA molecules have been available for many years, such as northern blotting, reverse-transcription polymerase chain reaction (RT-PCR), expressed sequence tags (ESTs), and serial analysis of gene expression (SAGE). However, high-throughput transcriptome profiling became a reality after the development of gene expression microarrays around the mid-1990s [31]. In 2003, the Affymetrix U133 Plus 2.0 Array was first used in a large cancer study [32]. At present, the Affymetrix GeneChip is the most widely used microarray platform.

Microarrays

Microarray technology enables the study of the effect of toxic compounds on thousands of genes simultaneously or on the whole genome of a cell in a single experiment. This provides comprehensive information about the gene expression profile under different experimental conditions. The main principle of a transcriptome microarray is that for each individual mRNA sequence, one or more probes are designed, consisting of complementary oligonucleotide sequences. Depending on the platform, these oligonucleotide sequences are either spotted or sequenced on a solid surface (Affymetrix platform), or attached to individual microspheres (Sentrix Illumina platform). The sequences of these oligonucleotides are usually based on a genome sequence or on a known or predicted open reading frame. Transcripts are extracted from samples of the cell or tissues to be investigated, labeled with fluorescent dyes (either one or two colors), and hybridized to the arrays. After the hybridization, unbound and noncoding sequences are washed off the chip. Thereafter, fluorescent or luminescent signals are detected by a scanner. Probes that correspond to transcribed RNA hybridize to their complementary target. Because transcripts are labeled with fluorescent dyes, light intensity can be used as a measure of gene expression. In the last decade, microarray transcriptome profiling has become common practice in almost any biomedical research laboratory. At present, the Gene Expression Omnibus (GEO), the largest repository for transcriptome data, contains more than 456,000 human, 135,000 murine, and 37,000 rat transcriptome experiments [33]. This vast amount of data has led to the development of a range of novel strategies for data analysis and experimental design. The quality of microarray data has improved dramatically, as compared to 10 years before. For instance, during that time different microarray platforms appeared to produce diverse results with the same samples [34], the reproducibility between different labs was limited [35], and ozone differentially degraded the fluorescent dyes [36]. Recognition of such biases by, for instance, the MicroArray Quality Control Consortium (MAQC), has led to the development of quality control standards to ensure an improved consistency of microarray experiments [37].

Until now, toxicogenomics studies that assessed the effects of immunotoxicants on the transcriptome have employed DNA microarrays to identify gene expression signatures, and employed quantitative real-time PCR (qRT-PCR) to validate differential gene expression. However, microarrays are somewhat limited in their readout. In the past, the role of mRNA splicing has also been assessed by employing splicing arrays. The disadvantage of these splicing arrays is that they require probes designed to be complementary to junctions, and these can therefore be generated only if the genes and the distinct isoforms produced from them are already known. Now, the role of mRNA splicing in the modulation of immune responses is becoming clearer, owing to recent advances in RNA sequencing techniques that have enabled interrogation of the spliceosome [38,39]. The relevance of RNA splicing for immunotoxicity still needs to be confirmed. Recent studies suggest that immunotoxic compounds can alter RNA splicing. For instance, a proteomics screen assessing the effects of the immunotoxic organotin compound TBTO in a murine T-cell line identified that the levels of two proteins involved in splicing of pre-mRNAs were decreased [40]—the splicing factor arginine/serine-rich 2 (2.6-fold), and chromodomain-helicase-DNA binding protein 4 (4.5-fold). Although this finding is no direct proof that TBTO alters RNA splicing itself, it does suggest that TBTO may affect this process, underscoring a role in immunotoxicity.

RNA sequencing

Meanwhile, a revolution in the analysis of RNA has come about through the development of tools for massively parallel transcriptome-wide sequencing of RNA by what is known as RNA-Seq [41]. Deep sequencing of RNA on Illumina's Genome Analyzer and HiSeq instruments and the Applied Biosystems' SOLiD instrument are now fast-developing alternatives for profiling the transcriptome [42].

Instead of using molecular hybridization to "capture" transcript molecules of interest, RNA-Seq samples transcripts present in the starting material by direct sequencing. Transcript sequences are then mapped back to a reference genome. Reads that map back to the reference are then counted to assess the level of gene expression, the number of mapped reads being the measure of expression level for that gene or genomic region. In this way, RNA-Seq has several advantages when compared to DNA microarrays. RNA-seq can simultaneously quantify the transcriptome and the miRNome, whereas DNA microarrays assess these separately; it provides direct access to the sequence, and junctions between exons can be assayed without prior knowledge of the gene structure. Thus, RNA editing and splicing events can be detected, knowledge of polymorphisms can provide direct measurements of allele-specific expression, and individual transcript isoforms can be quantified [39]. Recently, RNA-seq data have revealed that the human transcriptome undergoes extensive RNA editing [43], and that RNA editing potentially plays important roles in human complex diseases [44]. At present, the relatively low cost of microarrays is a great advantage, as compared with RNA-seq, but the cost differences between microarrays and RNA-seq are decreasing continuously owing to the increasing output of RNA-seq data, and, in the near future, will decrease further owing to label-free techniques such as nanopore sequencing [45].

3.3.5 Bioinformatics

Bioinformatics is defined as the application of statistics and computer science to the field of molecular biology. Therefore, this discipline has many applications, ranging from the design and development of software tools for devising PCR and sequencing primers, short interfering RNA (siRNA) molecules, probes for DNA/locked nucleic acid (LNA) microarrays, and DNA constructs for generating genetically modified organisms (GMOs), to the normalization and functional analyses of proteome, transcriptome, metabolome, and miRNome data. In toxicogenomics studies, bioinformatics is mainly used for the identification of effect markers and unraveling mechanisms of toxicity. Therefore these facets of bioinformatics will be highlighted.

Quality control

Regarding microarray data, bioinformatics starts with quality control (QC) of the scanned images, often TIFF files, and includes intra- and inter-array analyses of the background and spot intensities, respectively. These analyses enable the identification of outlier microarrays, or outlier spots on the arrays that can subsequently be filtered out. Several software packages are available for this. The most popular open source package that is applied for this purpose is Bioconductor (http://www.bioconductor.org), which is written in R [46]. This package can be used for most common microarray platforms, including Affymetrix Genechips, Agilent oligonucleotide microarrays, and Illumina Beadarrays. Furthermore, regarding the Affymetrix chip platform, unreliable probes can be filtered out by re-annotating the chip definition files (CDF) by employing Brainarray in Bioconductor (http://brainarray.mbni.med.umich.edu/Brainarray/) or by using the CDF-Merger algorithm [47]. Upon removing the outlier probes, one can proceed with the data normalization, which can be performed within the arrays (intra-array), and between the arrays (inter-array). Intra-array normalization is particularly important for dual color microarray data, such as miRNA microarray data from the Agilent or Exiqon platforms. This is because in dual color microarrays the signal intensity distribution of the two colors is mostly non-linear and must be corrected, for instance by using the Loess algorithm [48]. The normalization algorithms that are most commonly applied to single color microarray platforms, such as the Affymetrix Genechips and the Illumina Beadarrays, include the robust multichip average (RMA) and GCRMA [49,50]. These methods similarly correct the probe intensities for background and non-specific binding, and GCRMA also corrects non-specific signals due to high GC contents of the probes. At present, the normalization methods for mRNA deep sequence (RNA-seq) data have not matured as much as those methods currently available for microarray data, although good progress is being made. For instance, the Quantile normalization method, which corrects for inter-sample differences in read-counts, seems to be an effective way of normalizing RNA-seq data [51].

Upon removal of the outlier probes and genes from the microarray data and the subsequent data normalization, the next step is the identification of outlier samples. The employment of principle component analysis (PCA) and hierarchical clustering has proven to be effective for identifying and visualizing outlier samples in microarray data [46,52] and RNA-seq samples [53].

Statistics

Considering transcriptome, microRNome, or proteome data, the next step after normalization is the statistical data analysis, which serves at least three main purposes. These include the identification of (1) individual genes or proteins that are differentially expressed between two or more treatment groups, (2) a classifier gene set that can potentially distinguish between two or more treatment groups or could be predictive for exposure to certain toxicants or for specific endpoints of toxicity, and (3) canonical pathways or upstream transcription factors affected by a toxicant.

The statistical methods that can be employed for these three purposes mostly incorporate corrections for multiple testing errors, better known as the false discovery rate (FDR). The FDR is defined as the random chance of getting false-positive results upon statistical analysis. When subjected to statistical testing, high-content data have the inherent disadvantage that the number of false-positive outcomes depends on the number of statistical tests, which is equal to the number of genes in transcriptome data. As an example, we can compare the statistical analyses of two data sets, comprising 20 genes and 25,000 genes, respectively. When choosing a P-value threshold of 0.05 for dismissing the zero hypotheses, the random chance of getting false-positive outcomes is 5% of the number of genes being tested, equaling a single and 1250 false-positive genes for the data sets of 20 and 25,000 genes, respectively. In order to correct for these false positives, several multiple testing correction methods have been developed, including the Monte Carlo approach [54] and the more popular Benjamini–Hochberg FDR method [55]. Within each of these methods, one can vary the degree of conservativeness of the FDR correction. When choosing a conservative setting with a low FDR, one limits the chance of getting type I errors (false positives), but increases the chance of getting type II errors (false negatives), and vice versa. In addition, the FDR depends on various factors, including biological and technical variation, the data set size, the number of biological replicates, and the read-out method. Therefore, an empirical approach has to be followed in order to choose the proper FDR setting.

For the identification of individual mRNA or miRNA molecules that are differentially expressed between different treatments with toxicants, the most popular tools are Limma [56], the Cyber T-test [57], and significance analysis of microarrays (SAM) [58], respectively. Limma is based on linear models for microarray data [56], the Cyber T-test utilizes a Bayesian error model to estimate standard deviations (SDs) before conducting a regulated T-test [57], and SAM employs a non-parametric method [59]. These statistical tests can be applied to high-content data from microarrays or from RNA-seq [57,59], and can also generate two-class or multiple-class outcomes.

Classifier analyses

In the last decade, several algorithms have been employed for the identification of marker genes. These algorithms include support vector machines (SVM) [60]; SVM in combination with recursive cluster elimination (RCE) [61]; K-nearest neighbors (KNN) [62]; random forest (RF) [63]; and nearest shrunken centroids [64], applied in prediction analysis for microarrays (PAM) [65]. The application of RF and SVM has led to the identification of mRNA-based effect markers for immunotoxicity in vitro [66], and of markers that potentially distinguish between skin-sensitizing and skin-irritating compounds in vitro [67]. Thus, the RF and SVM algorithms are particularly useful for toxicogenomics applications when one needs to identify markers that distinguish between two classes of compounds. In theory, these classification algorithms could also be applied for the identification of markers that distinguish between more than two classes of toxicants or toxicity endpoints, although in practice this has not yet been done.

When a classifier analysis has led to the identification of one or more candidate effect markers for certain toxicities, the predictive value of this classifier set needs to be verified. First, the predictive value of the candidate effect marker set is verified internally by a suitable quantification technique, such as quantitative real-time PCR, deep sequencing, dedicated bead arrays, gene chips, or microarrays. Thereafter, one proceeds with external validation studies, in which the predictive value of the marker set is verified by exposing the biological system to compounds or external stimuli that have not been used for the initial identification of the marker set. With respect to clinical studies, the external validation of potential biomarkers is performed in a separate group of subjects.

The predictive value of a biomarker set for a toxicity endpoint might be limited when this set is based on a single molecular level, like mRNA, whereas the endpoint could be regulated at different molecular levels, such as miRNA or protein. This could particularly be the case when subsets of toxic compounds with similar toxicity endpoints have modes of action at different molecular levels, like miRNA or protein. Different strategies have been developed in order to overcome this potential problem, and to increase the predictive value of potential biomarker sets. These strategies include (1) the incorporation of 'omics data from different sources, such as mRNA, miRNA, and proteins, and (2) combining 'omics data with in silico quantitative—activity relationship (QSAR) models. Such combined approaches have not yet led to the identification of improved effect markers for toxicity. However, the strategy of combining data from different 'omics sources has led to the identification of biomarkers with improved predictivities for endpoints of complex multifactorial diseases in humans, such as diabetic nephropathy (DN) [62], triple-negative breast cancer [68], and psychiatric disorders [63]. Furthermore, a tool has recently become publicly available called Resource, which can be used to identify biomarkers on the basis of multi-'omics data in an automated and standardized fashion [65]. Therefore, the integration of toxicogenomics data from more than one source, including the miRNome, transcriptome, epigenome, proteome, and phosphoproteome, is a potentially promising strategy with respect to the discovery of better effect markers for toxicity.

Pathway analysis

Several pathway analysis tools are available as open source applications, or as commercial packages. These tools employ genomic enrichment analyses of genes from existing metabolic and cellular pathways from, for example, Biocarta (http://cgap.nci.nih.gov/Pathways/BioCarta_Pathways), KEGG [69], or GenMAPP [70], from the gene ontology (GO) database (http://www.geneontology.org) [71], or from eukaryotic transcription factor databases, such as TRANSFAC, TRRD, and COMPEL [51,72]. The most popular open source pathway analysis tools include DAVID [73], gene set enrichment analysis (GSEA) [74], Mappfinder [75], molecular signature database (MySigDB) [76], pathway explorer [77], SAM [58], PAM [65], and Webgestalt [78]. Some of these tools, including GSEA, PAM, SAM, and whole-genome rVista [79], can provide insights about transcription factors that could control the mRNA expression levels of co-regulated genes. For this purpose, these tools employ databases that sort genes according to known and putative transcription factor binding sites (TFBS) that are conserved across the genomes of multiple mammalian species—human, mouse, rat, and dog for GSEA, PAM, and SAM [58], and human and mouse, or human and chicken for whole-genome rVista [79].

Many of these tools, including DAVID, Mappfinder, and Webgestalt, use Fisher's exact test to identify those pathways, GO terms, or upstream transcription factors that have increased relative abundances among groups of co-regulated genes, as compared to the complete data set or the whole genome. In contrast, the GSEA method is based on identifying differences in ranking of up- or down-regulation based on the Kolmogorov–Smirnov test [74]. In the past, concerns have been raised about the validity of the GSEA approach, because false positives may result when genes within a set of co-regulated genes have different variances than the genes outside this set [80,81]. At present, other methods are available that do account for different variances: functional class scoring [82] and parametric gene set enrichment (PAGE) [46]. At the moment, GSEA, SAM, and PAM are among the most useful pathway analysis tools of toxicogenomics transcriptome data. The application of these tools has expanded our knowledge about the MOA of various toxic compounds. For instance, the application of GSEA has led to the identification of the MOA of immunotoxic compounds [66,83–87], the application of SAM has led to the identification of radiation response genes in lymphoblastic cell lines in vitro [58], and the employment of PAM has led to the identification of putative effect marker genes for potential non-genotoxic hepatocarcinogenicity in rats in vivo [79].

In the near future, these pathway analysis tools could be succeeded by novel tools that potentially perform better owing to the application of more refined algorithms, such as the adaptive rank truncated product (ARTP) method [88], PAGE, and functional class scoring.

In general, the major disadvantage of these pathway analysis tools is that their outcomes are based on prior knowledge that is present within the fixed GO, pathway, and transcription factor databases that these tools employ. Therefore, for a given set of co-regulated genes, such tools will generally not provide insights into novel or poorly described pathways, cellular functions, or putative upstream transcription factors. An alternative approach, which has been developed to overcome this shortcoming, is the generation of gene networks based on text mining. Text mining is based on the principle of scanning databases containing scientific experiments or literature for gene pairs occurring within the text bodies or within the same experiment, respectively. Tools that employ text-mining methods include Anni [89], Ingenuity [90], Genomatix [91], Metacore [59], open mutation miner (OMM) [60], resource manager [65], and Segmine [48]. With respect to transcriptome data analyses, these text-mining tools can be employed to identify interactions between genes and miRNAs; cellular proteins, including receptors and signaling nodes; and putative upstream transcription factors. Some of these tools, like Ingenuity and Metacore, can simultaneously identify these different interactions, whereas others, like resource manager or OMM, specifically focus on identifying interactions with miRNAs or mutations, respectively.

Text mining methods have two inherent drawbacks regarding the biological or toxicological interpretation of their output. The first drawback is that a significant part of the connections within a gene network that is based on a text-mining method might not be functionally relevant. For example, in the scientific literature, the genes that are present within the same text body of a publication might not have a functional relationship. To overcome this problem, two of these tools, Ingenuity and Metacore, implement the strategy of manually verifying the functional relevance of each co-citation before incorporating it into an internal database. Such manual curation has the drawback of being labor intensive and time-consuming. Another downside of text-mining methods is the publication bias [55]. For instance, within text-mining results for gene interactions, the co-citations will be over-represented for well-described genes, such as the transcription factor p53. Thus, text-mining results are inherently biased towards the best-described data, which is a problem that has not yet been resolved.

Another method consists of comparing the co-regulated genes or proteins between your own experiments and other toxicogenomics experiments available in public high-content data repositories, such as the transcriptome data repositories GEO [33], Arrayexpress [92], RNA-seq Atlas [62], the Connectivity Map [93], and the Comparative Toxicogenomics Database (CTD) [67]. Typical outcomes of such cross-database analyses provide correlations between expression signatures and exposure conditions, diseases, or genetic alterations such as mutations and single-nucleotide polymorphisms (SNPs), respectively. Cross-database analysis can be done manually by importing publicly available data into a private database, but this has the disadvantage of being labor intensive. Therefore, several tools have recently been developed that can perform cross-database comparisons of high-content data in an automated fashion, including DvD [94], SPIED [64], MARQ [68], and ProfileChaser [63]. The Drug versus Disease (DvD) and the Searchable Platform-Independent Expression Database (SPIED) tools both compare drug- and disease-related expression profiles from several public microarray repositories, whereas the MARQ and ProfileChaser tools mine data from the public transcriptome database GEO. These tools have not yet been frequently applied in the analysis of toxicogenomics data, as they are relatively new. Thus, particularly the tools that can simultaneously cross-compare databases, such as DvD and SPIED, are promising with respect to gaining insights into the MOA of toxic compounds and the identification of toxicity-related markers, respectively.

Toxicogenomics data infrastructure

The open source packages Cytoscape [95,96] and VANTED [97] include powerful functionalities that enable integrated analysis of high-content data from the genome, metabolome, and proteome, thereby enabling a systems toxicology approach in data analysis. In addition, the ToxPI graphical user interface is a powerful tool to visualize toxicity data from various sources [98].

With the goal of integrating experimental toxicogenomics data from internal and external sources, the FDA has developed the Arraytrack database [67]. Arraytrack consists of three main components: MicroarrayDB, TOOL, and LIB. MicroarrayDB stores the data, whereas TOOL normalizes, analyzes, and visualizes the data, and LIB can store information from public repositories, including gene annotation, protein function, and pathways. Arraytrack has the advantage that the database environment is stable, user-friendly, and flexible in terms of storing experiment annotation data. However, Arraytrack has the drawback that it is mainly focused on storage and analysis of microarray data and cannot fully integrate high-content data from other sources, such as the metabolome and the proteome. In the future, multi-'omics data integration within Arraytrack might be enabled via implementation of a toxicoinformatics integrated system (TIS) [67]. A true multi-'omics data integration infrastructure has not yet been realized, although the need for such an infrastructure still remains. Within the EU, a bioinformatics project has been proposed with the aim to realize such a system, named the Data Infrastructure for Chemical Safety (diXa) project (see http://www.dixa-fp7.eu). In addition, diXa aims to enable integration with experimental and modeling data from toxicokinetics and toxicodynamics studies. If the aims of the diXa project are achieved, then this will facilitate the systems toxicology approach by enabling a true integration of high-content data from the genome, epigenome, metabolome, and the proteome.

3.3.6 Immunotoxicogenomics studies: state of the art

Within the scope of direct immunotoxicity, toxicogenomics studies can serve different purposes. They can be employed for identifying marker genes or proteins for exposure to immunotoxicants, or for identifying or confirming mechanisms of immunotoxicity. Genomic or proteomic markers for exposure level can potentially be useful for biomonitoring purposes, whereas markers that correlate with immunotoxic endpoints can be employed for identification or characterization of hazard for direct immunotoxicity, for instance during regulatory safety testing of chemicals or drugs.

Human in vivo toxicogenomics studies

By the time of publication of this book, the number of toxicogenomics studies that have assessed genomic effects of immunotoxicant exposure in humans in vivo is still limited to three. An overview of these studies is given in Table 3.3.1.

Human in vivo studies are particularly useful for the identification of effect markers for immunotoxicants. For instance, human population studies can be designed to identify effect markers for environmental immunotoxicants. Examples of in vivo toxicogenomics studies that have led to the identification of potential effect markers for immunotoxicants are those by McHale et al. [99], and Jetten et al. [100].

The study by McHale assessed the effects of 2,3,7,8-tetrachlorodibenzo-p-dioxin (TCDD) exposure on the peripheral blood mononuclear cell (PBMC) transcriptome in a cohort of adult women from the 1976 Seveso disaster [99]. During this disaster, thousands of residents were exposed to very high levels of TCDD, and during the following two decades the local population suffered from increased incidence of hematopoietic neoplasms [101] and the skin condition chloracne [102]. In order to identify gene expression changes upon chronic exposure to high levels of TCDD, 20 years after the disaster, the PBMC transcriptomes were compared between women with high and low plasma TCDD levels. This comparison led to the identification of 135 genes that were differentially

Table 3.3.1 Human In Vivo Immunotoxicogenomics Studies

Study	Subjects and Sampling	Exposures, Sampling	Pathology/ Phenotype	Results, Transcriptome Changes
[100]	N = 5 ♂ N = 2 ♀, aged 48.4 ± 4.5 years Whole blood (transcriptome, miRNome), urine metabolome	APAP; single oral, 0.5, 2, or 4 g Blood sampled 0, 1, 7, and 25 h post oral dose Urine samples 0 and 24 h	APAP metabolites linked to hepatotoxicity and oxidative APAP metabolism	**Enriched GO-terms mRNA analyses:** 2 and 4 g (25 h): immune response, lipid and drug metabolism, cholesterol transport, apoptosis and survival/ oxidative stress/DNA damage **miRNA–mRNA correlation analyses:** 2 and 4 g (25 h): multiple reciprocal correlations observed. Correlating mRNAs enriched in the same GO terms as all APAP responding genes, but excluding drug metabolism
[105]	Neonates: N = 45 ♂ N = 66 ♀ Transcriptome umbilical-cord-blood-derived PBMC	PCBs and TCDD, cumulative maternal exposure estimated by FFQ PBMC sampled postpartum from cord vein	Measles vaccination response at age of three, determined by circulating Ig levels	**Maternal chemical exposure–mRNA correlations:** ♀ and ♂: many genes correlated well. In ♀ best correlations found for PCB169: N = 484 genes (+313, −171), and in ♂ for TCDD: N = 399 genes (+235; −164) **Ig levels measles vaccination–mRNA correlations:** ♀ N = 2890 genes (+1848, −1042), ♂ N = 1750 genes: (+755, −995). Genes with − correlations to chemical exposures generally showed + correlations with Ig levels **Enriched GO-terms genes correlating to exposures:** ♀ and ♂: immune-related genes
[99]	Seveso cohort ♀ aged 24–49 years: N = 13 exposed, N = 13 controls Transcriptome PBMC	Plasma TCDD high (24–268 ppt), or low (4.5–7.9 ppt) Peripheral blood sampled 20 years post-exposure.	Chloracne	**High versus low plasma TCDD:** N = 22 genes up, including N = 4 histone genes (*HIST1H3H* confirmed by real-time PCR). N = 113 genes down. Gene networks involving cancer, cell death, cellular growth and proliferation, replication, recombination, and repair of DNA **Chloracne versus no chloracne:** N = 23 genes up, including N = 5 hemoglobins; N = 9 genes down. Networks involving cell death, proliferation, immunological and hematological disease

expressed between high and low plasma TCDD levels. The majority of these genes had a lowered mRNA expression in the high-plasma-TCDD group. At the individual gene level, the observed effect size of TCDD was very small, as only 20 genes were modulated by more than 1.5-fold. At pathway level, several processes were affected that could be indicative of immunosuppression, such as cell death, and cellular growth and proliferation. In conclusion, it is questionable whether these 135 genes can be used as effect markers for chronic dioxin exposure in vivo, because the small effect size in relation to the dioxin exposure level would probably limit the sensitivity of a future assay based on these genes. Furthermore, it is questionable whether plasma TCDD levels can be used as an accurate measure for the internal dose of TCDD, because plasma TCDD levels formed the basis for the identification of these effect markers.

Jetten et al. assessed the acute effects (1–25 hours) of a single oral dose of acetaminophen (APAP) on the transcriptome, microRNome, and metabolome of seven healthy human volunteers [100]. This study led to the identification of mRNA and miRNA molecules that correlated with the APAP exposure level. Among these APAP-responsive mRNA and miRNA molecules, the levels of 11 miRNAs inversely correlated with 89 mRNAs. Furthermore, GO analysis indicated that APAP had immune-activating properties. The latter finding seems at variance with results from rodent studies that point out immunosuppressive properties of APAP [103,104]. However, in these rodent studies, APAP only induced immunosuppression upon chronic administration of high doses that were linked to hepatotoxicity [103]. Therefore, it is still a matter of debate whether APAP is directly immunotoxic, or hepatotoxic with secondary immunotoxic properties.

Toxicogenomics can also be employed in humans in vivo to identify marker genes or proteins for endpoints of immunotoxicity. In order to identify such markers for immunotoxic endpoints, toxicogenomics and internal exposure data need to be combined with an in vivo parameter for immune functionality, such as an antibody response upon vaccination. An example is the study by Hochstenbach et al. [105], which led to the identification of potential markers for reduced immune functionality upon exposure to immunotoxicants in humans in vivo. This study interrogated the transcriptomes of umbilical-cord-blood-derived PBMC from 111 newborns of the Norwegian BraMat cohort. Earlier studies on this cohort found that maternal exposure to polychlorinated biphenyls (PCBs) and TCDD, based on validated food-frequency questionnaires (FFQs), correlated with an increased risk of infections in the offspring after 1 year [106] and reduced measles vaccination responses at the age of 3 [107]. The toxicogenomics study by Hochstenbach et al. found that the mRNA levels of different gene sets correlated with the estimated exposure levels of different PCBs and TCDD, respectively. Furthermore, those genes that had positive correlations with antibody levels induced upon measles vaccination were found to have negative correlations with PCB and TCDD exposure in general, and vice versa. In addition, these genes were known to be involved in immune-related processes. Altogether, these findings suggest that these genes could potentially be used as markers for immune functionality in umbilical-cord blood of neonates. Clinical validation studies will be required to determine the predictive value of these genes for immune functionality in vivo. In addition to these findings, this study led to the identification of gender differences in the respective transcriptome responses to PCB and TCDD and to measles vaccination. The female neonates were found to be more responsive than the males in the number of genes for which the mRNA levels correlated with the chemical exposure and the immunogenicity of the measles vaccination, respectively. Thus, these findings suggest that females are more sensitive to the immunotoxic properties of aryl hydrocarbon receptor (Ahr) agonists than males. Evidence from the literature suggests that Ahr-independent effects play roles in these gender-related sensitivity differences to PCBs and TCDD, such as estrogenicity [108,109]. Further research is required to more completely characterize the molecular basis for these gender-related differences.

Rodent in vivo toxicogenomics studies

In risk assessment of direct immunotoxicity, the application of toxicogenomics in rodent models in vivo is particularly useful for risk characterization purposes, because the outcomes provide mechanistic information on the potential MOA at the molecular level. Such detailed information will not be provided at the molecular level when one only evaluates phenotypic changes of the immune system upon exposure to the test compound.

Several rodent in vivo studies, in which direct immunotoxicity had been evaluated by combining toxicogenomics with phenotypic endpoint of the immune system, have provided in-depth information on the MOA of direct immunotoxicants (see Table 3.3.2). In addition, these in vivo studies all employed transcriptome data analysis on target immune cells or tissues of animals that had been exposed to different levels or durations of the immunotoxicants. This was done in order to determine the relationships between the level and duration of exposure with potentially different mechanisms of immunotoxicity. In two murine toxicogenomics studies, the MOA of immunotoxicants was compared with that of immunosuppressive drugs. This was done to establish the degree of diversity in the potential mechanisms of direct immunotoxicity. For instance, Baken et al. compared the MOA of the immunotoxicants APAP and benzo(a)pyrene (B(a)P) with the immunosuppressive drug cyclosporin A (CsA) using splenocytes [104], whereas Frawley et al. compared the MOA of TCDD with three different immunosuppressive drugs using thymocytes [110]. In both of these studies, all of the immunomodulating test compounds induced structural changes in the thymus, which included macroscopic and histological changes, and growth reduction. Baken et al. mainly found that the different immunotoxicants had overlapping mechanisms of action: B(a)P, CsA and bis(tri-n-butyltin) oxide (TBTO) all showed gene expression signatures for cell cycle arrest. Frawley also identified overlapping MOAs between the different immunotoxicants that she tested: (1) altered T-cell receptor and CD28 signaling by cyclophosphamide (CPS), diethylstilbestrol (DES), and dexamethasone (DEX), and (2) induction of antigen presentation and dendritic cell (DC) maturation by DES, DEX, and TCDD. For TCDD, its effects on the thymocyte transcriptome, which indicated immune activation, seem at variance with its effects on the immunopathology of the murine thymus and the transcriptome of human PBMC in vivo, both indicative of immunosuppression [99,105].

The studies by van Kol et al. [85] and McMillan et al. [111] employed transcriptomics on murine thymocytes in order assess immunotoxic mechanisms of single compounds, i.e. the mycotoxin deoxynivalenol (DON) and the dioxin TCDD, respectively. These studies identified different mechanisms of immunotoxicity for DON and TCDD. van Kol et al. found that DON induced genes involved in T-cell activation (TCA) and the negative selection of thymocytes that recognize "self-antigens," and down-regulated genes encoding mitochondrial and ribosomal proteins, regulators of cell proliferation and protein synthesis, and markers for early CD4+/CD8+ double positive (DP) precursor thymocytes. These data suggest that DON potentially induces thymic TCA, thereby evoking an increased negative selection and depletion of thymocytes. In this study, the TCA properties of DON were confirmed by additional experiments with murine thymocytes in vitro, in which DON evoked calcium release from the endoplasmatic reticulum (ER), and nuclear translocation of the proinflammatory transcription factor NFAT.

Table 3.3.2 Rodent In Vivo Immunotoxicogenomics Studies

Study	Sex, Strain, Species	Compounds, Exposure Route, Dosing (mg/kg BW/Day),	Organs, Exposure Time and Group Size	Pathology/Phenotype	Transcriptome Changes
[110]	Female B6C3F1 mice	CPS IP 0, 5, 50 DES SC 0, 0.8, 8 DEX IP 0, 0.5, 5 TCDD OG: 0, 0.3, 3	Thymus, 6 days, N = 4	Thymic atrophy, and altered ratios CD3+, CD4+, and CD8+ cells	CPS, DES, DEX: TCR and CD28 signaling; TCDD, DES, DEX: (induction) antigen presentation and DC maturation
[85]	Male C57BL6 mice	DON OG 0, 5, 10, 25	Thymus, 3, 6, and 24 h, N = 3 pools of N = 2 thymuses	Thymic atrophy: 24-h exposure to highest dose reduced thymus weights by 33% relative to BW	Up: inflammatory- and immune-response genes; down: cell-cycle-regulating, ribosome, and protein-translation-related genes
[104, 120]	Male C57BL6 mice, male Wistar rats	**Mice:** APAP AL 0, 5000 ppm B[a]P OG 0, 13 mg/kg BW 3 × weekly CsA AL 0, 300 ppm TBTO AL 0, 5–300 ppm; **Rats:** TBTO AL 0, 5–80 ppm	Spleen and thymus, 3, 7, 14 days; rats also 28 days, N = 4 pools of N = 2 organs	TBTO: atrophy of spleen (mice) and thymus, and altered histology (7/8 mice); B[a]P and CsA (mice): atrophy spleens; APAP: no atrophy spleen	**Mice:** B[a]P, CsA, TBTO: cell division (anti-proliferative), DNA replication, and metabolism of xenobiotics **Rats:** Thymus TBTO: N = 2 genes; spleen N = 1 gene
[111]	Male mice: several lines	N = 2: TCDD 0 (vehicle), and 100 μg/kg BW single IP	Thymus, 48 h	Induction novel CD3−, CD4−, CD8− triple-negative thymocyte phenotype with intermediate expression CD44 and CD25	Involvement KLF2 induction in the TCDD-mediated differentiation block thymocytes towards CD4/CD8 DP T-cell precursors

The murine toxicogenomics study by MacMillan et al. identified the involvement of a gene set involved in blocking thymocyte precursor maturation upon induction by TCDD [111]. This gene set included the zinc finger transcription factor KLF2 and its target genes. Furthermore, KLF2 is normally only expressed at a more mature stage, and its physiological role is to enforce egress of thymocytes to the peripheral lymphoid organs [112,113] and to induce peripheral T-cell quiescence [114].

In summary, rodent in vivo toxicogenomics studies have mainly led to an increased mechanistic understanding of the different aspects of immunotoxic endpoints, but have not yet led to the identification of effect markers for immunotoxicity. Furthermore, the outcomes of these studies are difficult to translate to the human situation, owing to many factors, including interspecies differences in sensitivity and metabolism, and different exposure levels, routes, and durations.

For the majority of the immunotoxic test compounds, evidence of their immunomodulating properties is mainly based on rodent models for human toxicity, such as for the antipyretic and analgesic drug APAP [103,104,115], the polycyclic aromatic hydrocarbon B(a)P [104,116], the mycotoxin DON [117–119], the organotin compound TBTO [120–123], and dioxin [124–126]. Therefore, the relevance for human immunotoxicity needs to be further verified for these test compounds, including their associated modes of action. When such animal-to-human interspecies comparisons are made, one should also take into account the different sensitivities of the molecular targets of the immunotoxicants. Therefore, further research will be required to elucidate the MOA of additional direct immunotoxicants that form potential human health risks upon exposure though the environment [69,127], food [10,127], clinical treatment [1,89], or upon occupational exposure [61,74,128]. In addition, one should focus on the MOA of direct immunotoxicants to which children can be exposed, because the developing immune system seems to be particularly sensitive to immunotoxicants [61,129,130].

In vitro toxicogenomics studies

When comparing the potential applications of in vitro with in vivo employment of toxicogenomics, both approaches can be used for the identification of effect markers for immunotoxicants, and also for risk identification of direct immunotoxicity. In risk assessment of chemicals or drugs, toxicogenomics can also be employed in vitro for characterizing the potential risk or hazard for direct immunotoxicity, depending on whether bioavailability data are provided from in vivo studies.

Toxicogenomics studies employing human and rodent lymphocytes to assess genomic effects of immunotoxicants in vitro and ex vivo are summarized in Table 3.3.3.

Effect markers for direct immunotoxicity

The study by Hochstenbach et al. [66] is an example of employing toxicogenomics in vitro for the purpose of identifying effect markers for immunotoxicants. In this study, transcriptome data analysis led to the identification of an mRNA expression signature for immunotoxicants, consisting of 48 genes. Further pre-validation studies will be required to determine whether this gene expression signature can be employed as effect markers for human immunotoxicants in vivo. If so, then this set could potentially be employed for biomonitoring purposes.

Table 3.3.3 In Vitro and Ex Vivo Toxicogenomics Studies

Study, Species, and Cell Types	Study Type	Test Compounds	Exposure Time, Concentration	Transcriptome/Proteome Changes
[66], human, PBMC	In vitro transcriptome	AFB1, B[a]P, DON, EtOH, 4-HNE, MDA, PCB153, TCDD; controls AA, DMNA, IQ, PhIP	20h, 100, 10, and 1%	Identified mRNA expression signature for immunotoxicity of N = 48 genes. These genes are partly involved in immune-system-related processes such as acute-phase response, myeloid leukocyte activation, and cytokine production
[86,87], human, Jurkat, PBMC		DON, TBTO, EtOH (vehicle)	3, 6, 24h **Jurkat:** DON: 0.25, 0.5μM TBTO: 0.2, 0.5, 1μM **PBMC:** DON 2, 4μM	**DON and TBTO**-induced genes involved in apoptosis, ER-stress, NFAT-mediated TCA, signaling of AP-1 and NFκB, and own-regulated anti-apoptotic and immune response genes **DON:** induction of genes involved in regulating transcription and translation, and ribosome function **TBTO:** down-regulation of genes involved in oxidative phosphorylation, indicative of inhibiting mitochondrial activity
[133], mouse, thymocyte		TBTO, EtOH (vehicle)	3, 6, 11h 0.1, 0.5, 1.0, 2.0μM	Induction of genes involved in DNA damage response, ER stress, apoptosis, TCA, and NFκB and TNF pathways; TCA corroborated by nuclear translocation of NFAT
[167], rat, primary thymocyte		TBTO	3, 6h 0.3, 0.5, 1.0μM	**Low level (0.3uM):** expression of apoptosis-related genes both up and down, stimulation of glucocorticoid receptor signaling **High level (1.0uM):** inhibition of mitochondrial activity and immune cell activation

(Continued)

Table 3.3.3 In Vitro and Ex Vivo Toxicogenomics Studies (Continued)

Study, Species, and Cell Types	Study Type	Test Compounds	Exposure Time, Concentration	Transcriptome/Proteome Changes
[131], mouse, thymocyte	Ex vivo transcriptome	TCDD, anisole/peanut oil (vehicle)	24h; oral gavage, 0, 15 μg/kg BW	Induction of CCR9, GZMB, BLIMP-1, TGF-beta3, *Blimp1*, *Gzmb*, and genes of the IL12-Rb2 signaling pathway, including an enhanced phosphorylation of STAT4; inhibition IL-2 production through inhibition of *Gitr*; 2% of TCDD-treated CD4(+) cells expressed *Foxp3*, suggesting that *Ahr* does not rely on *Foxp3*-suppressive activity
[40,136], mouse, EL-4	In vitro phosphoproteome and proteome by nano-LC MS/MS and SDS/PAGE; phosphopeptides enriched by two-step TiO_2-column	TBTO, EtOH (vehicle)	6h, 0.5 μM	**Phosphoproteome:** altered energy balance (ACC, GFAT1, PDH), protein synthesis (lower RPS6KA1 level). Down-regulation cell cycle-regulators MAPK, MATR3, and RRM2 **Proteome:** down-regulation cytoskeleton proteins (myosin-9, SPNB2, and plectin 8), and proteins involved in RNA splicing (CHD4, SRSF2), translation (EIF4G1, S10), and cell proliferation (PTMA)
[135], human, Jurkat, RPMI1788	In vitro phosphoproteome; 2D-gel electrophoresis + MALDI-TOF MS/MS of tryptic peptides	DON, EtOH (vehicle)	24–54h 0.8, 1.7 μM	**Altered phosphorylation:** protein folding (HSPA8 ↓), signal transduction (GRB2 ↓, NME1 ↓), metabolism (MTHFD1 ↑, NDKA↑), translation (EIF3S1 ↑, EIF5A2 ↑)

ACC, acetyl-CoA carboxylase isoform 1; EIF3S1, eukaryotic translation initiation factor 3 subunit I; EIF4G1, eukaryotic translation factor 4 gamma 1; EIF5A2, eukaryotic elongation factor 2; EL-4, EL-4T-cell line; GFAT1, glutamine-fructose-6-phosphate amidotransferase; GRB2, growth factor receptor-bound protein 2; HSPA8, heat shock cognate 71kDa protein; Jurkat, Jurkat T-cell line clone E6-1; MATR3, matrin-3; MAPK, mitogen-activated protein kinase; MTHFD1, cytoplasmic C-1-tetrahydrofolate synthase; NFAT, nuclear translocation of transcription factor; NFκB, nuclear factor kappa B; NME1, nucleoside diphosphate kinase A; PBMC, peripheral blood monocyte cells (primary cells); PDH, pyruvate dehydrogenase; PTMA, prothymosin alpha; RPMI1788, RPMI1788 human B-cell line; RPS6KA1, ribosomal protein S6 kinase 1; RRM2, ribonucleoside-diphosphate reductase subunit M2; S10, ribosomal protein S10; SRSF2, serine/arginine-rich splicing factor 2; SPNB2, spectrin beta2; TCA, T-cell activation; TNF, tumor necrosis factor.

Molecular mechanisms of direct immunotoxicity

Toxicogenomics has also been applied ex vivo in order to elucidate molecular mechanisms of immunotoxicity upon a secondary exogenous antigenic or mitogenic stimulus. The study by Marshall et al. [131] is an example of an ex vivo study that was conducted for this purpose with respect to the immunosuppressive mode of action of TCDD. One crucial finding of this study was that the activation of the Ahr may represent a novel pathway for the induction of regulatory T cells by TCDD.

Studies employing toxicogenomics in vitro in order to identity potential mechanisms of direct immunotoxicity have been conducted at the transcriptome level by Baken et al. [132], Katika et al. [86,87], van Kol et al. [133], and Schmeits et al. [134], and at the proteome level by Nogueira et al. [135] and Osman et al. [40,136] (see Table 3.3.3). The two studies by Katika et al. individually focused on the mechanistic characterization of the mycotoxin DON [87] and the organotin TBTO [86] in human lymphocytes in vitro. These studies showed that DON and TBTO similarly induced genes involved in calcium and NFAT signaling, jun proto-oncogene (JUN) and nuclear factor kappa B (NFκB)-mediated transcription, and DNA-damage inducible transcript 3 (DDIT3)-mediated ER stress responses leading to T-cell activation, followed rapidly by apoptosis. DON specifically induced genes involved in ribotoxic stress responses. TBTO specifically down-regulated genes involved in oxidative phosphorylation, whereas DON had no effect on this process. Furthermore, DON elicited comparable transcriptome changes in the Jurkat cell line as in primary lymphocytes from human peripheral blood. These data indicate that DON more specifically induces ribotoxic stress, whereas TBTO is a more specific inhibitor of mitochondrial activity. A further in-depth characterization of the MOA of TBTO in vitro by Katika et al. [137] confirmed that this organotin compound induced oxidative stress, ER stress, and activated calcium signaling, at the same exposure conditions that applied for the transcriptome analyses. These findings confirmed earlier work in which tributyltin (TBT) was found to induce thymocyte apoptosis through a calcium-mediated induction of ER and ribotoxic stress [138]. The murine in vivo toxicogenomics study by van Kol et al. [85] also complemented the mechanistic in vitro data of Katika et al. [87], as in both studies DON altered the expression levels of genes involved in apoptosis, ER stress, and TCA. The genes involved in TCA were part of CD40 and chemokine signaling. In addition, in both these human and murine studies, the employment of qRT-PCR confirmed the induction of ER stress and TCA. In contrast with these data, Schmeits et al. observed that DON (6H, 2 μM) down-regulated mRNA expression of genes involved in TCA, ER stress, and apoptosis in CTLL-2 cells, a murine cytotoxic-T-cell line [134]. In this study, it was proposed that these conflicting findings in CTLL-2 cells may be explained by a partial lack of a signal transduction route that connects secondary cellular stress responses to ribotoxic stress.

The ER and ribotoxic stress response pathways are generally considered as being pro-apoptotic to lymphocytes [80,90,139], thereby indicating that the induction of lymphocyte apoptosis is a potential mechanism of direct immunotoxicity that the immunotoxicants DON and TBTO have in common. With respect to immunotoxicity-related gene expression changes that were elicited by DON, the transcriptome data confirmed earlier findings of functional characterization studies with leukocytes in vitro. In these studies, DON induced immune-stimulatory genes in sub-cytotoxic conditions, whereas in cytotoxic conditions DON induced ribotoxic stress and promoted apoptosis with concomitant immune suppression [139].

When comparing organotin-mediated changes of the lymphocyte transcriptomes of murine [133,134], rat [132], and human origin [86,140], respectively, the outcomes are strikingly similar.

At sub-cytotoxic levels, all of the toxicogenomics studies showed that organotin compounds primarily affected lipid metabolism, whereas at cytotoxic levels TBTO elicited more broad stress responses, including ER stress, which eventually led to the induction of apoptosis. However, one striking interspecies difference was observed. When cytotoxic conditions prevailed, both the rodent studies observed that TBTO induced genes involved in glucocorticoid receptor (GR) signaling [132,133], whereas no GR-target genes were induced in human T lymphocytes [86]. However, an important difference between these rodent and human studies is that thymocytes had been employed in the rodent studies, whereas PBMC and Jurkat T cells had been used in the human studies. Therefore, additional experiments are required to establish whether the differences in GR responsiveness between these studies were caused by an inherent difference between rodents and humans or between thymocytes and lymphocytes.

The in vitro toxicogenomics study of Shao et al. [140] aimed to identify mechanisms for direct immunotoxicity and to explore the degree of diversity among these mechanisms. For these purposes, Shao studied the effects of various direct immunotoxicants on the transcriptome of a human T-cell line in vitro, including different structural classes of immunotoxic chemicals and immunosuppressive drugs. This approach led to the identification of several pathways and processes that were transcriptionally modulated by distinct subsets of immunotoxicants with dissimilar structures. Shao et al. discovered that large sets of direct immunotoxicants transcriptionally modulated several cellular stress response pathways, including retinoic acid/X receptor (RAR/RXR) signaling, cholesterol biosynthesis, and Notch receptor signaling. Based on these findings, we hypothesize that these pathways play key roles in mediating direct immunotoxicity in T cells in vivo. Verification studies need to be conducted in order to confirm the potential mechanisms of direct immunotoxicity that have been identified in this study.

Nogueira et al. applied phosphoproteomics in order to assess the functional effects of DON on human B and T lymphocytes in vitro [135]. In this study, DON decreased the phosphorylation status of proteins involved in signal transduction and protein folding, and increased the phosphorylation of proteins that regulate metabolism and the translational machinery. These findings underline the observations that DON affects protein translation in human T lymphocytes, as has also been found by the transcriptome analyses of Katika [87]. With the current data available, it is difficult to determine the underlying factors that caused these different outcomes between the in vivo and in vitro studies that have assessed the MOA of DON. Additional studies will be required to determine which putative factors could explain these different outcomes, such as the differences in dosing and timing, and the human–mouse interspecies difference.

Two studies by Osman et al. employed proteomics and phosphoproteomics, respectively, in order to assess the functional effects of TBTO on a murine lymphoblastic T-lymphocyte cell line [40,136] (see Table 3.3.3). The proteomics study showed that TBTO mainly down-regulated the expression levels of cellular proteins involved in maintaining cytoskeleton structure and integrity, RNA splicing, protein translation, and cell proliferation. The phosphoproteomics study showed that TBTO altered the levels of phosphopeptides involved in the energy balance and protein translation, and down-regulated cell-cycle-regulating phosphopeptides. Comparing the outcomes of these two studies at the functional level shows overlaps between the proteome and phosphoproteome with respect to TBTO-mediated inhibition of cell proliferation and protein translation. In contrast, regulators of the energy balance seemed to be selectively regulated by dephosphorylation, whereas cytoskeletal proteins appeared to be selectively down-regulated without altering the phosphorylation status. Further research has to validate these overlaps and differences between the respective effects of TBTO on the proteome and phosphoproteome. When comparing the functional outcomes of these proteomics and phosphoproteomics studies with the in vitro transcriptomics studies by van Kol and Katika et al. [86,133], all three studies observed that TBTO affected the cell cycle. These outcomes also overlap with the observations by Baken et al. that TBTO down-regulated genes involved in cell division in the murine thymus [132]. An integrated network analysis of the genes, proteins, and phosphoproteins that are altered by TBTO could shed light on the molecular switches behind the cell-cycle-regulating properties of TBTO.

In summary, the in vitro application of toxicogenomics has led to the identification of effect markers for direct immunotoxicants, and of several signaling pathways that are potentially linked to mechanisms of direct immunotoxicity. These effect markers can potentially be employed for hazard identification purposes in vitro as well as in vivo, such as in rodent standard toxicity studies (STS), and in human population studies. At present, it is unclear whether effect markers for direct immunotoxicity that have been identified in vitro could be employed in vivo. In order to verify this, one could start a novel in vivo study, or employ in vitro–in vivo extrapolations on existing toxicogenomics data. Furthermore, for the majority of the pathways that have been identified by toxicogenomics in vitro, the potential roles in regulating immunotoxic endpoints need further validation. This could be established by employing transgenic rodent models in vivo, as well as by applying functional genomics to lymphocytes in vitro. Some of these potential mechanisms of direct immunotoxicity have already been confirmed. This has been done both in vitro, by applying biochemistry or immunocytochemistry, and in vivo, by immunophenotyping. Examples include those studies conducted by van Kol et al. [85,133] and Katika et al. [137], in which the immunotoxicants DON and TBTO could activate T cells and induced ER stress by activating Ca^{2+}-mediated signaling and eliciting nuclear translocation of the transcription factor NFAT, respectively. An example of in vivo confirmation experiments by immunophenotyping include those presented by McMillan [111], who showed that TCDD altered thymocyte development by activating the transcription factor KLF2 and its target genes in vivo.

In addition, in vitro toxicogenomics studies are providing the first clues as to why immune cells are specifically sensitive to the effects of these immunotoxicants. In lymphocytes, NFAT signaling seems to play a central role in mediating both the activation of T cells at a lower level and the induction of apoptosis at a higher level. Furthermore, certain responses of the human T-cell transcriptome, at the level of gene functions, appear to be unique to certain structural classes of immunotoxicants, whereas other responses overlap between different classes. These findings suggest that direct immunotoxicants can generally be ordered into specific functional classes. Additional confirmation studies are required to further characterize these different functional classes with respect to their interactions with the upstream signal transduction pathways involved in the transcriptional regulation of these processes. When such confirmation studies are done, the data can also be used to select sets of functional markers for these pathway-linked processes. These markers could eventually form the basis of a novel generation of in vitro screens for direct immunotoxicity. These in vitro screens can potentially refine, reduce, and replace test animals in standard toxicity screens of chemicals and pharmaceuticals in the preclinical stage.

3.3.7 Future directions

Mechanisms of direct immunotoxicity

Since the unraveling of the human genome, SNP and mutation analyses, and the in vivo and in vitro applications of gene-knockout models of human immunity, we have broadened our mechanistic knowledge about the interactions between signal transduction pathways, gene networks, and immune-regulating processes. In addition, we have identified a wide variety of chemical and biological structures that modulate immune functionality. In contrast, however, for only a minute fraction of the immunomodulatory compounds are we just beginning to understand the chemical interactions with signal transduction pathways and gene networks, for immunosuppressive drugs, aryl hydrocarbon receptor agonists, tributyltins, and some mycotoxins.

In the near future, by applying toxicogenomics we will more rapidly gain mechanistic insights into an expanding number of immunotoxicants. In this respect, true systems biology approaches will be required, mainly owing to the complexity of immune system functionality and the various ways in which immunotoxicants can interact with the different components of the immune system, at the levels of the microRNome, transcriptome, proteome, and metabolome. In addition, in order to understand all possible ways that direct immunotoxicants can interact with the immune system, integrated approaches will be required in which data from multi-'omics platforms, obtained from cell culture systems, animal models, and human clinical trials, will be integrated with other relevant parameters, such as in vivo pathology, immunophenotyping, and in vitro cytotoxicity dose–response data. The main challenges in such integrated systems biology approaches are of an organizational nature, as different research disciplines have to be brought together for longer periods in time. If these challenges are overcome, besides acquiring a true understanding of the interactions between an increasing number of chemicals and the immune system, the systems biology approach will also lead to the identification of novel markers. These markers can be applied in predictive in vitro assays that will, therefore, become more sensitive and specific than those currently available. These in vitro assays will first reduce, and thereafter eventually replace animal testing for immunotoxicity.

In vivo validation of in vitro markers for immunotoxicity

With respect to direct immunotoxicity, in vitro tests can employ markers for three distinct purposes: the assessment of immunotoxicant effects, the mode of action (MOA), and the functional or physical damage to the immune system. Biomarkers with predictive values for immunotoxic effects can potentially be used for hazard identification purposes; for instance, in an occupational or forensic setting. In vitro assays that have predictive values for either detecting the MOA or the resulting damaging effects to the immune system have the potential to replace, reduce, and refine the use of laboratory animals in immunotoxicity testing.

MOA and effect markers

Two human studies have addressed the effects of direct immunotoxicants on the transcriptome of human lymphocytes in vivo. These observational studies both assessed the effects of Ahr agonists [99,105], a single functional class of direct immunotoxicants. One intervention study has been performed with the cyclooxygenase (COX) inhibitor APAP [100]. However, APAP is not directly immunotoxic, as its immunotoxic potential is dependent on hepatotoxicity [103]. Regarding direct immunotoxicity, these observational studies have gained insights into the MOA of Ahr agonists in humans in vivo. However, for the majority of direct immunotoxicants, at present toxicogenomics studies on the effects on human lymphocytes in vivo are lacking. Thus, in vivo toxicogenomics data need to be generated by performing clinical trials with immunosuppressive drugs, in order to identify effect markers for direct immunotoxicity that overlap between the in vitro and in vivo situation. In addition, in these clinical trials, multi-'omics data need to be generated, instead of evaluating the transcriptome only, because immunotoxic effects are not only mediated at the transcriptional level, which implies that the predictive power of mRNA-based effect markers would be improved by complementation with, for instance, protein- or miRNA-based markers.

In vitro immunotoxicity testing for the registration of chemicals and drugs

At present, the immunotoxic potential of compounds, which is required to be evaluated for the registration of chemicals, pharmaceuticals, and biologicals, is embedded into general toxicity testing applying in vivo animal models. Regulatory guidelines do allow the application of in vitro assays, but only as tools for gaining mechanistic insights into the immunotoxic effects that have initially been observed in animal models in vivo. One exception is the human CFU-GM in vitro assay, which has been validated by ECVAM to replace a second non-rodent species for evaluating myelotoxicity [141]. Meanwhile, for the last few decades a paradigm shift has continued to materialize towards the reduction, refinement, and replacement of animal models in toxicity testing. There are, however, several reasons why in vitro assays for direct immunotoxicity are not yet employed as primary prediction models for humans in vivo. Firstly, within the current scientific advisory boards, more emphasis is put on the development of in vitro alternatives for other toxicity endpoints, such as for skin sensitization, owing to the 7th Amendment of the EU Cosmetics Directive that bans animal tests for personal care products [142], but also for carcinogenicity and reproductive toxicity testing [143]. Secondly, there is consensus that the current in vitro tests cannot yet predict human immunotoxicity at the level of accuracy that is required to replace animal testing [16,143]. The main reasons for this consensus are that when in vitro assays are applied, (1) information about bioavailability is lost, (2) xenobiotics are not biotransformed by lymphocytes, (3) the level of complexity is too low compared to the in vivo situation, and (4) the predictive biomarkers for direct immunotoxicity that have been identified in vitro [66,140] have not yet been validated in vivo. However, some recently developed alternative strategies draw optimism among those who advocate the future replacement of test animals. Firstly, with respect to predicting the absorption, distribution, metabolism, and excretion (ADME) patterns of xenobiotics for humans in vivo, the refinement of combined in silico–in vitro pharmacokinetics (PBPK) approaches is proceeding. These combined in vitro–in silico ADME approaches are also known as quantitative in vitro–in vivo extrapolation (QIVIVE). At present, QIVIVE options are already available for high-throughput settings and for more complicated full PBPK models as well [144]. Furthermore, QIVIVE-based predictions do not have to be perfect before they are to replace animal-based predictions, because the current animal models perform rather poorly in predicting human bioavailability in vivo [145].

Novel tiered in vitro approach for immunotoxicity risk assessment

When, at some point in the future, in vitro assays replace these animal models, we will be disposing of some basic drawbacks of animal-based STS, such as the required interspecies extrapolation

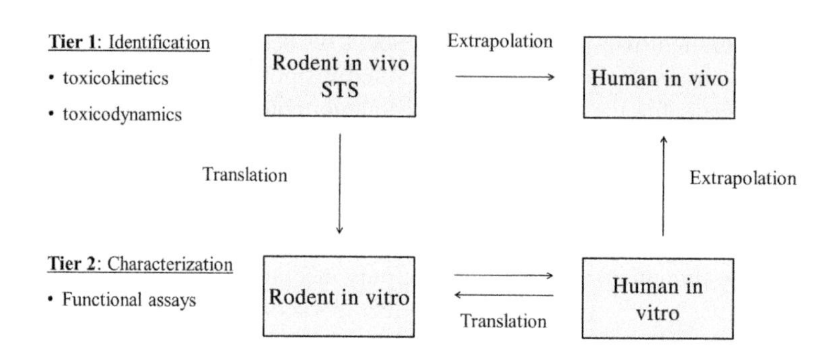

FIGURE 3.3.1

STS-Based Parallelogram Approach for Direct Immunotoxicity Testing.

of test outcomes towards the human situation, the larger amounts of time, labor, and money that are required, and the ethical concerns from society. When, at some point in time, immunotoxicity testing is solely assessed by applying in vitro assays, the current parallelogram approach, as depicted in Figure 3.3.1, will no longer apply. We therefore envisage a less complex tiered in vitro approach for immunotoxicity risk assessment of chemicals, pharmaceuticals, and biologicals (see Figure 3.3.2). This top-down in vitro approach will primarily be based on human cell systems, to avoid interspecies extrapolation towards the human situation. A battery of in vitro assays will have to be employed in order to predict different aspects of the immune system in vivo, as was proposed by Dean et al. in 1979 [14]. However, in our opinion, a tiered approach would be the most efficient setup because it will employ a minimum amount of in vitro assays, which is analogous to the current STS-based testing strategies for chemicals and drugs. In this respect, the functional in vitro assays will only verify those functions that are indicated to be affected in initial high-throughput screens (HTS). In brief, the in vitro approach that we propose would consist of three basic tiers. Tier 1 will generate toxicokinetics data for each test compound. These toxicokinetics data will be used as input for the immunotoxicity HTS in Tier 2. The potential immunomodulating properties of the test compounds, upon identification in Tier 2, will be further characterized in Tier 3. The in vitro assays that we envisage to be parts of the second and third tiers of our alternative approach are summarized in Table 3.3.4. We will now discuss each tier in more detail.

Tier 1: QIVIVE

In Tier 1, toxicokinetics data should be generated for each compound by conducting quantitative in vitro–in vivo extrapolation (QIVIVE). The QIVIVE approach integrates *in silico* pharmacokinetics models with data from in vitro systems [144]. The resulting toxicokinetics data, such as bioavailability, the internal exposure, and metabolism, should be used as inputs for the second tier. In our opinion, QIVIVE can be especially powerful when it also includes relatively complex cell-based absorption and biotransformation models that more closely resemble human physiology. Examples of such in vitro models that are promising are the gut-on-a-chip intestinal absorption model, which employs microfluidics and includes microbial flora [146], and the 3D HepaRG hepatic biotransformation model [147].

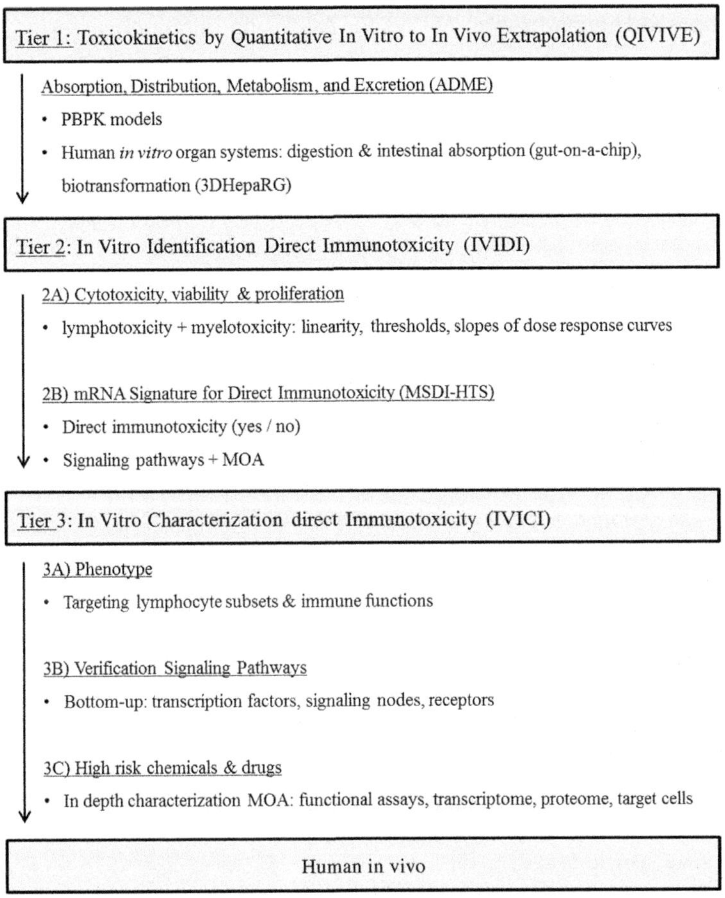

Tier 1: Toxicokinetics by Quantitative In Vitro to In Vivo Extrapolation (QIVIVE)

Absorption, Distribution, Metabolism, and Excretion (ADME)

- PBPK models
- Human *in vitro* organ systems: digestion & intestinal absorption (gut-on-a-chip), biotransformation (3DHepaRG)

Tier 2: In Vitro Identification Direct Immunotoxicity (IVIDI)

2A) Cytotoxicity, viability & proliferation

- lymphotoxicity + myelotoxicity: linearity, thresholds, slopes of dose response curves

2B) mRNA Signature for Direct Immunotoxicity (MSDI-HTS)

- Direct immunotoxicity (yes / no)
- Signaling pathways + MOA

Tier 3: In Vitro Characterization direct Immunotoxicity (IVICI)

3A) Phenotype

- Targeting lymphocyte subsets & immune functions

3B) Verification Signaling Pathways

- Bottom-up: transcription factors, signaling nodes, receptors

3C) High risk chemicals & drugs

- In depth characterization MOA: functional assays, transcriptome, proteome, target cells

Human in vivo

FIGURE 3.3.2

Future Tiered In Vitro Approach for Direct Immunotoxicity Testing as an Alternative to Test Animals.

Tier 2: HTS for in vitro identification of direct immunotoxicity

The second tier will consist of HTS for the in vitro identification of direct immunotoxicity (IVIDI). In practice, this tier will consist of two sub-tiers, Tier 2A and Tier 2B. Tier 2A will incorporate HTS for three basic endpoints: cytotoxicity, viability, and proliferation. Tier 2B will incorporate the HTS based on effect markers for direct immunotoxicity.

Within Tier 2A, the myelotoxic and lymphotoxic potentials of test compounds will be determined by employing these HTS on myeloblasts and lymphoblasts or lymphocytes, respectively. Besides the lymphotoxic potential, which is the predominant parameter for direct immunotoxicity, myelotoxicity is determined because a compound with myelotoxic potential is immunotoxic by definition [2]. This is, for example, the case for the organochlorine herbicide propanil [148]. The following parameters need to be generated within Tier 2A, in order to give an accurate estimate of the cytotoxic potential of the

Table 3.3.4 Proposed Human In Vitro Assays for Assessing Immunotoxicity, as Replacement for Animal Tests

Assays	References
Tier 2A: *in vitro identification direct immunotoxicity (IVIDI)* *High-throughput screens (HTS) for cytotoxicity and anti-proliferative effects*	
Lymphotoxicity: viability (ATP, dehydrogenases: WST-1), cytotoxicity (plasma membrane integrity: AK-leakage) Myelotoxicity: FMCA-GM HTS using cord blood CD34+ mononuclear progenitor cells	Cytotoxicity and viability: [150,168]; FMCA-GM HTS: [154]
Tier 2B: *IVIDI* *mRNA signature for direct immunotoxicity (MSDI), HTS*	
MSDI-HTS using Jurkat T cells or PBMC employing biomarkers for lymphocyte signaling pathways, stress responses, apoptosis, T-cell activation, chemotaxis, cholesterol homeostasis, and differentiation/lineage commitment	MSDI-HTS: patent pending, [87,137,140]
Tier 3A: *in vitro characterization of direct immunotoxicity (IVICI)* *Immunophenotyping subsets of lymphocytes (lymphotoxicity) or myeloblasts (myelotoxicity)*	
Flow cytometry to determine cytotoxicity using whole blood or PBMC: **T cells:** CD3+, CD4 (Th), CD8+ (cytotoxic), CD16/CD56 (NK), CD4+CD25+FOXP3+ (Treg) **B-cells:** CD19(+), CD19(+)CD25(+) activated B-cells, **DC, monocytes:** CD11c, CD14 **Apoptosis versus necrosis:** $[Ca^{2+}]i$ (Fluo-8 dye), PARP, cleaved casp-3 and -9, annexin-V, PI	[19,149,157]; in vivo: [128]
Cytokine release upon antigenic/mitogenic stimuli: whole-blood cytokine release assay: (Luminex/Meso scale/ELISAs): Th1 (IL-2, IFN- γ), Th2 (IL-3, IL-4, IL-5, IL-6, IL-10, IL-13), Th17 (IL-17), IL-12 (macrophage)	[2,169,170]
Antibody and proliferative responses of CD4+ and CD8+ T cells upon infection with influenza virus (human lymphocyte activation (HuLA) assay, and activation of primary B and/or T cells upon antigenic and/or mitogenic stimuli **MLR:** proliferation Jurkat upon allogeneic stimulus by PBMC	[19,155–157]
Cytolytic activities CD8+ or CD56+ NK T cells: PBMC on Jurkat T cells (CTL), or on K-562 cells (NK cell activity)	[156]
T-cell-dependent antibody response (TDAR): antibody (IgM + A + G) production tonsil-derived lymphocytes upon mitogenic or antigenic stimuli	[171–173]
T-cell chemotaxis: 2D transmembrane migration T-lymphocytes upon a chemotactic stimulus, such as CCL21, CXCL12, or S1P	[174,175]
Differentiation and proliferation: CFU-GM, IC90 and IC50: microscopic scoring colony-forming units Immunophenotyping myeloid lineage subsets using flow cytometry	[176]

(Continued)

Table 3.3.4 Proposed Human In Vitro Assays for Assessing Immunotoxicity, as Replacement for Animal Tests (Continued)

Assays	References
Tier 3B: *IVICI*	
Characterization signaling pathways by determining transcription factor activities	
Multiplex DNA-binding assays (Luminex) and Reporter assays: **Cytokine promoters:** fluorescent cell chip (IL-1β, IL-4) **T-cell lineage commitment:** CD4/CD8 (RUNX1/3, ZBTB7B), NK T (RUNX3), TCR (RUNX1/3, ZBTB7B), Th1/Th2 (GATA3, TBX21), Th17 (HIF1α, FOXP3, LXR, RUNX1, RORγT), Treg and γδT (FOXO1, RUNX1) **T-cell/thymocyte migration:** MTF1, FOXO1-KLF2-S1P_1 **Hormone disruption:** AHR, AR, ESR, GR, PR **Lipid metabolism:** RAR/RXR, SREBP **Stress responses:** ATF3, p53, JNK, AP-1, NFATC1, NF-κB, NR4A1, NRF2/KEAP1	[177–183]; AHR, AR, ESR, GR, and PR CALUX-assays: [184,185]
Tier 3C: *IVICI*	
In-depth characterization MOA high-tonnage and high-risk compounds: modulation TF activity, signaling transduction, receptor activities	
Diverse endpoints including the modulation of receptors, kinases/phosphatases (kinome arrays, phosphoproteomics), and transcription factors. Application functional genomics: silencing of signaling nodes using siRNA/shRNA, chemical inhibitors, or GMO	[137,186, 187]

IFN, interferon; IL, interleukin; NK, natural killer; Th, T-helper cell; Treg, regulatory T cell.

test compounds. At a 24-hour exposure time, dose–response curves will be used to determine those compound concentrations that reduce cell viability by 20% (CV80), and by 50% (CV50), and increase cytotoxicity by 20% (CC120) and by 50% (CC150), as compared to the vehicle controls. These values will be used to categorize the test compounds according to their cytotoxic potencies, and to determine the exposure concentrations for the HTS in Tier 2B. The degree of linearity and the slopes of the dose–cytotoxicity and –viability response curves also need to be determined by implementing curve-fitting algorithms. Together with the 24-hour CV80, CC120, CV50, and CC150 values, these parameters will help to determine for each test compound whether or not threshold concentrations for cytotoxicity exist, which is essential information for further risk assessment. In addition, it is necessary that cell density data be generated in parallel with the cytotoxicity and viability data, preferentially in multi-parameter format. This is essential because the cytotoxic potential of a test compound would be over- or underestimated if that compound specifically inhibited or stimulated cell proliferation, respectively. In addition, the inhibition of cell proliferation is a potential mechanism of immunosuppression. Therefore, in Tier 2A, proliferation assays should be undertaken for myeloblasts and lymphocytes or lymphoblasts. Recently, readout methods have greatly improved for in vitro cytotoxicity assays. There is now the option to use multi-parameter viability assays in high-throughput format. Such assays can simultaneously quantify parameters for mitochondrial activity (ATP, dehydrogenases), plasma membrane integrity (extracellular activity of kinases or proteases), and the induction of apoptosis (poly

(ADP-ribose) polymerase (PARP), caspase 3/7) by employing bioluminescence and/or fluorescence up to a 1536-well format [149,150]. In theory, it is also possible to extract multi-parameter viability data for different lymphocyte subsets within a single whole-blood sample when employing multicolor flow cytometry with specific antibodies for cell-surface markers of different lymphocyte subsets, such as B-, T-helper 1 (Th1), Th2, Th17, and regulatory T cells (Treg), respectively. This has, however, not yet been demonstrated. When this becomes reality, one can simultaneously identify and characterize the lymphotoxic properties of the test compounds in a single HTS.

In order to determine the cytotoxic potential in vitro, it is possible to use primary cells, such as PBMC-derived lymphocytes to quantify lymphotoxicity, and cord-blood-derived myeloblasts to determine myelotoxicity. However, when using primary cells, one has to take into account all the factors that potentially interfere with the responsiveness of the cells to the test compounds. For instance, in the literature there are reports that factors such as gender, diet, smoking and drinking habits, medication, and age are determinants in innate, humoral, and cell-mediated immune responses ([105]). When these factors are standardized, one can still apply primary human cells in these HTS. There would, for instance, be the need to generate separate cytotoxicity data for both genders. When cytotoxicity data are generated for both genders, this would also lead to a more sensitive screening method, because gender-specificity of putative immunotoxicants will also be detected. If inter-individual variation needs to be ruled out in these cytotoxicity assays, one can also employ cell lines to determine the lymphotoxic and myelotoxic potentials of the test compounds, such as Jurkat T cells, RPMI-1788 B-cells, and KG-1 myeloblasts.

Furthermore, if the compound is known to undergo biotransformation, it would be necessary to perform co-cultures with either microsomal S9-fractions, or with metabolically competent cells, such as cytochrome P450 transgenic HepG2 cells [151].

Tier 2B: Functional biomarkers for direct immunotoxicity

The cytotoxic and anti-proliferative potentials that have been determined in Tier 2A are required as input for Tier 2B, because the test compounds will be assessed at sub-cytotoxic and cytotoxic levels. With respect to the immunotoxic potential of test compounds, sub-cytotoxic conditions are required to identify early effect biomarkers, whereas cytotoxic conditions will provide additional information about the cytotoxic MOA. Based on a set of 10–40 of these functional biomarkers for direct immunotoxicity, we have developed a predictive in vitro assay, named Messenger RNA Signature for Direct Immunotoxicity in High-Throughput Screening format (MSDI-HTS; patent pending). The functional biomarkers for early effects have been developed under sub-cytotoxic conditions—a 6-hour exposure time to a 24-hour ≤ CV80 and ≤ CC120 [140]. Therefore, the MSDI-HTS will most probably be based on exposure times of 6 hours. The sub-cytotoxic concentrations will be based on 24-hour ≤ CV80 and ≤ CC120 values, whereas the cytotoxic conditions will be based on 24-hour CV50 and CC150 values.

In our previous toxicogenomics studies we have identified transcription factors (TFs) and TF-target genes that are involved in immune-regulating processes, such as such as T-cell differentiation and lineage commitment; cell cycle block; chemotaxis; cholesterol homeostasis; cytokine expression; antibody production; T-cell activation (TCA); and ER, oxidative, hypoxic, and ribotoxic stress responses [86,87,140]. In addition, quantitative PCR technologies have been downsized to the nanofluidics scale [152,153], thereby enabling the development of an HTS for direct immunotoxicity. Further pre-validation studies will have to prove whether the MSDI-HTS is accurate enough to replace the animal-based STS. If this is the case, then IVIDI can solely be based on the MSDI-HTS. In addition, the sensitivity of a future MSDI-HTS could be improved by adding functional biomarkers to detect those functional and structural subclasses of immunotoxicants that were not previously detected.

In line with the tiered approach recommended by ECVAM, pre-screening for direct immunotoxicity should also start with the evaluation of myelotoxicity, since all immune cells arise from hematopoietic stem cells in the bone marrow, and therefore a myelotoxic compound is, by definition, also immunotoxic. Thus, Tier 2 also includes an HTS for myelotoxicity. However, whether the immunotoxic compound is a direct immunotoxicant then still needs to be determined by the MSDI-HTS, and by further functional analyses in Tier 3. At present, the cytotoxicity-based Fluorimetric Multiculture Cytotoxicity Assay—granulocyte/macrophage (FMCA-GM) HTS, which was developed by Haglund, seems a promising in vitro alternative for myelotoxicity testing [154]. Unfortunately, the development of gene-expression-based biomarkers for myelotoxicity lags behind, when compared to the biomarkers for direct immunotoxicity. If additional functional mRNA- or miRNA-based biomarkers for myelotoxicity can be identified, their future implementation in an additional HTS would add functional information to the FMCA-GM test, and therefore give more refined outcomes of the myelotoxicity tests in Tier 2.

Tier 3: in vitro characterization direct immunotoxicity

When the outcomes of the HTS for direct immunotoxicity or myelotoxicity are positive in the second tier, a further in vitro characterization of immunotoxicity (IVICI) will be performed in the third tier.

Tier 3A: phenotypic characterization

The gene expression changes of specific functional biomarkers can be used to select those in vitro assays that can validate the specific immune function that was suggested by that particular biomarker. In order to better investigate a particular immunotoxic MOA, one can choose from an array of in vitro assays that will give a phenotypic characterization of the different components of immune functionality. This part of in vitro immunotoxicity testing is defined as Tier 3A. The classical ex vivo assays that are used for risk characterization comprise the whole-blood cytokine release assay, the lymphocyte proliferation assay, the mixed lymphocyte reaction (MLR), the cytotoxic T cell (CTL) activity assay, the natural killer (NK) cell activity assay, and the T-cell-dependent antibody response (TDAR) assay [2]. In vitro alternatives have been developed for these assays, and are based on human peripheral-blood mononuclear cells (hPBMC) from frozen blood [155–157]. The human in vitro CTL and NK cell activity assays determine the cytolytic activities of hPBMC and hPBMC-derived NK cells on the Jurkat lymphoblastic and K562 erythromyeloblastoid cell lines, respectively [156]. The human in vitro MLR assay determines the unidirectional proliferation response of hPBMC [156]. A potential human in vitro alternative for the ex vivo TDAR assay is the human lymphocyte activation (HuLa) assay [155]. This assay determines the proliferative and antibody responses of CD4+ and CD8+ T cells derived from hPBMC, upon infection with the influenza virus.

Tier 3B: verification of signaling pathways

When the application of one or more functional in vitro assays in Tier 3A has led to the identification of the target immune cell or component of a particular test compound, a further verification of the putative signaling pathway involved is performed in Tier 3B.

In mammalian lymphocytes, signal transduction cascades are generally hierarchical. Typically, after ligand-binding at the outer plasma membrane, the intracellular part of a receptor activates a cytosolic signaling cascade through protein phosphorylation by intracellular kinases. This leads to the dissociation of a protein–transcription factor complex, enabling the translocation of that transcription factor to the nucleus where it can activate the transcription of its target genes upon DNA-binding [158].

In our opinion, a bottom-up approach would be the most logical to identify the molecular target of the immunotoxic test compound within a signal transduction cascade. The rationale for this bottom-up approach is that the transcription factor (TF) or TF-target gene, being downstream of the signal transduction cascade, has already been identified in Tier 2. Therefore, starting with the TF, it would make sense to work our way up in that particular transduction pathway. Thus, step one of Tier 3B would be to confirm the compound's effects on TF activities by employing transactivation assays, such as (1) human TF-reporter cell lines, or (2) more laborious multiplex TF–DNA-binding ELISAs (enzyme-linked immunosorbent assays) on protein extracts of cellular nuclei. When these experiments show positive results, enough evidence has been gathered in Tiers 1–3B to determine the MOA of immunotoxicity, and at that point the evidence can be delivered to the risk assessors.

It should be emphasized that an individual immunotoxic compound could potentially affect different molecular pathways at different internal exposure doses and durations. For instance, toxicogenomics studies have assessed the dose– and time–effect relationships of the immunotoxicants DON and TBTO on lymphocytes in vitro [86,87], as well as in vivo [120,133]. These studies indicate that these immunotoxicants activate molecular pathways involved in T-cell activation at low doses, and induce cell cycle block and apoptosis at higher doses when cytotoxic conditions prevail. In addition, for these two immunotoxicants, time-dependent differences were also observed at the level of molecular pathways. Therefore, it is advisable that, for each test compound, different exposure doses and durations should be incorporated into Tiers 1–3B.

It might also be necessary to gather more in-depth information about the MOA of the so-called high-risk compounds in an additional tier (3C). These high-risk compounds include compounds (1) with a high risk/benefit ratio, (2) with a high tonnage of production, or (3) where a number of functionally or structurally related compounds have given positive test results in Tier 2 and 3A.

Tier 3C: in-depth characterization of mode of action of immunotoxicants

Additional in-depth experiments can be employed to identify the molecular target(s) of an immunotoxicant in the upstream signaling pathway, such as the application of functional genomics. These experiments are, however, more laborious and expensive, and sometimes more time-consuming. In general, transcriptome analysis of the target lymphocyte upon exposure to the test compound, both at sub-cytotoxic and cytotoxic levels, gains critical insights into immunotoxicity-related modes of action. This has, for instance, been shown for the organotin compound TBTO [86]. Such an approach could also be employed for assessing the proteome, because proteomics can also clarify the MOA of immunotoxicants, as has been done for the mycotoxin DON in human lymphocytes [135]. However, the application of proteomics is more labor intensive and time-consuming than using transcriptomics. Therefore, it would be pragmatic to commence Tier 3C with the employment of transcriptomics to assess the compound's effects under various conditions; for instance, during a time-course experiment with at least two compound concentrations, thereby generating both sub-cytotoxic and cytotoxic conditions. Based on the outcomes of the transcriptome analysis, one can then decide whether it is necessary to interrogate effects on specific signal transduction pathways. Furthermore, if the transcriptome analysis is inconclusive, it is more likely that the immunotoxicant has more specific effects at the translational or posttranslational level. In that case, the molecular target could be identified by means of a systems toxicology approach, to include analyses of the microRNome, proteome, or phosphoproteome. However, the employment of more targeted approaches would improve the chances of identifying the molecular target of the immunotoxicant. For instance, chemical–protein interactions can be identified by combing proteomics with in silico or in vitro pull-down assays, whereas the Chip-seq approach can be employed to identify chemical interactions with transcription factors and chromosomal DNA [159]. In addition, recently developed multiplex technologies employing antibodies, such as multicolor phospho-flow cytometry [160,161] and Luminex, MesoScale, and antibody and kinase arrays [162], are powerful and promising tools that could potentially lead to a more rapid identification of the molecular targets of immunotoxicants [163]. Furthermore, the technological advances in automated multi-parameter imaging [164]; droplet microfluidics, enabling multi-parameter high-throughput screens using single cells [165]; multiple applications of nanofluidics technology, such as multiplex qPCR profiling in single cells [166]; and deep sequencing with unlimited read lengths without the requirement of expensive dyes or cameras [45] will also greatly expand the possibilities of more a rapid identification of the MOA of immunotoxicants.

References

[1] Galbiati V, Mitjans M, Corsini E. J Immunotoxicol. 2010;7:255–67.
[2] Lankveld DP, van Loveren H, Baken KA, Vandebriel RJ. In vitro testing for direct immunotoxicity: state of the art. Methods Mol Biol 2010;598:401–23.
[3] Descotes J, Choquet-Kastylevsky G, van Ganse E, Vial T. Toxicol Pathol 2000;28:479–81.
[4] Campistol JM, Cuervas-Mons V, Manito N, Almenar L, Arias M, Casafont F, et al. Transplant Rev 2012;26:261–79.
[5] Vial T, Descotes J. Toxicology 2003;185:229–40.
[6] Descotes J. Toxicol Lett 2004;149:103–8.
[7] Rysavy NM, Maaetoft-Udsen K, Turner H. J Appl Toxicol 2013;33:1–8.
[8] Milicevic DR, Skrinjar M, Baltic T. Toxins 2010;2:572–92.
[9] Okoro HK, Fatoki OS, Adekola FA, Ximba BJ, Snyman RG, Opeolu B. Rev Environ Contam Toxicol 2011;213:27–54.
[10] Borchers A, Teuber SS, Keen CL, Gershwin ME. Clin Rev Allergy Immunol 2010;39:95–141.
[11] Food and Drug Administration Fed Regist 2006;71:19193–19194.
[12] OECD test no. 443: extended one-generation reproductive toxicity study, OECD guidelines for the testing of chemicals, section 4. : OECD Publishing; 2011. p. 25.
[13] Institoris L, Siroki O, Desi I, Lesznyak J, Serenyi P, Szekeres E, et al. Hum Exp Toxicol 1998;17:206–11.
[14] Dean JH, Padarathsingh ML, Jerrells TR. Drug Chem Toxicol 1979;2:5–17.
[15] Becking GC, Chen BH. Biol Trace Elem Res 1998;66:439–52.
[16] WHO, WHO library cataloguing-in-publication data. WHO; 2012.
[17] Sundwall A, Andersson B, Balls M, Dean J, Descotes J, Hammarström S, et al. Toxicol In Vitro 1994;8:1067–74.
[18] Gennari A, Ban M, Braun A, Casati S, Corsini E, Dastych J, et al. J Immunotoxicol 2005;2:61–83.
[19] Carfi M, Gennari A, Malerba I, Corsini E, Pallardy M, Pieters R, et al. In vitro tests to evaluate immunotoxicity: a preliminary study. Toxicology 2007;229:11–22.

[20] Sebastian K, Ott H, Zwadlo-Klarwasser G, Skazik-Voogt C, Marquardt Y, Czaja K, et al. Toxicol Appl Pharmacol 2012;262:283–92.
[21] Gotz C, Pfeiffer R, Tigges J, Ruwiedel K, Hubenthal U, Merk HF, et al. Exp Dermatol 2012;21:364–9.
[22] Kaluzhny Y, Kandarova H, Hayden P, Kubilus J, d'Argembeau-Thornton L, Klausner M. ATLA 2011;39:339–64.
[23] Costin GE, Raabe HA, Priston R, Evans E, Curren RD. ATLA 2011;39:317–37.
[24] Welss T, Basketter DA, Schroder KR. Toxicol In Vitro 2004;18:231–43.
[25] Pertea M. Genes 2012;3:344–60.
[26] Ma F, Xu S, Liu X, Zhang Q, Xu X, Liu M, et al. Nat Immunol 2011;12:861–9.
[27] Quinn SR, O'Neill LA. Int Immunol 2011;23:421–5.
[28] Chen Q, Wang H, Liu Y, Song Y, Lai L, Han Q, et al. PLoS ONE 2012;7:e42971.
[29] Contreras J, Rao DS. Leukemia 2012;26:404–13.
[30] Alexander RW, Eargle J, Luthey-Schulten Z. FEBS Lett 2010;584:376–86.
[31] Schena M, Shalon D, Davis RW, Brown PO. Science 1995;270:467–70.
[32] Shen-Ong GL, Feng Y, Troyer DA. Can Res 2003;63:3296–301.
[33] Barrett T, Troup DB, Wilhite SE, Ledoux P, Evangelista C, Kim IF, et al. Nucleic Acids Res 2011;39:D1005–D1010.
[34] Tan PK, Downey TJ, Spitznagel Jr. EL, Xu P, Fu D, Dimitrov DS, et al. Nucleic Acids Res 2003;31:5676–84.
[35] Irizarry RA, Warren D, Spencer F, Kim IF, Biswal S, Frank BC, et al. Nat Methods 2005;2:345–50.
[36] Fare TL, Coffey EM, Dai H, He YD, Kessler DA, Kilian KA, et al. Anal Chem 2003;75:4672–5.
[37] Shi L, Reid LH, Jones WD, Shippy R, Warrington JA, Baker SC, et al. Nat Biotechnol 2006;24:1151–61.
[38] Martinez NM, Pan Q, Cole BS, Yarosh CA, Babcock GA, Heyd F, et al. RNA 2012;18:1029–40.
[39] Agarwal A, Koppstein D, Rozowsky J, Sboner A, Habegger L, Hillier LW, et al. BMC Genomics 2010;11:383.
[40] Osman AM, van Kol S, Peijnenburg A, Blokland M, Pennings JL, Kleinjans JC, et al. J Immunotoxicol 2009;6:174–83.
[41] Wang Z, Gerstein M, Snyder M. Nat Rev Genet 2009;10:57–63.
[42] Bradford JR, Hey Y, Yates T, Li Y, Pepper SD, Miller CJ. BMC Genomics 2010;11:282.
[43] Peng Z, Cheng Y, Tan BC, Kang L, Tian Z, Zhu Y, et al. Nat Biotechnol 2012;30:253–60.
[44] Costa V, Aprile M, Esposito R, Ciccodicola A. Eur J Hum Genet 2013;21:134–42.
[45] Stranneheim H, Lundeberg J. Biotechnol J 2012;7:1063–73.
[46] Kim SY, Volsky DJ. BMC Bioinformatics 2005;6:144.
[47] de Leeuw WC, Rauwerda H, Jonker MJ, Breit TM. BMC Res Notes 2008;1:66.
[48] Podpecan V, Lavrac N, Mozetic I, Novak PK, Trajkovski I, Langohr L, et al. BMC Bioinformatics 2011;12:416.
[49] Harr B, Schlotterer C. Nucleic Acids Res 2006;34:e8.
[50] Schmid R, Baum P, Ittrich C, Fundel-Clemens K, Huber W, Brors B, et al. BMC Genomics 2010;11:349.
[51] Heinemeyer T, Wingender E, Reuter I, Hermjakob H, Kel AE, Kel OV, et al. Nucleic Acids Res 1998;26:362–7.
[52] Fierro AC, Vandenbussche F, Engelen K, van de Peer Y, Marchal K. Curr Genomics 2008;9:525–34.
[53] Raaum RL, Wang AB, Al-Meeri AM, Mulligan CJ. BioTechniques 2010;48:449–54.
[54] Lin DY. Bioinformatics 2005;21:781–7.
[55] Harmston N, Filsell W, Stumpf MP. Hum Genomics 2010;5:17–29.
[56] Masuya H, Makita Y, Kobayashi N, Nishikata K, Yoshida Y, Mochizuki Y, et al. Nucleic Acids Res 2011;39:D861–70.
[57] Gunther OP, Chen V, Freue GC, Balshaw RF, Tebbutt SJ, Hollander Z, et al. BMC Bioinformatics 2012;13:326.

[58] Xie X, Lu J, Kulbokas EJ, Golub TR, Mootha V, Lindblad-Toh K, et al. Nature 2005;434:338–45.
[59] Ekins S, Bugrim A, Brovold L, Kirillov E, Nikolsky Y, Rakhmatulin E, et al. Xenobiotica 2006;36:877–901.
[60] Naderi N, Witte R. BMC Genomics 2012;13(Suppl. 4):S10.
[61] Yousef M, Jung S, Showe LC, Showe MK. BMC Bioinformatics 2007;8:144.
[62] Krupp M, Marquardt JU, Sahin U, Galle PR, Castle J, Teufel A. Bioinformatics 2012;28:1184–5.
[63] Engreitz JM, Chen R, Morgan AA, Dudley JT, Mallelwar R, Butte AJ. Bioinformatics 2011;27:3317–8.
[64] Williams G. BMC Genomics 2012;13:12.
[65] Tilton SC, Tal TL, Scroggins SM, Franzosa JA, Peterson ES, Tanguay RL, et al. BMC Bioinformatics 2012;13:311.
[66] Hochstenbach K, van Leeuwen DM, Gmuender H, Stolevik SB, Nygaard UC, Lovik M, et al. Toxicol Sci 2010;118:19–30.
[67] Tong W, Cao X, Harris S, Sun H, Fang H, Fuscoe J, et al. Environ Health Perspect 2003;111:1819–26.
[68] Vazquez M, Nogales-Cadenas R, Arroyo J, Botias P, Garcia R, Carazo JM, et al. Nucleic Acids Res 2010;38:W228–32.
[69] Kanehisa M, Goto S. Nucleic Acids Res 2000;28:27–30.
[70] Salomonis N, Hanspers K, Zambon AC, Vranizan K, Lawlor SC, Dahlquist KD, et al. BMC Bioinformatics 2007;8:217.
[71] Lewis S, Ashburner M, Reese MG. Curr Opin Struct Biol 2000;10:349–54.
[72] Wingender E, Dietze P, Karas H, Knuppel R. Nucleic Acids Res 1996;24:238–41.
[73] Dennis Jr. G, Sherman BT, Hosack DA, Yang J, Gao W, Lane HC, et al. Genome Biol 2003;4:P3.
[74] Subramanian A, Tamayo P, Mootha VK, Mukherjee S, Ebert BL, Gillette MA, et al. Proc Natl Acad Sci USA 2005;102:15545–50.
[75] Doniger SW, Salomonis N, Dahlquist KD, Vranizan K, Lawlor SC, Conklin BR. Genome Biol 2003;4:R7.
[76] Liberzon A, Subramanian A, Pinchback R, Thorvaldsdottir H, Tamayo P, Mesirov JP. Bioinformatics 2011;27:1739–40.
[77] Mlecnik B, Scheideler M, Hackl H, Hartler J, Sanchez-Cabo F, Trajanoski Z. Nucleic Acids Res 2005;33:W633–7.
[78] Zhang B, Kirov S, Snoddy J. Nucleic Acids Res 2005;33:W741–8.
[79] Dubchak I, Ryaboy DV. Methods Mol Biol 2006;338:69–89.
[80] Damian D, Gorfine M. Nat Genet 2004;36:663. author reply 663.
[81] Wang L, Zhang B, Wolfinger RD, Chen X. PLoS Genet 2008;4:e1000115.
[82] Pavlidis P, Qin J, Arango V, Mann JJ, Sibille E. Neurochem Res 2004;29:1213–22.
[83] Li Z, Srivastava S, Yang X, Mittal S, Norton P, Resau J, et al. BMC Syst Biol 2007;1:21.
[84] De S, Ghosh S, Chatterjee R, Chen YQ, Moses L, Kesari A, et al. Environ Int 2010;36:907–17.
[85] van Kol SW, Hendriksen PJ, van Loveren H, Peijnenburg A. Toxicol Appl Pharmacol 2011;250:299–311.
[86] Katika MR, Hendriksen PJ, van Loveren H, Peijnenburg A. Toxicol Appl Pharmacol 2011;254:311–22.
[87] Katika MR, Hendriksen PJ, Shao J, van Loveren H, Peijnenburg A. Transcriptome analysis of the human T lymphocyte cell line Jurkat and human peripheral blood mononuclear cells exposed to deoxynivalenol (DON): new mechanistic insights. Toxicol Appl Pharmacol 2012;264:51–64.
[88] Evangelou M, Rendon A, Ouwehand WH, Wernisch L, Dudbridge F. PloS ONE 2012;7:e41018.
[89] Jelier R, Schuemie MJ, Veldhoven A, Dorssers LC, Jenster G, Kors JA. Genome Biol 2008;9:R96.
[90] Raponi M, Belly RT, Karp JE, Lancet JE, Atkins D, Wang Y. BMC Cancer 2004;4:56.
[91] Scherf M, Epple A, Werner T. Brief Bioinform 2005;6:287–97.
[92] Liu F, White JA, Antonescu C, Gusenleitner D, Quackenbush J. BMC Bioinformatics 2011;12:46.
[93] Numata K, Yoshida R, Nagasaki M, Saito A, Imoto S, Miyano S. BMC Bioinformatics 2008;9:494.
[94] Pacini C, Iorio F, Goncalves E, Iskar M, Klabunde T, Bork P, et al. Bioinformatics 2013;29:132–4.
[95] Shannon P, Markiel A, Ozier O, Baliga NS, Wang JT, Ramage D, et al. Genome Res 2003;13:2498–504.
[96] Saito R, Smoot ME, Ono K, Ruscheinski J, Wang PL, Lotia S, et al. Nat Methods 2012;9:1069–76.

[97] Rohn H, Junker A, Hartmann A, Grafahrend-Belau E, Treutler H, Klapperstuck M, et al. BMC Syst Biol 2012;6:139.
[98] D.M. Reif, M. Sypa, E.F. Lock, F.A. Wright, A. Wilson, T. Cathey, et al., Bioinformatics, 2013;29:402–3.
[99] McHale CM, Zhang L, Hubbard AE, Zhao X, Baccarelli A, Pesatori AC, et al. Toxicology 2007;229:101–13.
[100] Jetten MJ, Gaj S, Ruiz-Aracama A, de Kok TM, van Delft JH, Lommen A, et al. Toxicol Appl Pharmacol 2012;259:320–8.
[101] Pesatori AC, Consonni D, Rubagotti M, Grillo P, Bertazzi PA. Environ Health 2009;8:39.
[102] Baccarelli A, Pesatori AC, Consonni D, Mocarelli P, Patterson Jr. DG, Caporaso NE, et al. Br J Dermatol 2005;152:459–65.
[103] Masson MJ, Peterson RA, Chung CJ, Graf ML, Carpenter LD, Ambroso JL, et al. Chem Res Toxicol 2007;20:20–6.
[104] Baken KA, Pennings JL, Jonker MJ, Schaap MM, de Vries A, van Steeg H, et al. Toxicol Appl Pharmacol 2008;226:46–59.
[105] Hochstenbach K, van Leeuwen DM, Gmuender H, Gottschalk RW, Stolevik SB, Nygaard UC, et al. Toxicol Sci 2012;129:315–24.
[106] Mocchegiani E, Giacconi R, Cipriano C, Malavolta M. J Clin Immunol 2009;29:416–25.
[107] Yin X, Knecht DA, Lynes MA. BMC Immunol 2005:6.
[108] Swedenborg E, Kotka M, Seifert M, Kanno J, Pongratz I, Ruegg J. Mol Cell Endocrinol 2012;362:39–47.
[109] Yoshioka H, Hiromori Y, Aoki A, Kimura T, Fujii-Kuriyama Y, Nagase H, et al. Toxicol Lett 2012;211:257–65.
[110] Frawley R, White Jr K, Brown R, Musgrove D, Walker N, Germolec D. Environ Health Perspect 2011;119:371–6.
[111] McMillan BJ, McMillan SN, Glover E, Bradfield CA. J Biol Chem 2007;282:12590–12597.
[112] Carlson CM, Endrizzi BT, Wu J, Ding X, Weinreich MA, Walsh ER, et al. Nature 2006;442:299–302.
[113] Bai A, Hu H, Yeung M, Chen J. J Immunol 2007;178:7632–9.
[114] Bista P, Mele DA, Baez DV, Huber BT. Mol Immunol 2008;45:3618–23.
[115] Oteiza PI. Free Radic Biol Med 2012;53:1748–59.
[116] Lee WW, Cui D, Czesnikiewicz-Guzik M, Vencio RZ, Shmulevich I, Aderem A, et al. Rejuvenation Res 2008;11:1001–11.
[117] Liu J, Liu Y, Habeebu SS, Klaassen CD. Toxicol Appl Pharmacol 1999;159:98–108.
[118] Ranaldi G, Caprini V, Sambuy Y, Perozzi G, Murgia C. Toxicol In Vitro 2009;23:1516–21.
[119] Rotter BA, Prelusky DB, Pestka JJ. J Toxicol Environ Health 1996;48:1–34.
[120] Baken KA, Pennings JL, de Vries A, Breit TM, van Steeg H, van Loveren H. J Immunotoxicol 2006;3:227–44.
[121] Vos JG, De Klerk A, Krajnc EI, van Loveren H, Rozing J. Toxicol Appl Pharmacol 1990;105:144–55.
[122] Verdier F, Virat M, Schweinfurth H, Descotes J. J Toxicol Environ Health 1991;32:307–17.
[123] Penninks AH. Food Addit Contam 1993;10:351–61.
[124] Pillet S, Rooney AA, Bouquegneau JM, Cyr DG, Fournier M. Toxicology 2005;209:289–301.
[125] Kimura T, Itoh N, Takehara M, Oguro I, Ishizaki JI, Nakanishi T, et al. Biochem Biophys Res Commun 2001;280:358–62.
[126] Singh RP, Kumar S, Nada R, Prasad R. Mol Cell Biochem 2006;282:13–21.
[127] Klaassen CD, Liu J, Diwan BA. Toxicol Appl Pharmacol 2009;238:215–20.
[128] Biro A, Fodor Z, Major J, Tompa A. Immunotoxicity monitoring of hospital staff occupationally exposed to cytostatic drugs. Pathol Oncol Res 2011;17:301–8.
[129] Markou A, Strati A, Malamos N, Georgoulias V, Lianidou ES. Clin Chem 2011;57:421–30.
[130] Git A, Spiteri I, Blenkiron C, Dunning MJ, Pole JC, Chin SF, et al. Breast Cancer Res 2008;10:R54.
[131] Marshall NB, Vorachek WR, Steppan LB, Mourich DV, Kerkvliet NI. J Immunol 2008;181:2382–91.
[132] Baken KA, Arkusz J, Pennings JL, Vandebriel RJ, van Loveren H. Toxicology 2007;237:35–48.
[133] van Kol SW, Hendriksen PJ, van Loveren H, Peijnenburg A. Toxicology 2012;296:37–47.

[134] Schmeits PC, Volger OL, Zandvliet ET, van Loveren H, Peijnenburg AA, Hendriksen PJ. Toxicol Lett 2013;217:1–13.
[135] Nogueira da Costa A, Keen JN, Wild CP, Findlay JB. Biochim Biophys Acta 2011;1814:850–7.
[136] Osman AM, van Loveren H. Toxicol Sci 2012;126:84–100.
[137] Katika MR, Hendriksen PJ, de Ruijter NC, van Loveren H, Peijnenburg A. Immunocytological and biochemical analysis of the mode of action of bis (tri-n-butyltin) tri-oxide (TBTO) in Jurkat cells. Toxicol Lett 2012;212:126–36.
[138] Kass GE, Orrenius S. Environ Health Perspect 1999;107(Suppl. 1):25–35.
[139] Pestka JJ. Food Addit Contam Part A Chem Anal Control Expo Risk Assess 2008;25:1128–40.
[140] Shao J, Katika M, Schmeits P, Hendriksen P, van Loveren H, Peijnenburg A, et al. Toxicol Sci 2013;135:328–42.
[141] Pessina A, Bonomi A, Cavicchini L, Albella B, Cerrato L, Parent-Massin D, et al. ATLA 2010;38:105–17.
[142] Schumann R. ATLA 2002;30(Suppl. 2):213–4.
[143] Basketter DA, Clewell H, Kimber I, Rossi A, Blaauboer B, Burrier R, et al. Altex 2012;29:3–91.
[144] Yoon M, Campbell JL, Andersen ME, Clewell HJ. Crit Rev Toxicol 2012;42:633–52.
[145] Shanks N, Greek R, Greek J. Philos Ethics Humanit Med 2009;4:2.
[146] Kim HJ, Huh D, Hamilton G, Ingber DE. Lab Chip 2012;12:2165–74.
[147] S.B. Leite, I. Wilk-Zasadna, J.M. Zaldivar Comenges, E. Airola, M.A. Reis-Fernandes, M. Mennecozzi, et al., Toxicol Sci, 2012.
[148] Blyler G, Landreth KS, Lillis T, Schafer R, Theus SA, Gandy J, et al. Fundam Appl Toxicol 1994;22:505–10.
[149] Luu YK, Rana P, Duensing TD, Black C, Will Y. Profiling of toxicity and identification of distinct apoptosis profiles using a 384-well high-throughput flow cytometry screening platform. J Biomol Screen 2012;17:806–12.
[150] Niles AL, Moravec RA, Riss TL. In vitro viability and cytotoxicity testing and same-well multi-parametric combinations for high throughput screening. Curr Chem Genomics 2009;3:33–41.
[151] Tolosa L, Donato MT, Perez-Cataldo G, Castell JV, Gomez-Lechon MJ. Toxicol In Vitro 2011
[152] Devonshire AS, Sanders R, Wilkes TM, Taylor MS, Foy CA, Huggett JF. Methods 2012
[153] Spurgeon SL, Jones RC, Ramakrishnan R. PloS ONE 2008;3:e1662.
[154] Haglund C, Aleskog A, Hakansson LD, Hoglund M, Jacobsson S, Larsson R, et al. The FMCA-GM assays, high-throughput non-clonogenic alternatives to CFU-GM in preclinical hematotoxicity testing. Toxicol Lett 2010;194:102–7.
[155] Collinge M, Cole SH, Schneider PA, Donovan CB, Kamperschroer C, Kawabata TT. Human lymphocyte activation assay: an in vitro method for predictive immunotoxicity testing. J Immunotoxicol 2010;7:357–66.
[156] Lebrec H, Roger R, Blot C, Burleson GR, Bohuon C, Pallardy M. Immunotoxicological investigation using pharmaceutical drugs: in vitro evaluation of immune effects using rodent or human immune cells. Toxicology 1995;96:147–56.
[157] Alam I, Goldeck D, Larbi A, Pawelec G. Flow cytometric lymphocyte subset analysis using material from frozen whole blood. J Immunoassay Immunochem 2012;33:128–39.
[158] Mosenden R, Tasken K. Cell Signal 2011;23:1009–16.
[159] Northrup DL, Zhao K. Immunity 2011;34:830–42.
[160] Frischbutter S, Schultheis K, Patzel M, Radbruch A, Baumgrass R. Cytometry A 2012.
[161] Oberprieler NG, Tasken K. Cell Signal 2011;23:14–18.
[162] Hilhorst R, Houkes L, van den Berg A, Ruijtenbeek R. Anal Biochem 2009;387:150–61.
[163] Duramad P, Holland NT. Int J Environ Res Public Health 2011;8:1388–401.
[164] Shariff A, Kangas J, Coelho LP, Quinn S, Murphy RF. J Biomol Screen 2010;15:726–34.
[165] Brouzes E, Medkova M, Savenelli N, Marran D, Twardowski M, Hutchison JB, et al. Proc Natl Acad Sci USA 2009;106:14195–200.
[166] Citri A, Pang ZP, Sudhof TC, Wernig M, Malenka RC. Nat Protoc 2012;7:118–27.

[167] Baken KA, Vandebriel RJ, Pennings JLA, Kleinjans JC, van Loveren H. Methods 2007;41:132–41.
[168] Cho MH, Niles A, Huang R, Inglese J, Austin CP, Riss T, et al. A bioluminescent cytotoxicity assay for assessment of membrane integrity using a proteolytic biomarker. Toxicol In Vitro 2008;22:1099–106.
[169] Kooijman R, Devos S, Hooghe-Peters E. Inhibition of in vitro cytokine production by human peripheral blood mononuclear cells treated with xenobiotics: implications for the prediction of general toxicity and immunotoxicity. Toxicol In Vitro 2010;24:1782–9.
[170] Langezaal I, Hoffmann S, Hartung T, Coecke S. Evaluation and prevalidation of an immunotoxicity test based on human whole-blood cytokine release. Altern Lab Anim 2002;30:581–95.
[171] Herzyk DJ, Holsapple M. Immunotoxicity evaluation by immune function tests: focus on the T-dependent antibody response (TDAR). J Immunotoxicol 2007;4:143–7.
[172] Ladics GS. Primary immune response to sheep red blood cells (SRBC) as the conventional T-cell dependent antibody response (TDAR) test. J Immunotoxicol 2007;4:149–52.
[173] Wood SC, Karras JG, Holsapple MP. Integration of the human lymphocyte into immunotoxicological investigations. Fundam Appl Toxicol 1992;18:450–9.
[174] Entschladen F, Drell IV TL, Lang K, Masur K, Palm D, Bastian P, et al. Analysis methods of human cell migration. Exp Cell Res 2005;307:418–26.
[175] Goetzl EJ, Schwartz JB, Huang MC. Defective T cell chemotaxis to sphingosine 1-phosphate and chemokine CCL21 in idiopathic T lymphocytopenia. J Clin Immunol 2011;31:744–51.
[176] Gorczyca W, Sun ZY, Cronin W, Li X, Mau S, Tugulea S. Immunophenotypic pattern of myeloid populations by flow cytometry analysis. Methods Cell Biol 2011;103:221–66.
[177] Ulleras E, Trzaska D, Arkusz J, Ringerike T, Adamczewska V, Olszewski M, et al. Development of the "Cell Chip": a new in vitro alternative technique for immunotoxicity testing. Toxicology 2005;206:245–56.
[178] Behnisch PA, Hosoe K, Brouwer A, Sakai S. Screening of dioxin-like toxicity equivalents for various matrices with wildtype and recombinant rat hepatoma H4IIE cells. Toxicol Sci 2002;69:125–30.
[179] Oppl B, Kofler A, Schwarz S, Rainer J, Kofler R. Establishing a sensitive and specific assay for determination of glucocorticoid bioactivity. Wien Klin Wochenschr 2011;123:222–9.
[180] Cui G, Qin X, Wu L, Zhang Y, Sheng X, Yu Q, et al. Liver X receptor (LXR) mediates negative regulation of mouse and human Th17 differentiation. J Clin Invest 2011;121:658–70.
[181] Tsun A, Chen Z, Li B. Romance of the three kingdoms: RORgammat allies with HIF1alpha against FoxP3 in regulating T cell metabolism and differentiation. Protein Cell 2011;2:778–81.
[182] Kerdiles YM, Stone EL, Beisner DR, McGargill MA, Ch'en IL, Stockmann C, et al. Foxo transcription factors control regulatory T cell development and function. Immunity 2010;33:890–904.
[183] Dejean AS, Hedrick SM, Kerdiles YM. Highly specialized role of forkhead box O transcription factors in the immune system. Antioxid Redox Signal 2011;14:663–74.
[184] Sonneveld E, Pieterse B, Schoonen WG, van der Burg B. Validation of in vitro screening models for progestagenic activities: inter-assay comparison and correlation with in vivo activity in rabbits. Toxicol In Vitro 2011;25:545–54.
[185] Sonneveld E, Riteco JA, Jansen HJ, Pieterse B, Brouwer A, Schoonen WG, et al. Comparison of in vitro and in vivo screening models for androgenic and estrogenic activities. Toxicol Sci 2006;89:173–87.
[186] Wang Y, Liu J, Cui J, Xing L, Wang J, Yan X, et al. ERK and p38 MAPK signaling pathways are involved in ochratoxin A-induced G2 phase arrest in human gastric epithelium cells. Toxicol Lett 2012;209:186–92.
[187] Khan NM, Poduval TB. Immunomodulatory and immunotoxic effects of bilirubin: molecular mechanisms. J Leukoc Biol 2011;90:997–1015.

SECTION

Reproduction Toxicity

4

CHAPTER

Implementation of Transcriptomics in the Zebrafish Embryotoxicity Test

4.1

Sanne A.B. Hermsen[*,**,†] and Aldert H. Piersma[**,†]

[*]*Department of Toxicogenomics (TGX), Maastricht University, Maastricht, The Netherlands,* [**]*Center for Health Protection Research (GZB), National Institute for Public Health and the Environment (RIVM), Bilthoven, The Netherlands,* [†]*Institute for Risk Assessment Sciences (IRAS), Utrecht University, Utrecht, The Netherlands*

4.1.1 The zebrafish embryo as alternative test model for developmental toxicity testing

During recent decades, many alternative assays for developmental toxicity testing have been developed and studied. Alternative methods vary from cell-based tests, such as the embryonic stem cell test (EST), to assays using complete embryos—for instance, the rat postimplantation whole-embryo culture (WEC) and the zebrafish embryotoxicity test (ZET). The zebrafish embryo (*Danio rerio*) holds great promise as a model for developmental toxicity testing. In both the ZET and WEC, the possible developmentally toxic properties of a compound can be evaluated in a complete developing vertebrate embryo, in contrast to cell-based tests; thereby providing the possibility to obtain information on specific types of effect that may occur in vivo as well. One advantage of the ZET compared to WEC is the developmental window of the embryos; the WEC covers only part of the developmental window, ranging from gestation day 10 to 12 in the rat [1,2], whereas for the ZET the developmental period from fertilization to hatching and early larval stage can be used. During these stages, the embryos are not defined as protected under European legislation, as they are yolk-dependent and thus non-free-feeding [3,4]. Moreover, zebrafish are oviparous, i.e. fertilization and development occur externally, meaning no euthanasia of parental animals is necessary to obtain embryos. Development of the zebrafish is well defined and, in combination with their transparency, this allows easy monitoring and evaluation of morphology [5,6]. Their small size, relatively cheap maintenance, high fecundity, and rapid development make the zebrafish embryo a suitable model organism for high-throughput assays [7].

4.1.2 The zebrafish embryotoxicity test—a variety of methods

Generally, the ZET comprises the exposure of a zebrafish embryo to a compound of interest by adding this substance to water containing a recently fertilized zebrafish egg. After exposure, the embryo

Toxicogenomics-Based Cellular Models.

is evaluated for diverse endpoints to assess its development. One of the first reports of the use of the zebrafish embryo as a test organism for teratogenic effects dates back to the 1950s. Hisaoka tested exposure to 2-acetylaminofluorene, a known carcinogen, during different stages of development, and described morphological effects and investigated histochemical parameters [8,9]. However, only in recent decades has the popularity of the zebrafish embryo as model organism for compound screening increased. In environmental risk assessment, the use of the embryo was promoted as an alternative for the acute fish toxicity test [10,11]. A draft version of test guidelines for chemical testing in the fish embryo, with emphasis on the zebrafish, is currently being developed and validated [12].

Most studies use morphological evaluation as the endpoint of the test system after a certain time period of exposure. However, there is still no standardized or validated protocol in use. Currently described procedures may vary in different parameters, such as evaluated endpoints and exposure duration. In the method described by Nagel et al. [10], embryos with an intact chorion are exposed from the four-cell stage onwards up to 48 hours post-fertilization (hpf), with final evaluation of several lethal, sublethal, and teratogenic endpoints. This exposure period covers most of the early development of the zebrafish embryo, including cleavage and segmentation. However, fin development and hatching will only become apparent in the next 24 hours; therefore, are not taken into account in this protocol. In addition, morphological assessment is more difficult at this pre-hatching stage compared to hatched embryos.

To improve the existing methods, more elaborate and extensive protocols have been developed, though differences are still present. For instance, Brannen et al. used dechorionated embryos from 4–6 hpf for their test, with a 120-h exposure window and no medium changes, in contrast to Padilla et al., who exposed 6–8-hpf embryos within their chorion with daily test solution changes and evaluation at 144 hpf [13,14]. Also, the evaluation method differed between the two research groups. Padilla et al. assigned scores to the monitored endpoints, such as head malformation and cardiovascular function, ranging from 0 to 4 points increasing with severity of malformation or developmental delay. Similarly, Brannen et al. assigned scores ranging from 0.5 to 5, but increasing with normal development. Furthermore, body length and head–trunk angle were measured, the distance between eye and otic vesicle was estimated, and somite pairs were counted, which makes this method relatively labor intensive.

Hermsen et al. developed a semi-quantitative scoring method to evaluate embryos on readily observable developmental hallmarks in time [15]. A general morphology score (GMS) was developed analogous to the one described for WEC, although with fewer endpoints and limited score levels. In this protocol, treatment is initiated at the 4–32-cell stage with embryo evaluation at 72 hpf. As shown in Figure 4.1.1, a maximum score of 15 points can be obtained for a normally developed embryo at this time-point. All deviations from normal development will result in reduced scores. In addition to the GMS evaluation, malformations, such as pericardial edema or scoliosis, are recorded as well. This approach allows the monitoring of developmental delay as well as teratogenicity.

4.1.3 Developmental toxicity prediction using the zebrafish embryo

Even though all methods vary and are not yet validated, concordance of zebrafish results with mammalian data is high. Brannen et al. developed a prediction model to classify compounds as developmentally toxic or developmentally nontoxic. They evaluated 31 compounds in the zebrafish embryo

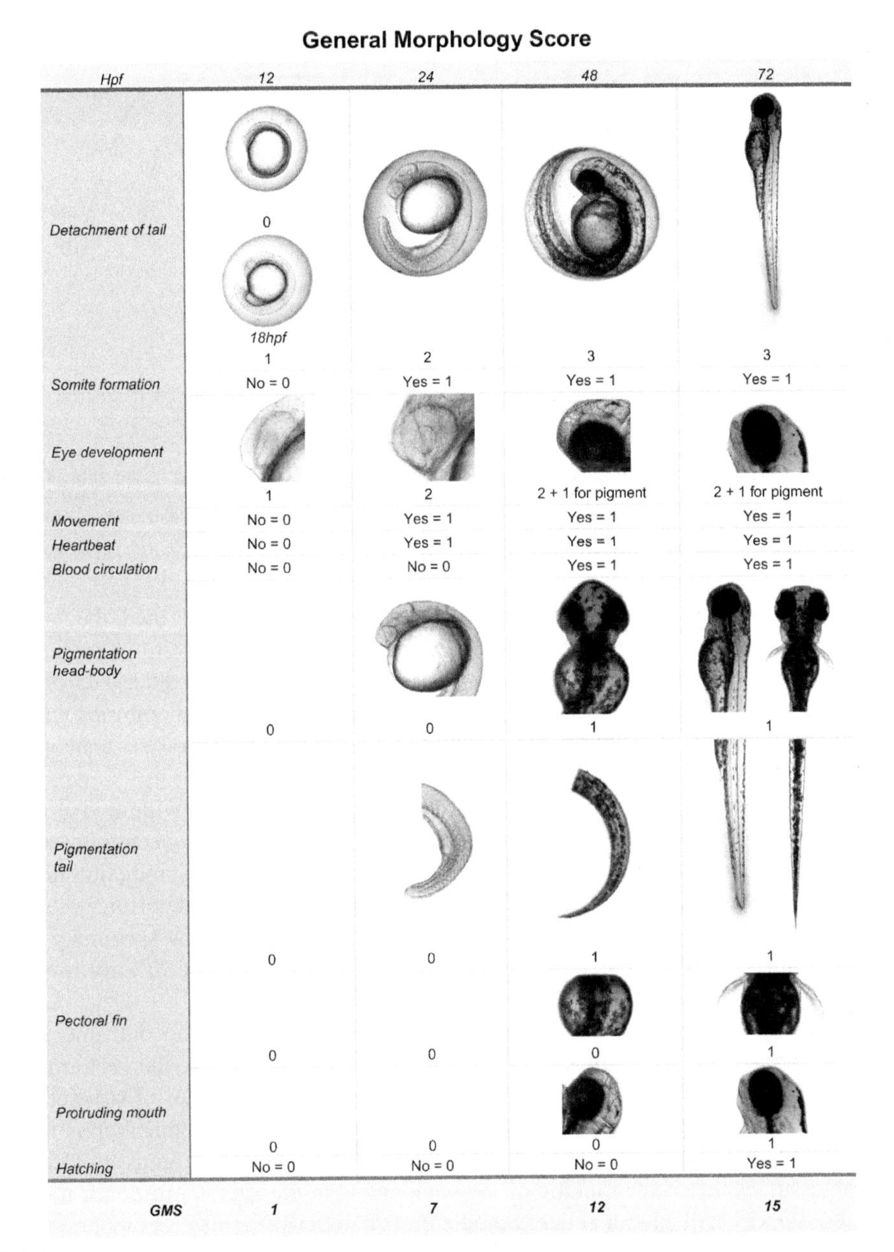

FIGURE 4.1.1

General morphology scoring system showing normal development of a zebrafish embryo up to 72 hours post-fertilization with different scores assigned to specific developmental endpoints in time.

–Reprinted from [15], with permission from Elsevier

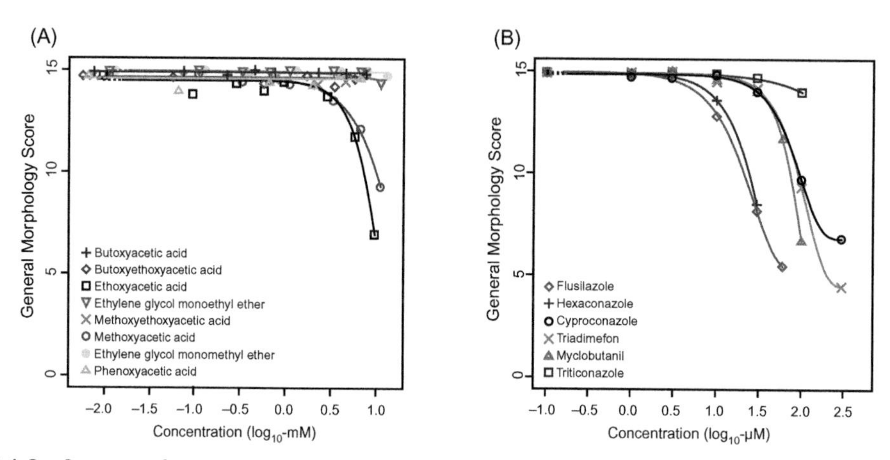

FIGURE 4.1.2 Concentration–response curves of general morphology score evaluated in the zebrafish embryo. Evaluations at 72 hours post-fertilization for (A) glycol ether metabolites and (B) 1,2,4-triazole antifungals.

–Adapted from [15], with permission from Elsevier

and determined the lethal concentration for 25% of the embryos (LC25) and the no-observed-adverse-effect level (NOAEL). With the LC25/NOAEL ratio, compounds were classified as teratogenic or non-teratogenic, giving an overall concordance with in vivo mammalian data of 87%. This protocol was further optimized by Gustafson et al. and tested at different laboratories with 20 compounds. The inter-laboratory assessment before optimization resulted in 60–70% concordance, and using the optimized protocol, a total concordance of 85% was achieved.

Selderslaghs et al. obtained a concordance of 74–81% for 27 compounds with in vivo animal data using the teratogenic index (TI), which is the ratio between the 50% lethal concentration (LC50) and the 50% effect concentration (EC50)[16]. To obtain this high concordance, a prediction model with a TI >2 as cutoff value was applied. With a smaller set of 15 compounds, van den Bulck et al. achieved only 60% overall concordance with mammalian data, mostly owing to a low specificity because of a high number of false-positive predictions in their assay [17]. Sensitivity of 75% for this study was reported.

Results from these studies showed a good prediction based on a yes/no outcome in terms of developmental toxicity with only limited sets of compounds. To characterize the performance of the zebrafish embryo within a class of chemicals, Hermsen et al. tested two series of structurally related compounds, glycol ethers and 1,2,4-triazoles. Compounds from the same chemical class were ranked on their relative embryotoxicity using the GMS and compared to the in vivo ranking [15]. This category approach assumes that if the ranking of the compounds in the ZET corresponds to the in vivo ranking, the test system will give a reliable prediction of toxicity for new compounds within the same class [18,19]. Using a 5% decrease in GMS, benchmark concentrations were calculated based on the concentration–response curve (Figure 4.1.2) for each compound within the chemical class (BMC_{GMS}). Benchmark concentrations based on teratogenicity (BMC_T), defined as a 5% decrease in the number of embryos with one or more teratogenic effect, were calculated as well. Using these BMCs, a ranking of embryotoxic potency of compounds was established, and ZET results were compared to the ranking of in vivo mammalian developmental toxicity data. For the tested classes,

Table 4.1.1 Benchmark Concentrations for the Endpoints GMS and Teratogenicity for Different Triazoles in the ZET

Triazoles	BMC_{GMS}[a] (μM)	BMC_{T}[b] (μM)	dLEL[c] (μmol/kg bw/day)
FLU	4.8 (4.3–5.4)	8.1 (5.4–11.3)	1.3
HEX	7.0 (6.1–7.9)	10.1 (7.1–19.0)	8.0
CYP	27.7 (22.3–34.7)	19.8 (8.4–29.7)	41.1
TDF	29.2 (23.1–37.5)	6.6 (3.5–12.3)	170.2
MYC	30.2 (28.0–32.5)	51.4 (25.8–53.8)	1083.9
TTC	80.5 (66.7–101.5[d])	40.0 (16.2–96.2)	3146.5

[a]*Benchmark concentration for general morphology score at a 5% benchmark response.*
[b]*Benchmark concentration for teratogenicity at a 5% benchmark response.*
[c]*dLEL: lowest effect level for any developmental effect derived from the ToxRefDB.*
[d]*Exceeding highest concentration tested.*
bw, body weight; CYP, cyproconazole; FLU, flusilazole; HEX, hexaconazole; MYC, myclobutanil; TDF, triadimefon; TTC, triticonsazole.

potency ranking in the ZET showed high concordance with in vivo data, as presented for the triazoles in Table 4.1.1[15].

Morphological evaluation appears to be a good outcome measurement for predictive toxicology using the zebrafish embryo; however, it is, for the greater part, subjective and subtle non-visible effects may remain unnoticed. Furthermore, morphological effects provide little discrimination between the specific effects of toxic compounds. To this end, improvements on the endpoint of the test regarding the sensitivity, objectivity, and predictability would be useful.

4.1.4 ZET and toxicogenomics

Recent developments in improving in vitro alternatives for developmental toxicity testing include the implementation of molecular techniques to measure gene transcription profiles. These transcriptomic techniques may enable the detection of sub-morphological effects and could increase sensitivity and predictability of the ZET. With the application of transcriptomics in toxicology, commonly referred to as toxicogenomics, efforts have been made to identify specific biomarkers for the prediction of toxicological effects. One of the first studies implementing gene expression evaluation as an outcome measure for compound exposure in the developing zebrafish embryo was based on a marker for endocrine disruption [20]. The vitellogenin 1 gene (*vtg1*), generally present in oviparous animals, was measured using quantitative real-time polymerase chain reaction (PCR). This gene encodes the calcium-binding phosphoglycoprotein vitellogenin, which is synthesized in the liver in fish under the influence of the female sex steroid estradiol. Therefore, *vtg1* is often used to indicate estrogenicity and subsequent endocrine disruption [20]. By exposure of zebrafish embryos to different compounds, including bisphenol A, 4-nonylphenol, atrazine, cyproconazole, 17β-estradiol, and 17α-ethinylestradiol, the results indicated that *vtg1* is a good marker in the zebrafish embryo to measure estrogenic potential [21]. However, for further characterization of toxic effects of compound exposure it is necessary to measure more endpoints than estrogenicity alone. Besides the *vtg1* gene, in a subsequent study *mt2*, *cyp1a1*, and *rag1* expression levels were determined as markers for metal toxicity, polycyclic aromatic hydrocarbon toxicity, and immune disruption, respectively. Results showed these genes to be suitable markers and the zebrafish embryo to be a promising model for toxicity prediction using molecular markers [22].

Whereas in these studies only small groups of genes were evaluated with real-time PCR, the development of microarray analysis has enabled the high-throughput screen of many transcripts at the same time. With a DNA microarray containing only a selected subset of 230 genes, the effect of 4-nonylphenol on 48-h.p.f. zebrafish embryos was analyzed as a proof-of-principle experiment to detect toxic effects. Nine robustly regulated genes were identified that were associated with the toxic response of the compound, showing the potential use of microarray for predictive toxicology in the zebrafish embryo [23]. Using genome-wide microarrays with more than 16,000 genes present, Yang et al. were able to predict with high probability the identity of the tested compounds in their study based on barcode-like gene expression profiles. More recently, discrimination between two chemical classes based on gene expression profiles was demonstrated [24]. Exposure to compounds from one chemical class, either glycol ethers or 1,2,4-triazoles, resulted in different transcript profiles for each class specifically, but within the chemical classes expression profiles for the individual compounds were very alike. Similarly, in other alternative models for developmental toxicity testing, these observations have been corroborated [25,26].

In addition, both ZET studies showed that transcriptomics readout is more sensitive than morphological assessment; at no-effect concentrations of morphology, effects on gene expression were present. Given the sensitivity of this endpoint, gene expression may provide a way to explore the low end of the concentration–response curve at which no morphological effects are observed [27].

4.1.5 Concentration-dependent gene expression

As a well-known principle in the field of toxicology, it is very important to consider the concentration of the compound, to establish potency as well as cause-and-effect relationships. Morphological effect size and the extent of gene expression changes are both related to dose [28,29]. In the field of developmental toxicity, this has been demonstrated for several alternative test models, such as the EST and WEC [30,31].

In zebrafish embryos, with only a few concentrations of the tested compounds, Yang et al. showed differences in gene expression for each concentration. A ZET study including microarray analysis at 24 hpf was performed with eight concentrations of flusilazole (FLU), a triazole antifungal, to investigate the concentration dependency of gene expression [32]. Results revealed a number of genes and processes to be regulated in a concentration-dependent way in terms of fold change in 24-hpf embryos (Figure 4.1.3). The actual number of genes significantly differentially expressed increased with increasing concentration, as shown in Figure 4.1.3C. Functional analysis revealed the enrichment of several development-related processes, such as retinol metabolism and transcription, as well as processes corresponding to the antifungal mechanism of action, steroid biosynthesis, and fatty acid metabolism. Moreover, the development-related processes were already significantly altered in the absence of morphological effects at both 24 and 72 hpf. This indicates that gene expression is more sensitive than morphological readout, but that it may also precede the morphological effects. An increase in concentration was also positively associated with an increase in magnitude of expression

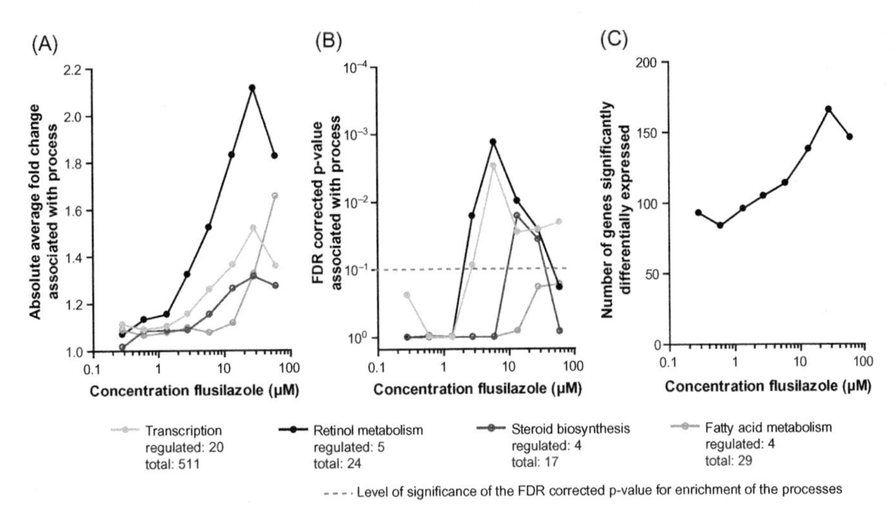

FIGURE 4.1.3 Results of a ZET study with microarray analysis to assess concentration dependency of gene expression.

(A) Concentration-related effect of FLU exposure in 24-hpf zebrafish embryos on absolute average fold change of a subset of enriched processes. Numbers underneath indicate the number of genes significantly regulated and total number of genes in the process. (B) Concentration-related effect of FLU exposure in 24-hpf zebrafish embryos on false discovery rate (FDR)-corrected p-value of a subset of processes, indicating at which concentration a process is enriched. Level of significance is depicted by the dotted line. (C) Number of significantly regulated genes compared with control after t-test ($p < 0.05$), which was used for determination of level of significance of enrichment.

–Reprinted from [32], with permission from Oxford University Press

for individual genes within these functional processes. However, in the presence of severe morphological effects at the time-point of sampling for gene expression, the transcriptional regulation seemed to have collapsed. Several studies have described similar phenomena in other model systems of developmental and reproductive toxicity [30,31,33]. Careful selection of compound concentration is therefore essential for the prediction of developmental toxicity by gene expression analysis.

4.1.6 Relative embryotoxicity using gene expression data

Concentration–response modeling may be a useful approach for determining potency ranking of compounds in the ZET using toxicogenomics. Developmental toxicants are structurally very diverse and may induce different types of developmental effects. Even though similar developmental effects can be induced by different types of compounds, the underlying mechanism that causes these effect may differ. The category approach can be used to create a relative potency ranking of compounds, to increase the predictive value of the test within classes of compounds. Chemicals with a similar structure are expected to have similar physicochemical and biological properties and, therefore, similar mechanisms of toxicity. If a class of chemicals has been correctly predicted in the in vitro test, the test system is likely to give a reliable prediction of toxicity for new compounds within the same class and, moreover, embryotoxic potency can be assessed using a read-across approach [34]. In addition, this will give information about the applicability domain of the test, since this may vary with the class of compounds tested [35]. In various model systems for developmental toxicity, the category approach has demonstrated promising results [18,36,37].

Potency ranking using the zebrafish embryo has been demonstrated with pyrethroid insecticides by DeMicco et al. [38]. Based on neurotoxicity data, ranking of these compounds was very similar to that using toxicity data from adult rats. For developmental toxicity, Hermsen et al. showed good correlation between the ranking in the ZET based on GMS and in vivo developmental toxicity data as described above [15].

As potency ranking was successfully used in the ZET with classical outcome measurement [15], implementation of toxicogenomics might allow additional assessment of specific mechanisms. Gene expression profiles have already been demonstrated to be chemical-class-specific in several in vitro systems, including EST, WEC, and ZET [24–26]. In studies using other in vitro models for developmental toxicity, it has been demonstrated that transcriptomic-based assessment can be used for potency ranking. In the WEC, for a group of phthalate esters, the application of gene-expression-based assessments showed differences in potency [39]. For the same group of toxicants, using concentration–response analysis for differential gene expression in the EST, the transcriptomics-based ranking mimicked the in vivo situation [40]. In the ZET, a group of six triazole antifungals was tested, of which one used in a complete-concentration–response design and the others tested at equipotent concentrations based on morphological outcome. Gene expression profiles for all compounds generally appeared very similar, although the extent of gene expression regulation varied among the tested compounds. Gene expression data allowed a detailed assessment of the chemical effects in terms of functional gene pathways for the discrimination of relative potencies. By comparing the degree of regulation of the triazoles' proposed developmentally toxic mechanism of action, retinol metabolism, and the antifungal mechanism, steroid biosynthesis, relative potencies between the compounds were indicated [41]. As depicted in Figure 4.1.4, cyproconazole (CYP) showed a high regulation of retinol metabolism, in contrast to triticonazole (TTC), which regulated this process to a lesser extent. However, for steroid biosynthesis, highest regulation was for TTC, and CYP showed the lowest regulation compared to the other compounds. Gene expression findings were in line with morphological assessment, as TTC was also the least potent compound based on GMS. Although these first results of a ranking approach in the ZET look very promising, further research is needed to study the applicability domain of the model.

4.1.7 Identification of adaptive and adverse responses using transcriptomics

Identification of responses using toxicogenomics-based assessment was shown to be more sensitive than using morphological outcome parameters. Although regulation of genes and functional gene groups, such as pathways, appears to be a far more sensitive endpoint, the question remains of when to consider these responses as adaptive or adverse.

As shown in the transcriptomic ZET study with multiple concentrations of FLU, regulation of specific pathways, such as transcription and retinol metabolism, occurred before the onset of morphological effects [32]. Whereas regulation of these pathways was clearly present at low concentrations, it is difficult to distinguish adaptive versus adverse responses.

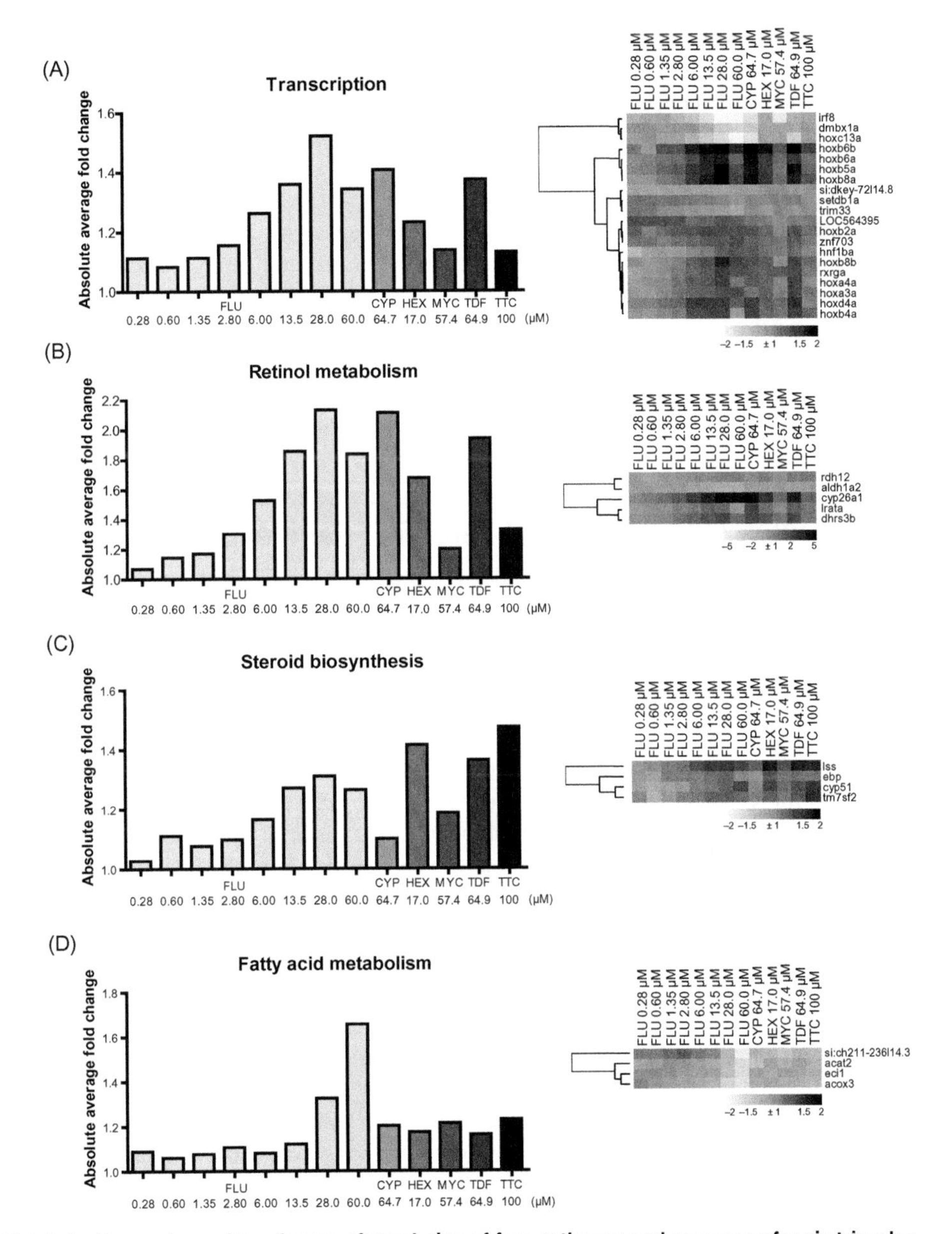

FIGURE 4.1.4 Comparison of the degree of regulation of four pathways and processes for six triazoles.

Absolute average fold change for the genes present in the processes or pathways for (A) transcription, (B) retinol metabolism, (C) steroid biosynthesis, and (D) fatty acid metabolism, calculated for each exposure group, with heat-maps showing individual gene expression (fold change) of the genes within the process or pathway.

–Reprinted from [41], with permission from Elsevier

In another study, with valproic acid, retinoic acid, carbamazepine, and caffeine tested in multiple concentrations, the benchmark concentration approach was applied to identify sensitively regulated genes [42]. The benchmark concentration for selected genes was calculated and compared to the benchmark concentration from morphology, to give an indication about the sensitivity of the genes. Some genes were found to be already regulated at very low concentrations relative to the concentrations needed to cause morphological effects.

Overall, regulation of genes and processes does not necessarily represent adverse outcomes. Under normal development, genes might be differentially regulated, indicating a possible homeostatic response to prevent deregulation of pathways eventually causing effects. To discern adaptive from adverse effects using gene expression, the level of regulation needs to be defined in order to interpret gene expression findings for hazard and risk assessment [43]. It is generally accepted that reproductive and developmental toxicants have a threshold of adversity [44]. Defining the threshold between adaptive and adverse responses at the gene expression level poses a challenge for future investigations.

4.1.8 Interspecies extrapolation of zebrafish gene expression data

The zebrafish embryo is a commonly used model for developmental biology. As stated before, its development is well characterized at the level of morphology. In addition, at the transcriptome level, different stages of embryonic development are being unraveled [45,46]. Cross-comparison of effects at transcriptomic level between different species may facilitate and enhance the identification of biomarkers for developmental toxicity and, furthermore, aid in the understanding of specific compound responses. Given the conservation of many molecular pathways between zebrafish and human [47], gene expression profiling provides the possibility to extrapolate the observed mechanisms of toxicity to the human situation. In recent studies, it has been demonstrated that zebrafish embryos are suitable for detection of known mechanisms of toxicity of compounds, for instance the triazoles. With transcriptomic analyses, the specific antifungal mechanism of triazoles, which is *cyp51* inhibition as reflected in regulation of steroid biosynthesis, was detected. The embryotoxic mechanism of action, modulation of *cyp26a1* expression and subsequent disruption of retinol metabolism, was also found to be highly regulated in the zebrafish embryos [32,41] and was in accordance with in vitro and in vivo data [48].

Nevertheless, to be able to extrapolate gene expression findings between species, the specific phylotypic stage of development needs to be taken into account. This provides information about developmental conservation across species during embryogenesis. In a recent paper, Irie and Kuratani used quantitative comparative transcriptomics to reveal the similarity between embryonic stages across different vertebrate species [49]. The pharyngula stage showed the highest similarity, corresponding to the 24-h.p.f. stage for the zebrafish embryo. Current transcriptomics studies in the zebrafish embryo are mostly performed covering this phylotypic stage, thereby facilitating extrapolation to the human situation.

4.1.9 Future perspectives

Improving the ZET—toxicokinetics

To even further improve toxicity prediction and extrapolation to the human situation, certain parameters might be considered for inclusion in the test. An important issue to consider for improvement of the ZET is the kinetics of the compound. Inclusion of kinetic parameters on absorption, distribution, metabolism, and excretion would improve interspecies extrapolation and enhance risk assessment, since they help to identify potential exposure duration and concentration. In both in vivo and in vitro studies, the toxicity prediction has been shown to improve when these parameters of the test compounds were included [50,51]. In the transcriptomic studies with external exposure described in this chapter, the zebrafish embryo is protected by a protein membrane, the chorion. Even though it is still unclear whether the chorion acts as a barrier for the uptake of compounds, pores of about 0.17 μm^2 are present in this membrane and may allow penetration of chemicals based on their size [52]. This size-dependent uptake has been reported for fluorescent dextrans greater than 3 kDa [53] and for a cationic polymer, Luviquat HM 522, which was only toxic in dechorionated or hatched embryos [54]. But the chorion acted as a barrier not only for large compounds, it also functioned as weak protection for some smaller compounds [55]. However, some groups suggest that the chorion has only a limited influence on compound delivery to the embryo and that the uptake of the compound by the embryo itself is more significant [56,57]. Several studies reported the amount or uptake of compounds in the embryo and used this information for their toxicity prediction [17,58]. As shown, if the compound is not or is only poorly absorbed by the embryo, this may give rise to false-negative results. Furthermore, relative potencies of compounds could be determined in more detail, which, in addition, would give more information about the applicability domain of the test. Toxicity prediction would, therefore, benefit from additional information on kinetics of the compound, and this needs to be studied in more detail.

Regulatory implementation

Current risk assessment for developmental and reproductive toxicology is mostly based on animal studies using rodents and rabbits. In view of the three Rs, to replace, reduce, and refine the use of animals in current testing methods, implementation of alternative tests such as the ZET would be very valuable. Using an intelligent testing strategy, including alternative assays, to assess toxicological profiles of compounds for hazard assessment, animal numbers and costs can be reduced. Intelligent testing strategies use a stepwise approach for toxicity testing, by which the test models used become more complex with every step, thereby evaluating the information obtained in each step to have a weighed decision on the need for further testing [59,60]. A battery of alternative assays included in a testing strategy, each one covering different parts of development, may be used to prioritize and select compounds for further in vivo testing of developmental toxicity [35]. Combining the alternative assays, such as the ZET, with transcriptional readouts and computational modeling may give detailed information on the toxicity of the compound and its mode of action. These transcriptional approaches will produce large data sets, which poses a challenge for the identification of toxicity, but supports the optimization and validation of alternative assays, such as the ZET. To optimize and validate the ZET, several critical factors need to be taken into account, including the applicability domain, interspecies extrapolation, and kinetics, as discussed in the previous sections.

The results presented in this chapter provide promising insight into the toxic effects of compounds on the zebrafish embryo; however, further research on the applicability of the ZET with transcriptomics is needed. As a next step forward, toxic responses at the biological pathway level causing subsequent effects on the embryo have to be identified. Many studies using transcriptomic approaches for predicting adverse effects focus on the regulation of individual genes as a biomarker. However, this method is not always successful, since compounds do not necessarily alter the regulation of the same genes even if the morphological outcome is similar [42]. A recently proposed approach is to identify and study pathways that are expected to initiate the toxic response that can lead to developmental effects [61,62]. Such an adverse-outcome pathway is a concept that combines existing knowledge concerning the linkage between a direct molecular initiating event and an adverse outcome, for instance developmental effects, at a biological level of organization relevant to risk assessment [61]. The use of transcriptomics data may lead to the discovery of adverse-outcome or toxicity pathways that, when sufficiently regulated, are likely to disturb normal development. Future research should be directed at the definition of these pathways and the level of dysregulation needed to initiate the developmentally toxic effects.

The research presented in this chapter represents the progress made in the implementation of transcriptomics in the ZET and provides a foundation for further study of the model and its use in developmental toxicity testing.

References

[1] Piersma AH. Whole embryo culture and toxicity testing. Toxicol In Vitro 1993;7:763–8.

[2] Piersma AH, Genschow E, Verhoef A, Spanjersberg MQ, Brown NA, Brady M, et al. Validation of the post-implantation rat whole-embryo culture test in the international ECVAM validation study on three in vitro embryotoxicity tests. Altern Lab Anim 2004;32:275–307.

[3] EU Directive 2010/63/EU of the European parliament and of the council of 22 September 2010 on the protection of animals used for scientific purposes. Off J EU 2010;L276:33–79.

[4] Strahle U, Scholz S, Geisler R, Greiner P, Hollert H, Rastegar S, et al. Zebrafish embryos as an alternative to animal experiments: A commentary on the definition of the onset of protected life stages in animal welfare regulations. Reprod Toxicol 2011

[5] Kimmel CB, Ballard WW, Kimmel SR, Ullmann B, Schilling TF. Stages of embryonic development of the zebrafish. Dev Dyn 1995;203:253–310.

[6] Gilbert SF. Developmental biology, 6th ed: Sinauer Associates; 2000.

[7] Hill AJ, Teraoka H, Heideman W, Peterson RE. Zebrafish as a model vertebrate for investigating chemical toxicity. Toxicol Sci 2005;86:6–19.

[8] Hisaoka KK. The effects of 2-acetylaminofluorene on the embryonic development of the zebrafish: I. Morphological studies. Cancer Res 1958;18:527–35.

[9] Hisaoka KK. The effects of 2-acetylaminofluorene on the embryonic development of the zebrafish: II. Histochemical studies. Cancer Res 1958;18:664–7.

[10] Nagel R. DarT: the embryo test with the zebrafish *Danio rerio*: a general model in ecotoxicology and toxicology. Altex 2002;19(Suppl 1):38–48.

[11] Braunbeck T, Lammer E. Background paper on fish embryo toxicity assays. Prepared for German Federal Environment Agency UBA Contract No. 203 85 122. This can be found at: <http://www.oecd.org/chemicalsafety/testing/36817242.pdf>; 2006.

[12] Organisation for Economic Co-operation and Development. Draft test guideline: Fish embryo toxicity (FET) test. OECD Guideline for the testing of chemicals. 1–11. This can be found at: <http://www.oecd.org/env/ehs/testing/2012-07-09_Draft_FET_TG_v8_FINAL.pdf >; 2006.

[13] Brannen KC, Panzica-Kelly JM, Danberry TL, Augustine-Rauch KA. Development of a zebrafish embryo teratogenicity assay and quantitative prediction model. Birth Defects Res B Dev Reprod Toxicol 2010;89:66–77.

[14] Padilla S, Corum D, Padnos B, Hunter DL, Beam A, Houck KA, et al. Zebrafish developmental screening of the ToxCast Phase I chemical library. Reprod Toxicol 2012;33:174–87.
[15] Hermsen SAB, van den Brandhof E-J, van der Ven LTM, Piersma AH. Relative embryotoxicity of two classes of chemicals in a modified zebrafish embryotoxicity test and comparison with their in vivo potencies. Toxicol In Vitro 2011;25:745–53.
[16] Selderslaghs IW, Blust R, Witters HE. Feasibility study of the zebrafish assay as an alternative method to screen for developmental toxicity and embryotoxicity using a training set of 27 compounds. Reprod Toxicol 2012;33:142–54.
[17] van den Bulck K, Hill A, Mesens N, Diekman H, De Schaepdrijver L, Lammens L. Zebrafish developmental toxicity assay: A fishy solution to reproductive toxicity screening, or just a red herring? Reprod Toxicol 2011;32:213–9.
[18] de Jong E, Louisse J, Verwei M, Blaauboer BJ, van de Sandt JJM, Woutersen RA, et al. Relative developmental toxicity of glycol ether alkoxy acid metabolites in the embryonic stem cell test as compared with the in vivo potency of their parent compounds. Toxicol Sci 2009;110:117–24.
[19] US Environmental Protection Agency. Development of chemical categories in the HPV challenge program. US Environmental Protection Agency; 1999.
[20] Muncke J, Eggen RI. Vitellogenin 1 mRNA as an early molecular biomarker for endocrine disruption in developing zebrafish (*Danio rerio*). Environ Toxicol Chem 2006;25:2734–41.
[21] Muncke J, Junghans M, Eggen RI. Testing estrogenicity of known and novel (xeno-)estrogens in the MolDarT using developing zebrafish (*Danio rerio*). Environ Toxicol 2007;22:185–93.
[22] Liedtke A, Muncke J, Rufenacht K, Eggen RI. Molecular multi-effect screening of environmental pollutants using the MolDarT. Environ Toxicol 2008;23:59–67.
[23] Hoyt PR, Doktycz MJ, Beattie KL, Greeley Jr. MS. DNA microarrays detect 4-nonylphenol-induced alterations in gene expression during zebrafish early development. Ecotoxicology 2003;12:469–74.
[24] Hermsen SAB, Pronk TE, van den Brandhof EJ, van der Ven LTM, Piersma AH. Chemical-class-specific gene expression changes in the zebrafish embryo after exposure to glycol ether alkoxy acids and 1,2,4-triazole antifungals. Reprod Toxicol 2011;32:245–52.
[25] Robinson JF, Tonk EC, Verhoef A, Piersma AH. Triazole-induced concentration-related gene signatures in rat whole-embryo culture. Reprod Toxicol 2012;34:275–83.
[26] van Dartel DAM, Pennings JLA, Robinson JF, Kleinjans JCS, Piersma AH. Discriminating classes of developmental toxicants using gene expression profiling in the embryonic stem cell test. Toxicol Lett 2011;201:143–51.
[27] Daston GP, Naciff JM. Predicting developmental toxicity through toxicogenomics. Birth Defects Res Part C: Embryo Today: Rev 2010;90:110–7.
[28] Andersen ME, Clewell HJ, Bermudez E, Willson GA, Thomas RS. Genomic signatures and dose-dependent transitions in nasal epithelial responses to inhaled formaldehyde in the rat. Toxicol Sci 2008;105:368–83.
[29] Goetz AK, Dix DJ. Mode of action for reproductive and hepatic toxicity inferred from a genomic study of triazole antifungals. Toxicol Sci 2009;110:449–62.
[30] van Dartel DAM, Pennings JLA, de la Fonteyne LJJ, Brauers KJJ, Claessen S, van Delft JH, et al. Concentration-dependent gene expression responses to flusilazole in embryonic stem cell differentiation cultures. Toxicol Appl Pharmacol 2011;251:110–8.
[31] Robinson JF, Guerrette Z, Yu X, Hong S, Faustman EM. A systems-based approach to investigate dose- and time-dependent methylmercury-induced gene expression response in C57BL/6 mouse embryos undergoing neurulation. Birth Defects Res B: Dev Reprod Toxicol 2010;89:188–200.
[32] Hermsen SAB, Pronk TE, van den Brandhof EJ, van der Ven LTM, Piersma AH. Concentration-response analysis of differential gene expression in the zebrafish embryotoxicity test following flusilazole exposure. Toxicol Sci 2012;127:303–12.

[33] Daston GP, Naciff JM. Gene expression changes related to growth and differentiation in the fetal and juvenile reproductive system of the female rat: Evaluation of microarray results. Reprod Toxicol 2005;19:381–94.
[34] Combes R, Barratt M, Balls M. An overall strategy for the testing of chemicals for human hazard and risk assessment under the EU REACH system. Altern Lab Anim 2003;31:7–19.
[35] Piersma AH. Alternative methods for developmental toxicity testing. Basic Clin Pharmacol Toxicol 2006;98:427–31.
[36] de Jong E, Doedee AM, Reis-Fernandes MA, Nau H, Piersma AH. Potency ranking of valproic acid analogues as to inhibition of cardiac differentiation of embryonic stem cells in comparison to their in vivo embryotoxicity. Reprod Toxicol 2011;31:375–82.
[37] Janer G, Verhoef A, Gilsing HD, Piersma AH. Use of the rat postimplantation embryo culture to assess the embryotoxic potency within a chemical category and to identify toxic metabolites. Toxicol In Vitro 2008;22:1797–805.
[38] DeMicco A, Cooper KR, Richardson JR, White LA. Developmental neurotoxicity of pyrethroid insecticides in zebrafish embryos. Toxicol Sci 2010;113:177–86.
[39] Robinson JF, Verhoef A, van Beelen VA, Pennings JL, Piersma AH. Dose-response analysis of phthalate effects on gene expression in rat whole-embryo culture. Toxicol Appl Pharmacol 2012;264:32–41.
[40] Schulpen SH, Robinson JF, Pennings JL, van Dartel DA, Piersma AH. Dose response analysis of monophthalates in the murine embryonic stem cell test assessed by cardiomyocyte differentiation and gene expression. Reprod Toxicol 2013;35:81–8.
[41] Hermsen SAB, Pronk TE, van den Brandhof EJ, van der Ven LT, Piersma AH. Triazole-induced gene expression changes in the zebrafish embryo. Reprod Toxicol 2012;34:216–24.
[42] Hermsen SAB, Pronk TE, van den Brandhof EJ, van der Ven LT, Piersma AH. Transcriptomic analysis in the developing zebrafish embryo after compound exposure: Individual gene expression and pathway regulation. Toxicol Appl Pharmacol 2013;272:161–71.
[43] Boekelheide K, Andersen ME. A mechanistic redefinition of adverse effects: A key step in the toxicity testing paradigm shift. ALTEX 2010;27:243–52.
[44] Piersma AH, Hernandez LG, van Benthem J, Muller JJ, van Leeuwen FX, Vermeire TG, et al. Reproductive toxicants have a threshold of adversity. Crit Rev Toxicol 2011;41:545–54.
[45] Mathavan S, Lee SG, Mak A, Miller LD, Murthy KR, Govindarajan KR, et al. Transcriptome analysis of zebrafish embryogenesis using microarrays. PLoS Genet 2005;1:260–76.
[46] Vesterlund L, Jiao H, Unneberg P, Hovatta O, Kere J. The zebrafish transcriptome during early development. BMC. Dev Biol 2011;11:30.
[47] Zon LI, Peterson RT. In vivo drug discovery in the zebrafish. Nat Rev Drug Discov 2005;4:35–44.
[48] Menegola E, Broccia ML, Di Renzo F, Giavini E. Postulated pathogenic pathway in triazole-fungicide-induced dysmorphogenic effects. Reprod Toxicol 2006;22:186–95.
[49] Irie N, Kuratani S. Comparative transcriptome analysis reveals vertebrate phylotypic period during organogenesis. Nat Commun 2011;2:248.
[50] Verwei M, van Burgsteden JA, Krul CA, van de Sandt JJ, Freidig AP. Prediction of in vivo embryotoxic effect levels with a combination of in vitro studies and PBPK modelling. Toxicol Lett 2006;165:79–87.
[51] Louisse J, de Jong E, van de Sandt JJ, Blaauboer BJ, Woutersen RA, Piersma AH, et al. The use of in vitro toxicity data and physiologically based kinetic modeling to predict dose-response curves for in vivo developmental toxicity of glycol ethers in rat and man. Toxicol Sci 2010;118:470–84.
[52] Cheng J, Flahaut E, Cheng SH. Effect of carbon nanotubes on developing zebrafish (*Danio rerio*) embryos. Environ Toxicol Chem 2007;26:708–16.
[53] Creton R. The calcium pump of the endoplasmic reticulum plays a role in midline signaling during early zebrafish development. Dev Brain Res 2004;151:33–41.
[54] Henn K, Braunbeck T. Dechorionation as a tool to improve the fish embryo toxicity test (FET) with the zebrafish (*Danio rerio*). Comp Biochem Physiol Part C: Toxicol Pharmacol 2011;153:91–8.

[55] Braunbeck T, Boettcher M, Hollert H, Kosmehl T, Lammer E, Leist E, et al. Towards an alternative for the acute fish LC(50) test in chemical assessment: the fish embryo toxicity test goes multi-species—an update. Altex 2005;22:87–102.

[56] Augustine-Rauch K, Zhang CX, Panzica-Kelly JM. In vitro developmental toxicology assays: a review of the state of the science of rodent and zebrafish whole-embryo culture and embryonic stem cell assays. Birth Defects Res C Embryo Today 2010;90:87–98.

[57] Eimon PM, Rubinstein AL. The use of in vivo zebrafish assays in drug toxicity screening. Expert Opin Drug Metab Toxicol 2009;5:393–401.

[58] Berghmans S, Butler P, Goldsmith P, Waldron G, Gardner I, Golder Z, et al. Zebrafish-based assays for the assessment of cardiac, visual and gut function: potential safety screens for early drug discovery. J Pharmacol Toxicol Methods 2008;58:59–68.

[59] van der Burg B, Kroese E, Piersma AH. Towards a pragmatic alternative testing strategy for the detection of reproductive toxicants. Reprod Toxicol 2011;31:558–61.

[60] van Leeuwen CJ, Patlewicz GY, Worth AP. Intelligent testing strategies. van Leeuwen CJ, Vermeire TG, editors. In risk assessment of chemicals. Netherlands: Springer; 2007. p. 467–509.

[61] Ankley GT, Bennett RS, Erickson RJ, Hoff DJ, Hornung MW, Johnson RD, et al. Adverse outcome pathways: a conceptual framework to support ecotoxicology research and risk assessment. Environ Toxicol Chem 2010;29:730–41.

[62] Krewski D, Acosta Jr. D, Andersen M, Anderson H, Bailar III JC, Boekelheide K, et al. Toxicity testing in the 21st century: a vision and a strategy. J Toxicol Environ Health B Crit Rev 2010;13:51–138.

CHAPTER

4.2 Transcriptomic Approaches in In Vitro Developmental Toxicity Testing

Elisa C.M. Tonk[*,**] and Aldert H. Piersma[**,†]

[*]Department of Toxicogenomics (TGX), Maastricht University, Maastricht, The Netherlands, [**]Center for Health Protection Research (GZB), National Institute for Public Health and the Environment (RIVM), Bilthoven, The Netherlands, [†]Institute for Risk Assessment Sciences (IRAS), Utrecht University, Utrecht, The Netherlands

4.2.1 Introduction to developmental toxicity testing

In vivo regulatory toxicity studies are expensive, time consuming, and require large numbers of animals. The majority of current regulatory toxicological testing for chemical safety is devoted to the assessment of reproductive and developmental health [1]. The implementation of the European REACH (Registration, Evaluation, Authorisation, and Restriction of Chemicals) legislation for chemical safety and additional societal pressure have resulted in a focus on the reduction of animal usage for toxicology studies. Still, the experimental paradigms for developmental toxicity testing have changed little over the past 40 years and rely entirely on the use of in vivo animal models.

Even though they are not implemented in current regulatory safety evaluations, in vitro developmental models have been important in advancing the understanding of mechanisms of embryogenesis and teratogenesis. These assays also aid the screening of compounds for further development as drug candidates and the prioritization of chemicals for animal testing. However, none of the assays is ready as yet to be utilized as a replacement for in vivo animal studies.

Recent technological advancements have enabled investigators to study thousands of molecular endpoints in a single assay. Currently available state-of-the-art techniques can be used to study the whole transcriptome, proteome, epigenome, or metabolome of a biological system. Transcriptomics, but also other so-called 'omics, can be used to identify molecular mechanisms underlying toxicity and to search for biomarkers to identify toxicity. Therefore, transcriptomics should be a useful new tool to characterize, classify, and potentially predict developmental toxicants. Multiple studies have shown that developmental-toxicant-induced gene expression alterations precede morphological changes that can be detected as part of the classical readout in these developmental in vitro models [2–4]. The use of transcriptomics may, therefore, shorten assay duration, which results in a higher throughput. Also, transcriptomics enables an objective endpoint evaluation. At the same time, the massive amount of data and complexity of statistical analysis make transcriptomic studies challenging. This chapter gives an overview of transcriptomic applications, in combination with various assay models, in the field of in vitro developmental toxicity testing.

Toxicogenomics-Based Cellular Models.

4.2.2 Alternative models for developmental toxicity testing

Whole-embryo culture

The rat whole-embryo culture (WEC) technique was first introduced by Denis New [5]. In this method, postimplantation rat embryos are typically explanted on gestational day (GD) 10, moved to an in vitro environment, and cultured for 48 h [6]. The culture medium consists primarily of rat serum or a mixture of rat and bovine serum, and test compounds can be added to the serum during the culture period. The endpoints, examined after 48 h of culture, usually include heartbeat and yolk sac circulation as an indication of viability and crown–rump length as a growth measure. The overall development is scored by the total morphological score first developed by Brown and Fabro [7] to quantitate embryonic growth and development, which was refined in a European Center for the Validation of Alternative Methods (ECVAM) protocol by Piersma [8] based on a more elaborate scoring method by van Maele-Fabry et al. [9,10]. The scoring method assesses the embryo's progression through neurulation and early organogenesis, including early formation of the brain, craniofacial structures, ears, eyes, heart, and limbs.

A major advantage of WEC is the use of intact embryos, which allows morphogenesis to proceed as it would in vivo. Additionally, the development of the embryo over the 48-h period in vitro parallels the in vivo development over the same period [11]. Furthermore, WEC offers the ability to determine the toxic potential of compounds as well as to perform more in-depth mechanistic studies. Also, WEC can be performed using rat, mouse, and rabbit embryos, which enables evaluation of species differences and the use of sensitive and resistant strains or species [12].

The relatively short culture duration and, therefore, limited developmental period of WEC can be a drawback. However, this developmental window represents a key period for organogenesis and provides a sensitive period for teratogenic insult associated with a majority of known teratogens [13]. These insults, however, may not manifest during the culture period using the classical scoring system. To detect subtle changes, a closer examination or alternative examination methods are required, such as the incorporation of gene expression profiling [3].

Zebrafish

Zebrafish as a model for developmental toxicity are gaining popularity. The zebrafish embryotoxicity test is a unique alternative that enables the study of a complete and well-characterized developmental period of a vertebrate embryo [14,15]. Hermsen et al. [16] described a general morphological scoring (GMS) system to assess the development of the zebrafish embryo based on specific endpoints in time. Embryos are treated from the 4- to 32-cell stage to 72 h post-fertilization (hpf) without removing the chorion. At 72 h.p.f., morphological evaluation of the embryos is performed using a GMS system similar to the scoring system established by Brown and Fabro [7] for rodent WEC. Evaluated endpoints include eye development, tail detachment, somite formation, movement, heartbeat, blood circulation, pigmentation development, pectoral fin development, mouth/jaw development, and tail detachment [16]. Malformations or other teratogenic effects are separately recorded as present or absent. The GMS scores can be analyzed using a benchmark dose (BMD) approach [17] in which a benchmark concentration (BMC) is calculated using a fitted dose–response curve. Endpoints include the BMC for the GMS (BMC_{GMS}), which relates to a 5% decrease in GMS, and the BMC that increases the percentage of embryos with one or more teratogenic effects by 5% (BMC_T).

The zebrafish as a model for developmental toxicity testing has a number of advantages, the major advantage being that, under European law, zebrafish embryos are considered a non-animal model of research up to 120 hpf, as they are not free-feeding up to that stage of development [18]. In general, zebrafish have a high fecundity, and the embryos develop independently of the maternal fish and are simply kept in water. The rapid development of the embryos allows the inclusion of all developmental stages in the assay. Also, evaluation of the transparent embryo permits clear observation of the early stages of development, which are similar to those of mammals. Still, applicability of the data derived from the zebrafish embryotoxicity test to mammalian systems has to be determined. Furthermore, more information is needed on uptake and metabolism of compounds in the zebrafish embryo and the role of the chorion as a potential barrier.

Embryonic stem cells

Blastocyst-derived murine pluripotent embryonic stem cells (ESCs) are able to self-renew as well as to differentiate in culture into a wide variety of cell types, including cardiomyocytes. The basic protocol for the mouse embryonic stem cell test (mEST) was first described by Spielmann et al. [19]. The cardiomyocyte differentiation pathway was selected because it allowed for a visual identification of contracting cells in differentiated ESC cultures. Briefly, mESTs from the D3 cell line are allowed to aggregate into embryoid bodies using the hanging drop culture technique for 3 days. These embryoid bodies resemble the egg-cylinder stage of a 5-day-old embryo and are able to differentiate into cells from each of the three germ layers, endoderm, ectoderm, and mesoderm. The embryoid bodies are then cultured in suspension culture for 2 days before transferring them to 24-well culture dish, where the cells adhere and differentiate for an additional 5 days. After a total culture time of 10 days, the percentage of wells with beating cardiomyocytes can be scored microscopically.

The mEST requires no live animals and uses only commercially available cell lines. In addition, it has been standardized and validated by ECVAM. However, the scoring of beating cardiomyocytes requires experience and is subject to observer bias. The introduction of molecular endpoints such as transcriptomics could allow for a more objective and high-throughput evaluation. Differentiation into a single lineage has been viewed as a disadvantage of the mEST; however, altered culture conditions allow for differentiation into other cell types derived from any of the three germ layers. Theunissen et al. [20] developed a protocol for neural differentiation of ESCs and de Jong et al. [21] developed a protocol for osteoblast differentiation, allowing for improvement of the predictive value of the mEST through incorporation of additional differentiation endpoints.

4.2.3 Application of transcriptomics in in vitro developmental toxicity assessments

Characterization

In vivo and in vitro models differ in biological complexity, structure, and observed toxic response. However, gene expression responses observed in vitro may correlate with response observed in vivo and can provide a comparable endpoint when in vitro studies are used to predict in vivo developmental toxicity. In WEC, transcriptomic responses associated with caffeine, methylmercury, monobutyl

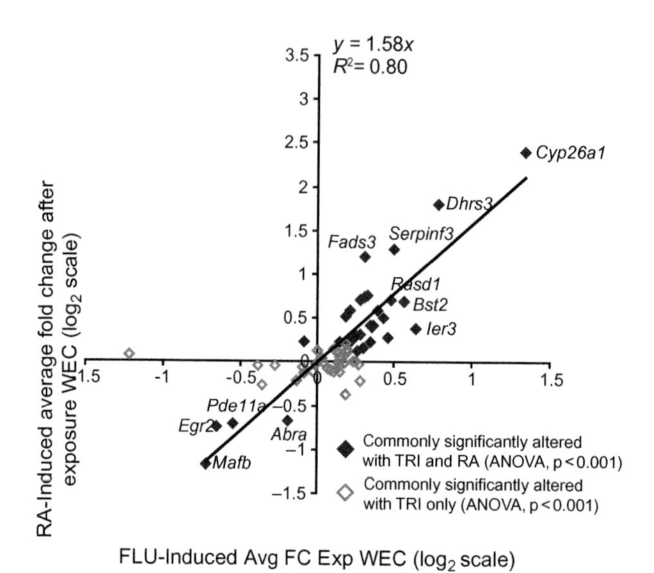

FIGURE 4.2.1 Direct comparison of the effects of flusilazole and retinoic acid on gene expression.

Cross-comparison based on the total number of genes commonly altered between flusilazole (FLU, 60 µg/ml) and retinoic acid (RA, 0.5 µg/ml) in WEC. Darkly shaded diamonds represent genes identified to be significantly altered by both flusilazole and retinoic acid (analysis of variance (ANOVA), $p < 0.001$).

–Reprinted from [22], with permission from Elsevier.

phthalate, and methoxyacetic acid correlated with previously hypothesized modes of developmental toxicity [3]. Multiple compounds induced common morphological effects; however, gene expression analysis revealed differences in biological processes affected in accordance with the modes of action of the compounds. Transcriptomics can also be used to investigate similar modes of action. A comparison of retinoic-acid- and flusilazole-induced effects in WEC was used to further investigate a suggested link between triazole exposure and altered levels of retinoic acid in association with teratogenicity [22]. The commonly regulated genes between flusilazole and retinoic acid exposure showed a similar directionality and ranking of response, further supporting a role for alteration of retinoic acid metabolism as a mode of developmental toxicity for triazoles (Figure 4.2.1).

Time-dependent effects

A developing organism is highly complex and the concept of developmental time is of great importance for toxicological testing. Transcriptomics can be used to describe the time-dependent nature of molecular changes during normal development and following exposure. Robinson et al. [23] exposed Wistar rat embryos to all-trans-retinoic acid (RA) both in utero and in WEC. The RA-induced gene expression changes at 2–48 h were identified to be time dependent (Figure 4.2.2). The in utero RA exposure showed a peak response in gene expression at the 6-h time-point, while a peak response for WEC was less distinct.

This time relationship was observed for the majority of the genes altered and could partly be explained by the kinetics of RA in both systems. However, some biological processes exhibited

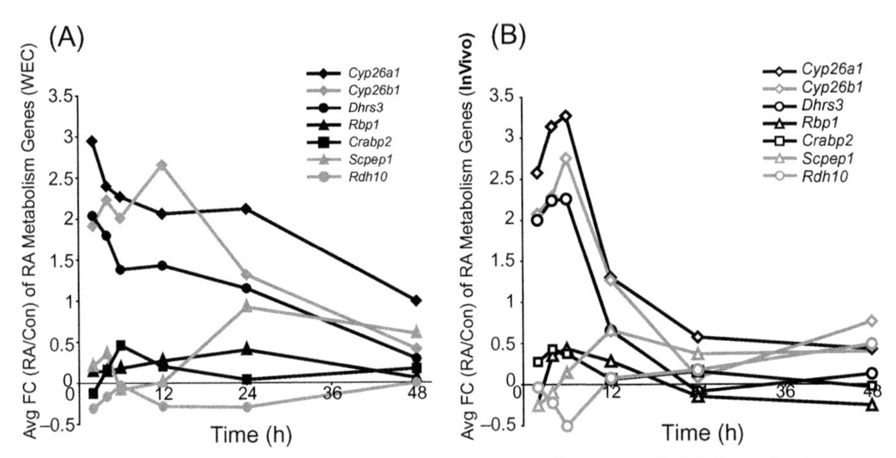

FIGURE 4.2.2 Time-dependent expression of retinoic acid metabolism genes in WEC and in vivo.

Retinoic-acid-induced changes in gene expression (average fold change (FC)) in genes associated with retinoic acid (or retinol) metabolism in (A) WEC and (B) in vivo ($\log_2$ scale).

–Reprinted from [23], with permission from Oxford University Press

different expression patterns; for instance, genes related to neural crest development and ear development were altered most at earlier time-points (<12 h), while genes related to glycolysis were altered at later time-points (>12 h). Similar time-dependent trends have been observed in both cardiomyocyte and neural EST [24,25]. In addition, the timing of exposure in differentiating ESCs was shown to affect the response in terms of sensitivity and type of genes regulated.

Concentration–response

Concentration–response assessment of gene expression is of great importance for the correct understanding of toxic properties of compounds. Studies have shown that the morphological effect size and the extent of gene expression changes, namely the magnitude of gene expression and the number of significantly regulated genes, were interrelated in models for developmental toxicity [22,26–29].

van Dartel et al. [29] investigated a concentration–response of flusilazole in EST (Figure 4.2.3). The assessment of the concentration-dependent gene expression modulation allowed for identification of the relative sensitivity of enriched processes. Using the differentiation track approach and previously defined gene sets, the degree of compound-induced ESC differentiation inhibition and gene ontology (GO) analysis of gene expression alterations correlated with the significance of deviation from the differentiation track. Therefore, the choice of test concentrations proved essential in view of the predictability of the EST for the detection of developmental toxicants.

Hermsen et al. [26] evaluated a concentration–response of differential gene expression in the zebrafish embryotoxicity test (ZET) following flusilazole exposure. Genes significantly altered in a monotonous or biphasic concentration–response fashion were identified and used in K-means clustering and term- or process-enrichment analysis. Steroid biosynthesis and fatty acid metabolism showed a concentration–response similar to the morphological response, while effects on retinol metabolism and transcription were observed at lower concentrations.

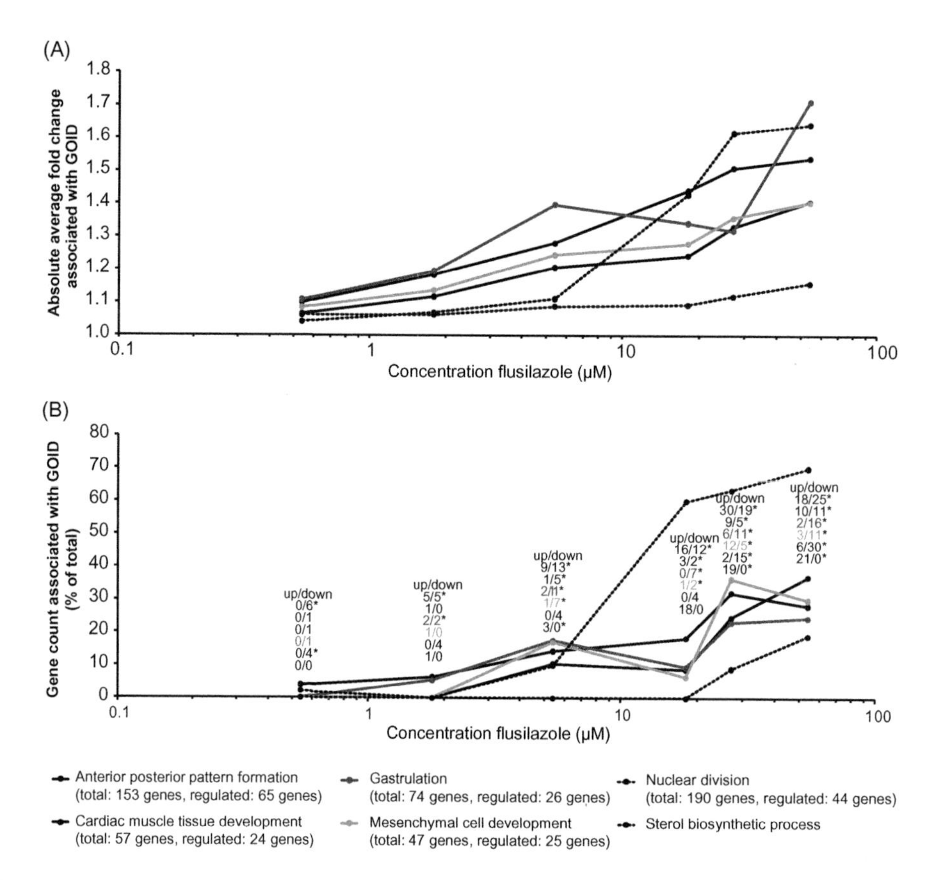

FIGURE 4.2.3 Concentration-dependent gene expression changes after flusilazole exposure in ESTc.

(A) Concentration-related effect of flusilazole exposure in embryonic stem cell test (cardiac) (ESTc) differentiation cultures on absolute average fold change of a subset of enriched gene ontology identifiers (GOIDs). For comparability, the absolute average fold change was calculated using the significantly differentially expressed genes that contributed to the enrichment in at least one of the experimental groups. (B) Concentration-related effect of flusilazole exposure in ESTc differentiation cultures on relative gene count associated with GOIDs. Continuous lines represent GOIDs related to development, dashed lines represent GOIDs related to cell division and lipid metabolism. "Total" refers to total number of genes annotated to that GOID, and "regulated" refers to total number of statistically significant genes that contributed to the enrichment of that GOID in at least one experimental group. Asterisks indicate statistically significantly enriched GOIDs.

–Reprinted from [29], with permission from Elsevier

Similar differential sensitivities were found evaluating triazole-induced gene signatures in WEC [22]. Significantly regulated genes included key members of hindbrain development, retinoic acid metabolism, neurogenesis, and steroid/lipid metabolism pathways. Flusilazole was tested at four concentrations, and benchmark doses representing a 10% shift in expression (BMD_{10}), were calculated for all significant genes within the selected enriched pathways. Using this BMD approach, hindbrain

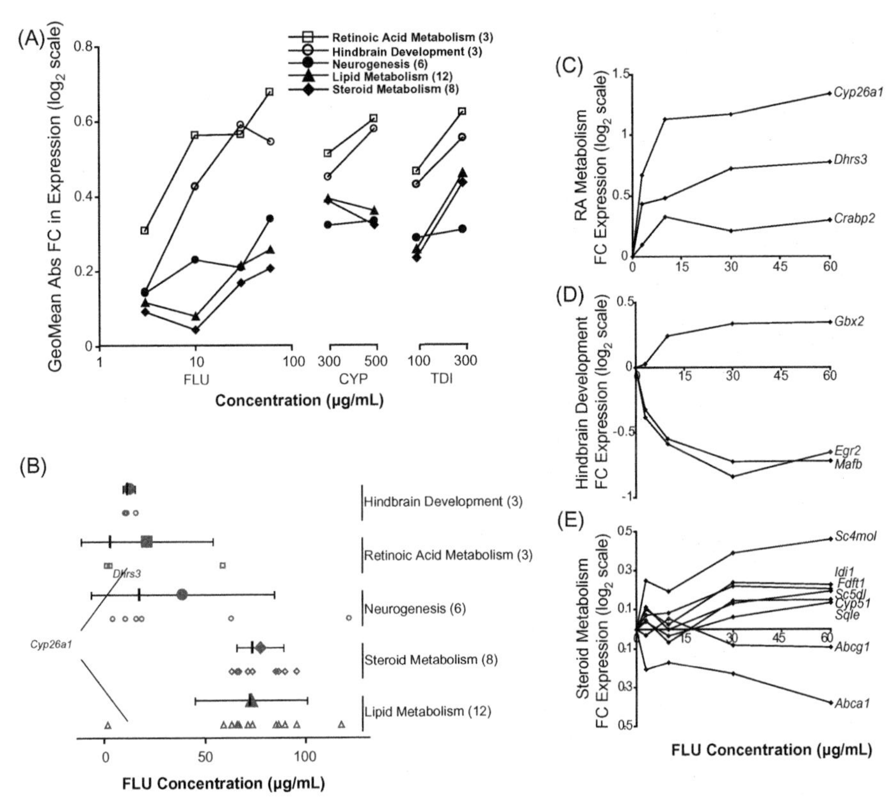

FIGURE 4.2.4 Functional analysis of dose-dependent effects of triazoles in WEC.

(A) Absolute average fold change response to triazole compounds in WEC of selected GO biological processes. The total number of genes altered by at least one of the three triazoles associated with each GO biological process is shown in parentheses (analysis of variance (ANOVA), $p < 0.001$). BMD analysis of significantly altered genes associated with enriched GO biological processes is shown. Smaller symbols are indicative of BMD_{10} values for independent genes. Single gene expression changes associated with (C) retinoic acid metabolism, (D) hindbrain development, and (E) steroid metabolism following flusilazole exposure. Abs FC, absolute average fold change; FLU, flusilazole; RA, retinoic acid.

–Reprinted from [22], with permission from Elsevier

development and retinoic acid metabolism appeared more sensitive to flusilazole as compared with steroid and lipid metabolism (Figure 4.2.4). Interestingly, using the same approach for phthalate-induced gene expression in WEC, regulation of the cholesterol/lipid/steroid metabolism pathways represented the most sensitive markers [27].

Based on these results, concentration–response assessment using transcriptomics allows for correlation of gene expression changes and dysmorphogenesis and can indicate potential mechanisms of toxicity.

Category approach

Transcriptomics can be used to distinguish between classes of compounds. van Dartel et al. [30] used transcriptomics in ESTc to differentiate between two classes of compounds, triazoles and phthalates, based on gene signatures using principle components analysis, enrichment of gene ontology biological processes using hierarchical clustering, and previously identified gene sets (Figure 4.2.5). The results show that compounds with similar modes of action can be identified using transcriptomics.

In the ZET, transcriptomics has also appeared promising for chemical classification and potency ranking. Hermsen et al. [2] identified specific gene signatures distinguishing exposures to glycol ethers and triazole compounds in ZET after 24-h exposure using principle components analysis of the genes commonly regulated per chemical class. In further studies, a concentration–response

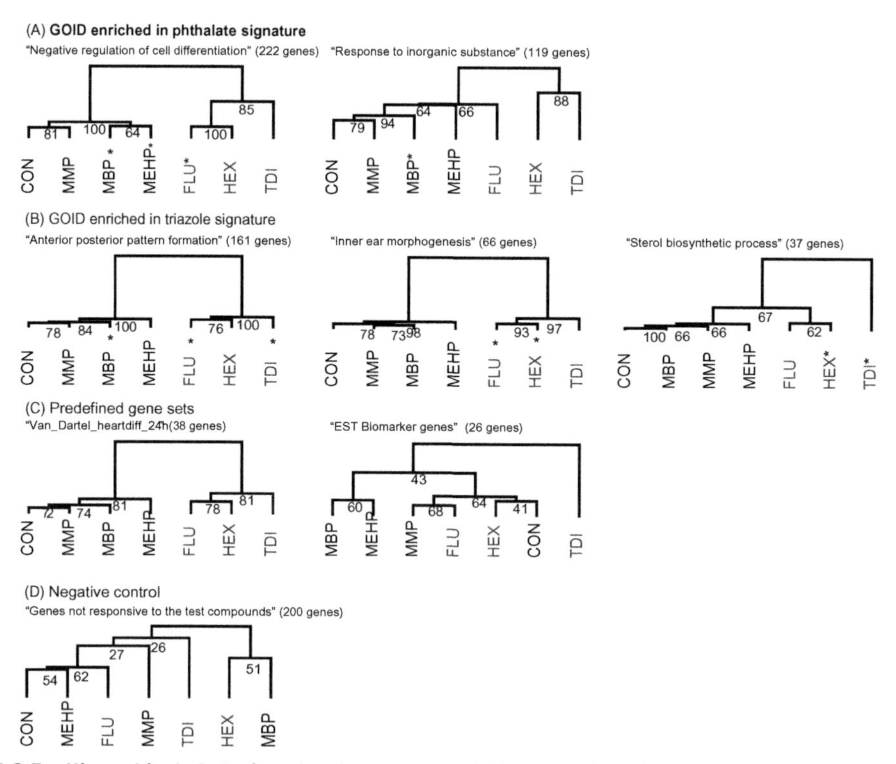

FIGURE 4.2.5 Hierarchical clustering of various compounds by transcriptomics.

Using all genes annotated to gene ontology identifiers (GOIDs) enriched in (A) the phthalate signature and (B) the triazole signature; using (C) predefined gene sets, and (D) a negative control gene set. Asterisks indicate significant GOID enrichment ($p < 0.05$) within the significantly regulated genes ($p < 0.001$) of that compound. The numbers indicate the confidence value (bootstrap value) for each node in the clustering tree. CON, control; FLU, flusilazole; HEX, hexaconazole; MBP, monobutyl phthalate; MEHP, monoethylhexyl phthalate; MMP, monomethyl phthalate; TDI, triadimefon.

–Reprinted from [30], with permission from Elsevier

assessment of flusilazole was used to identify triazole-specific pathways related to intended mechanisms and toxicological activity [26]. Subsequently, these data were used to discriminate between the relative potencies of multiple triazoles using specific gene groups [31]. A similar approach was used to evaluate gene expression alterations and relative potencies of triazole and phthalate exposures in WEC [22,27]. The directionality of gene expression regulation was observed to be similar across compound classes and the magnitude of effects on gene expression correlated with the degree of induced developmental toxicity.

Schulpen et al. [28] performed a potency ranking of four phthalates using concentration–response in the ESTc. Again, the same potency ranking was observed for morphology and gene expression alterations. The results of the concentration–response analysis in this study highlight an important aspect of the interpretation of in vitro assays, the threshold of adversity. Limited and stable gene expression alterations were observed at lower concentrations, with no toxic effects on morphology. Beyond a certain compound-specific concentration, gene expression greatly increased, followed closely by effects on morphology. The combination of a category approach and concentration–response analysis may therefore provide clues for the definition of adversity based on gene expression.

Classification

Due to their specificity, transcriptomic signatures can be used to classify compounds based on disruption of common genes or pathways. Implementation of this type of classification study in alternative developmental systems can possibly enhance the predictivity of these models. A series of studies published by van Dartel et al. and Pennings et al. [4,25,32–34] used a transcriptomic approach to classify and predict developmental toxic potential of compounds in ESTc. Using a principle-components-analysis-based differentiation track method, a gene set of 26 genes was able to discriminate between toxic and nontoxic exposures to 12 compounds with a prediction rate of 83% [33]. Further refinement using an integrated analysis led to the identification of 52 genes for which the data, combined with improvement of sample variability and effect size, suggested accuracies close to 100% might be reached [32].

Comparisons across in vivo and in vitro models

Conservation of gene expression responses across developmental model systems emphasizes the biological relevance of gene expression as a readout, especially in in vitro models [35]. To optimize the detection of developmental toxicants in vitro, increased insights into the biological domain, developmental stage, and complementary responses to chemicals within the various developmental in vitro assays are needed. These data can then be used to compose an effective combination of multiple complementary assays.

In vitro–in vivo comparisons

Comparisons of transcriptomic responses across in vivo and in vitro developmental models provide insights into the ability of in vitro models to correlate with in vivo hazard identification and toxicity prediction. Using transcriptomics, relevant molecular responses linked with developmental toxicity in vivo can be identified. Robinson et al. [23] used a transcriptomic approach to compare retinoic-acid-induced effects in Wistar rat embryos derived using WEC and in vivo. Across multiple time-points, strong similarities were observed in retinoic-acid-induced adverse morphological effects, including growth reduction, as well as alterations in neural tube, limb, branchial, and mandible development. Similar commonalities were observed on gene expression, both in directionality of response and significance. Subtle differences in gene expression alterations in WEC and in vivo could be linked to possible differential developmental progression and toxicokinetics, which were associated with specific differential morphological outcomes. Overall, the results supported the use of WEC to investigate morphological and gene expression responses relative to the in vivo situation. Similar observations were made for other in vitro developmental models. A comparison of available data assessing methylmercury-induced gene expression responses revealed a stronger similarity in the gene expression response between in vivo and in vitro developmental models as compared with the non-developing systems of liver, brain, and mouse embryonic fibroblasts [35] (Figure 4.2.6). Specific

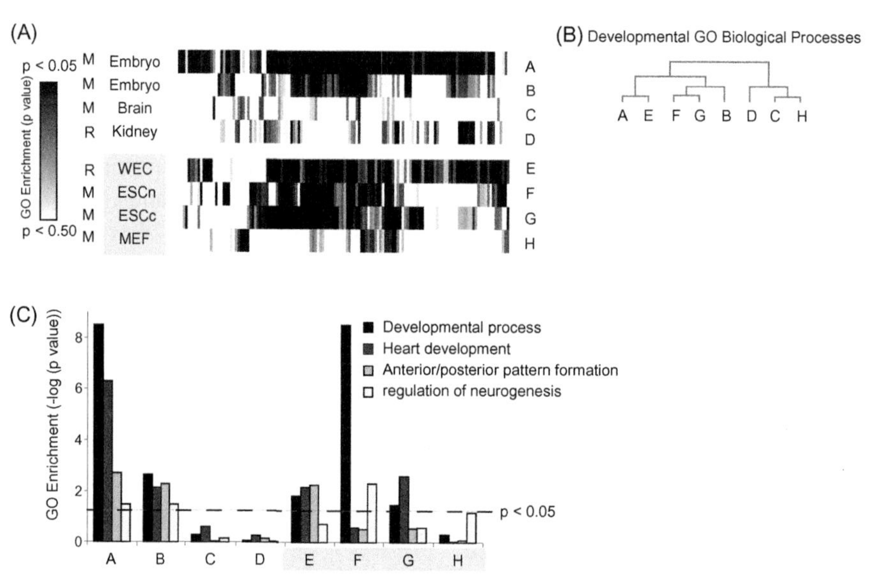

FIGURE 4.2.6 Enrichment of developmentally related GO biological processes across in vivo and in vitro models.

(A) Enrichment of GO biological processes related to development within the most prominent methylmercury-induced responses (top 1000 genes in each study) across eight in vitro and in vivo studies. Species (mouse (M) or rat (R)) and examined tissue/cell type are located to the left of the hierarchical clustering plot depicting the enrichment score (p value) of GO biological processes identified using DAVID. A total of 157 GO biological processes related to development identified to be significantly enriched in one of the eight data sets ($p < 0.05$, genes changes >5) are depicted by color intensity (columns). (B) Hierarchical clustering by study using the enrichment score (p value) of al 157 GO biological processes. (C) Examples of model-dependent enrichment of selected developmentally related GO biological processes.

–Reprinted from [35], with permission from Elsevier

methylmercury-induced gene expression alterations associated with developmental signaling and heart development were identified across WEC, ESTc, and in vivo systems, and the complementary nature of the ESTc and embryonic stem cell test (neural) (ESTn) was shown.

A comparison of phthalate-induced effects demonstrated an enrichment of cholesterol/lipid/steroid-metabolism-related genes across a variety of in vivo models, but only for the toxic phthalates. For in vitro systems, enrichment of cholesterol/lipid/steroid-metabolism-related genes was only found for WEC and not for ESTc or ESTn. Interestingly, the majority of these genes were down-regulated in vivo, while in WEC up-regulation was observed. The direction of regulation thus seems to vary dependent on dose, time, and model.

Overall, these data demonstrate the ability of in vitro developmental models to identify gene expression alterations relevant for developmental toxicity in vivo.

In vitro–in vitro comparisons

Robinson et al. [22] investigated the conservation of triazole-induced gene expression across in vitro developmental models derived from different species (Figure 4.2.7). The comparisons showed considerable overlap in terms of a significant induction of common genes. Furthermore, genes commonly regulated between studies showed a good correlation in directionality of response between WEC and ZET. Interestingly, commonly significantly altered genes in WEC and ESTc showed a similar directionality of induction; however, the degree of induction was different. Flusilazole induced the sterol-cholesterol-related *Cyp51* gene to a higher degree in the ESTc as compared to WEC, while the retinoic-acid-metabolism-related *Cyp26a1* showed an opposite pattern. This indicates that the extent of compound-induced effects on certain biological processes can differ between in vitro developmental toxicity assays.

Similar complementary responses were observed in comparisons of gene expression profiles of ESTc and ESTn after exposure to multiple compounds [36]. Although time-related gene expression responses showed similarities between the two EST systems, specific genes could be identified related to the different lineages of differentiation. Gene sets and biological processes found to be enriched were generally related to the morphological differentiation routes of the respective models. Therefore, a combination of transcriptomic data of both ESTc and ESTn model systems can provide complementary mechanistic information.

Combinations of toxicogenomic approaches

This chapter has focused on the application of transcriptomics in in vitro developmental toxicity testing. This is, however, only one aspect of the complexity of the biological system. Altered gene expression does not necessarily lead to protein alterations or effects at a tissue or organ level. Therefore, a combination of transcriptomics with other 'omics technologies is essential to fully understand toxicant-induced developmental adverse outcomes. An initial comparison of proteomic and mRNA profiles in phthalate-exposed ESCs suggests commonalities in protein expression and gene expression alterations but also illustrates the complexity of these types of comparisons [37]. As we move forward, combinations of toxicogenomic approaches can possibly bring new insight into mechanisms of developmental toxicity and simplify the differences between non-adverse, adaptive, and adverse effects.

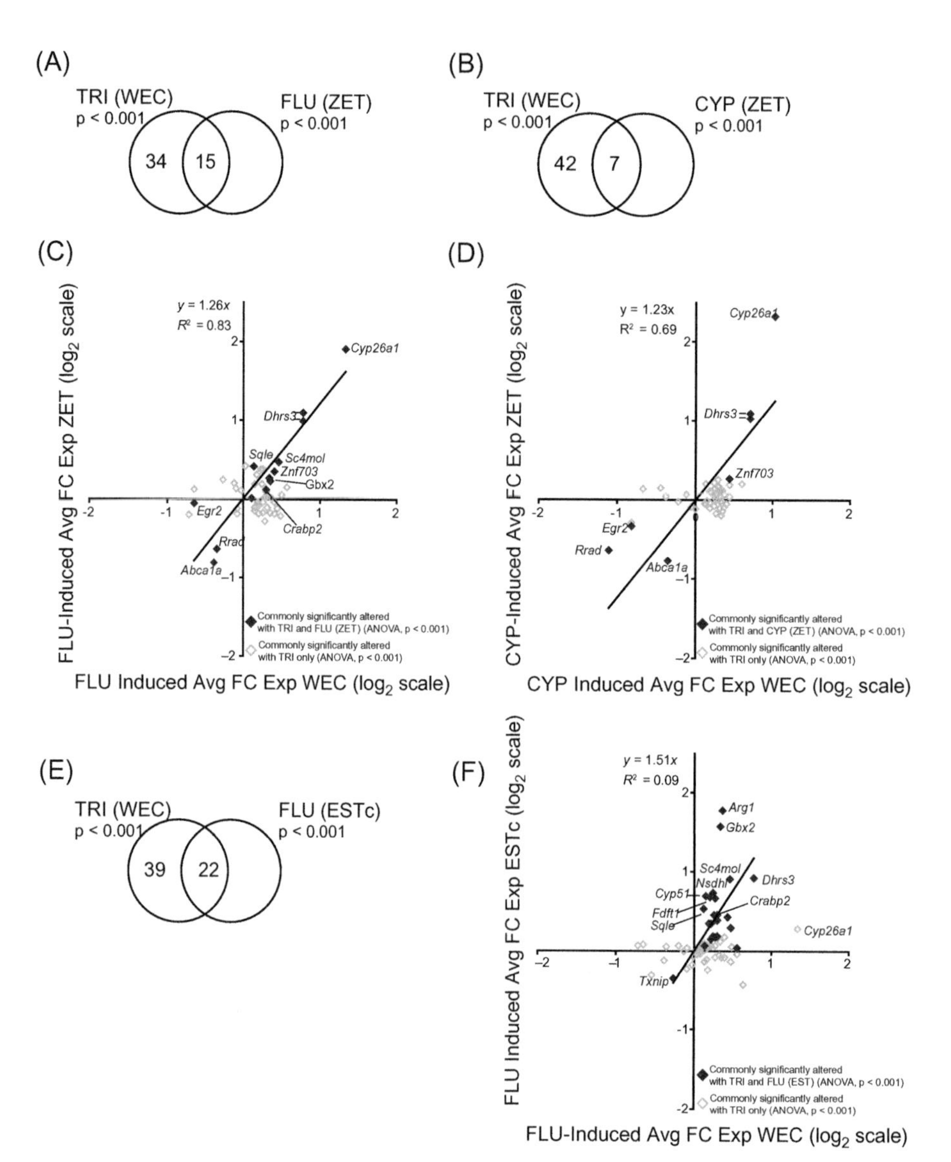

FIGURE 4.2.7 Triazole-induced impacts in WEC in comparison with ESTc and zebrafish developmental models.

(A, B) Venn diagrams of common gene expression alterations (analysis of variance (ANOVA), $p < 0.001$) between WEC exposed to triazoles (TRI) and zebrafish exposed to (A) flusilazole (FLU) or (B) cyproconazole (CYP). (C, D) Cross-comparison of gene expression alterations between WEC (FLU, 60 μg/ml) and (C) zebrafish (FLU, ±4 μg/ml) or (D) zebrafish (CYP, ±19 μg/ml). (E) Venn diagram of triazole-induced gene expression alterations (ANOVA, $p < 0.001$) between WEC and ESTc exposed to FLU. (F) Cross-comparison of gene expression alterations between WEC (FLU, 60 μg/ml) and ESTc (FLU, ±6 μg/ml).

–Reprinted from [22], with permission from Elsevier

4.2.4 Outlook

Recent technological advances in molecular techniques, such as transcriptomics, and bioinformatics, as well as novel mechanistic toxicological understanding, especially at the molecular level, have led to an increase in the amount and type of toxicological information available. This data is potentially useful for the optimization and validation of in vitro assays as models for the prediction of developmental toxicity in vivo. However, the massive amount of data emerging from this approach is challenging, and requires a targeted approach in a new dimension.

Traditionally, in transcriptomics, the regulation of individual genes has been studied. While an understanding of individual gene expression continues to be important, progressive insight has shown that a system-level understanding is needed for further advancement. Obtaining a system-level understanding, as advocated in a systems biology approach, requires a shift in our notion of "what to look for" in biology [38]. A systems-biology-based approach focuses on understanding and modeling the integrated, interacting networks of genes, proteins, biochemical reactions, or higher-order biological units. This way, it becomes possible to explore the integrated molecular responses in a biological system in the case of a toxic event. Drawing upon a systems biology perspective, adverse outcome pathways (AOPs) can be defined and used to gain insight into how molecular interactions can lead to toxicity. An AOP is a conceptual construct that portrays existing knowledge concerning the linkage between a direct molecular initiating event and an adverse outcome at a biological level of organization relevant to risk assessment [39]. The challenge clearly lies in the definition of AOPs and their level of detail in terms useful for employment in hazard assessment, and this should always be considered within the entire physiological framework of the biological system at hand.

For an AOP approach to be useful, AOPs must also be anchored to endpoints that are relevant to a specific hazard, in this case a developmental outcome. Advantages of models such as WEC and ZET are that they comprise whole embryos, potentially allowing the evaluation of initiating events, cellular responses, and adverse outcomes at the organism level. Although it is not possible to study adverse outcomes at the organism level in cell-based in vitro assays, they can play an important role in an AOP-based approach to risk assessment. By studying toxicity pathways, or cellular response pathways that, when sufficiently perturbed, are expected to result in adverse developmental outcomes, initiating events and proximal cellular responses can be evaluated in vitro. Transcriptomics, like many other techniques, can be used to obtain information and populate AOPs, thereby creating the possibility for the discovery of pathways that can be related to an adverse effect and have diagnostic value for the responsible initiating event [39]. Gene expression changes are not only important components of the cascade leading towards manifestations of toxicity, but also offer a way to explore the low end of the dose–response curve because they appear to be relatively sensitive [40]. Multiple studies have shown that gene expression changes can be used to investigate dose–response relationships [22,26–29]. Incorporation of dose–response analysis and dose metrics is important to be able to use in vitro models systems for hazard assessment purposes. To identify the lowest concentration of a chemical that causes adverse effects, a characterization of the dose–response relationships for several AOPs or pathways of toxicity must be performed. This way, the pathways most sensitive to a particular chemical can be identified, providing clues for mechanism of toxic action.

The transcriptomic approaches reviewed in this paper provide clues for further study towards the definition of pathways relevant for predicting developmental toxicity on the basis of gene expression changes in a variety of alternative assays.

References

[1] van der Jagt K, Munn S, Tørsløv, J, de Bruin, J. 2004 Alternative Approaches can Reduce the Use of Test Animals Under REACH. European Commission report EUR 21405EN.
[2] Hermsen SA, Pronk TE, van den Brandhof EJ, van der Ven LT, Piersma AH. Reprod Toxicol 2011; 32:245–52.
[3] Robinson JF, van Beelen VA, Verhoef A, Renkens MF, Luijten M, van Herwijnen MH, et al. Toxicol Sci 2010; 118:675–85.
[4] van Dartel DA, Pennings JL, Hendriksen PJ, van Schooten FJ, Piersma AH. Reprod Toxicol 2009; 27:93–102.
[5] New DA. Biol Rev Camb Philos Soc 1978;53:81–122.
[6] Piersma AH, Genschow E, Verhoef A, Spanjersberg MQ, Brown NA, Brady M, et al. ATLA 2004; 32:275–307.
[7] Brown NA, Fabro S. Teratology 1981;24:65–78.
[8] Piersma AH. Embryotoxicity Testing in Post-Implantation Whole Embryo Culture (WEC): Method of Piersma. ECVAM DB-ALM Protocol No. 123. 1999.
[9] van Maele-Fabry G, Delhaise F, Picard JJ. Toxicol In Vitro 1990;4:149–56.
[10] van Maele-Fabry G, Delhaise F, Picard JJ. Int J Dev Biol 1992;36:161–7.
[11] Robinson JF, Verhoef A, Piersma AH. Toxicol Sci 2012;126:255–66.
[12] Hansen JM, Carney EW, Harris C. Reprod Toxicol 1999;13:547–54.
[13] Augustine-Rauch K, Zhang CX, Panzica-Kelly JM. Birth Defects Res, Part C, Embryo Today: Rev 2010; 90:87–98.
[14] Hill AJ, Teraoka H, Heideman W, Peterson RE. Toxicol Sci 2005;86:6–19.
[15] Gilbert S. Developmental biology, 6th ed.: Sinauer Associates; 2000.
[16] Hermsen SA, van den Brandhof EJ, van der Ven LT, Piersma AH. Toxicol In Vitro 2011;25:745–53.
[17] Slob W. Toxicol Sci 2002;66:298–312.
[18] Strahle U, Scholz S, Geisler R, Greiner P, Hollert H, Rastegar S, et al. Reprod Toxicol 2012;33:128–32.
[19] Spielmann H, Pohl I, Döring B, Liebsch M, Moldenhauer F. In Vitr Mol Toxicol 1997;10:119–27.
[20] Theunissen PT, Schulpen SH, van Dartel DA, Hermsen SA, van Schooten FJ, Piersma AH. Reprod Toxicol 2010;29:383–92.
[21] de Jong E, van Beek L, Piersma AH. Toxicol In Vitro 2012;26:970–8.
[22] Robinson JF, Tonk EC, Verhoef A, Piersma AH. Reprod Toxicol 2012;34:275–83.
[23] Robinson JF, Verhoef A, Pennings JL, Pronk TE, Piersma AH. Toxicol Sci 2012;126:242–54.
[24] Theunissen PT, Pennings JL, Robinson JF, Claessen SM, Kleinjans JC, Piersma AH. Toxicol Sci 2011;122:437–47.
[25] van Dartel DA, Pennings JL, van Schooten FJ, Piersma AH. Toxicol Appl Pharmacol 2010;243:420–8.
[26] Hermsen SA, Pronk TE, van den Brandhof EJ, van der Ven LT, Piersma AH. Toxicol Sci 2012;127:303–12.
[27] Robinson JF, Verhoef A, van Beelen VA, Pennings JL, Piersma AH. Toxicol Appl Pharmacol 2012; 264:32–41.
[28] Schulpen SH, Robinson JF, Pennings JL, van Dartel DA, Piersma AH. Reprod Toxicol 2013;35:81–8.
[29] van Dartel DA, Pennings JL, de la Fonteyne LJ, Brauers KJ, Claessen S, van Delft JH, et al. Toxicol Appl Pharmacol 2011;251:110–8.
[30] van Dartel DA, Pennings JL, Robinson JF, Kleinjans JC, Piersma AH. Toxicol Lett 2011;201:143–51.
[31] Hermsen SA, Pronk TE, van den Brandhof EJ, van der Ven LT, Piersma AH. Reprod Toxicol 2012;34:216–24.
[32] Pennings JL, van Dartel DA, Robinson JF, Pronk TE, Piersma AH. Toxicology 2011;284:63–71.
[33] van Dartel DA, Pennings JL, de la Fonteyne LJ, Brauers KJ, Claessen S, van Delft JH, et al. Toxicol Sci 2011;119:126–34.

[34] van Dartel DA, Pennings JL, de la Fonteyne LJ, van Herwijnen MH, van Delft JH, van Schooten FJ, et al. Toxicol Sci 2010;116:130–9.
[35] Robinson JF, Theunissen PT, van Dartel DA, Pennings JL, Faustman EM, Piersma AH. Reprod Toxicol 2011;32:180–8.
[36] Theunissen P, Pennings J, van Dartel D, Robinson JF, Kleinjans J, Piersma A. Toxicol Sci 2013;132:118–30.
[37] Osman AM, van Dartel DA, Zwart E, Blokland M, Pennings JL, Piersma AH. Reprod Toxicol 2010;30:322–32.
[38] Kitano H. Science 2002;295:1662–4.
[39] Ankley GT, Bennett RS, Erickson RJ, Hoff DJ, Hornung MW, Johnson RD, et al. Environ Toxicol Chem 2010;29:730–41.
[40] Daston GP, Naciff JM. Birth Defects Res, Part C, Embryo Today: Rev 2010;90:110–7.

CHAPTER

Thyroid Toxicogenomics: A Multi-Organ Paradigm

4.3

Barae Jomaa

Division of Toxicology, Wageningen University, Wageningen, The Netherlands

4.3.1 Introduction

Disorders related to the thyroid have long been known to man owing to the visible enlargement of the gland known as goiter. This condition can be due to either an overactive or an underactive thyroid gland, with the former increasing and the latter decreasing the rate at which the body breaks down nutrients and produces cellular components necessary to sustain life [1]. The successful treatment of goiter with animal thyroid glands was reported in China as early as 643 CE [2]. That environmental contaminants, particularly lead, are associated with goiter had already been proposed by Paracelsus in the sixteenth century. In the nineteenth century, the essential element iodine that is a basic constituent of thyroid hormones (THs) was discovered [3,4]. Thyroid hormones were characterized in the twentieth century, starting with tetraiodothyronine (T4) in 1915, and later triiodothyronine (T3) in 1952 (Figure 4.3.1) [5,6].

The twentieth century saw a chain of events that raised the level of public concern over compounds that affect the thyroid, and ultimately culminated in the inclusion of thyroid disruption endpoints by the Organisation for Economic Co-operation and Development (OECD) in guidelines for the testing of chemicals (Figure 4.3.2). The "cranberry scare" of 1959 brought a herbicide, aminotriazole, and its effect on the thyroid into US politics, and toxicology to the forefront of public life. At the very core of the debate was a rat study that showed a high incidence of thyroid cancer upon exposure to high amounts of the pesticide. The idea that chemicals affecting rodents at high doses will affect humans at moderate doses, also known as the "mouse-as-little-man" principle, was so entrenched in people's minds at the time that it formed part of an amendment to the Federal Food, Drug, and Cosmetic Act known as the Delaney clause. To ease the newly created "chemophobia," the then-vice-president Richard Nixon ate cranberry sauce during his presidential election campaign [7]. A decade later, Nixon became president and proposed the establishment of the US Environmental Protection Agency (USEPA), which was approved by congress. Today, the jury is still out on the human relevance of the adverse effect of aminotriazole that was detected in the rat study, even after a retrospective cohort study of exposed Swedish railroad workers found it to be a "suspicious" carcinogen [8]. On the one hand, epidemiological data are often mired with confounders such

Toxicogenomics-Based Cellular Models.

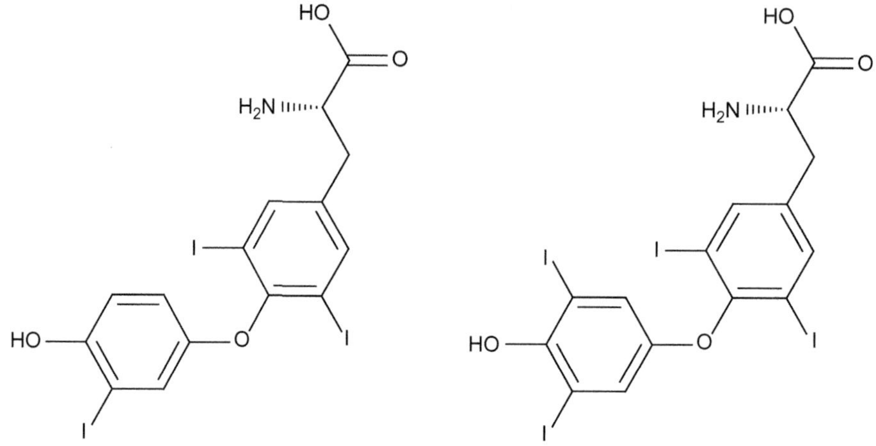

FIGURE 4.3.1

Chemical structures of T3 (left) and T4 (right).

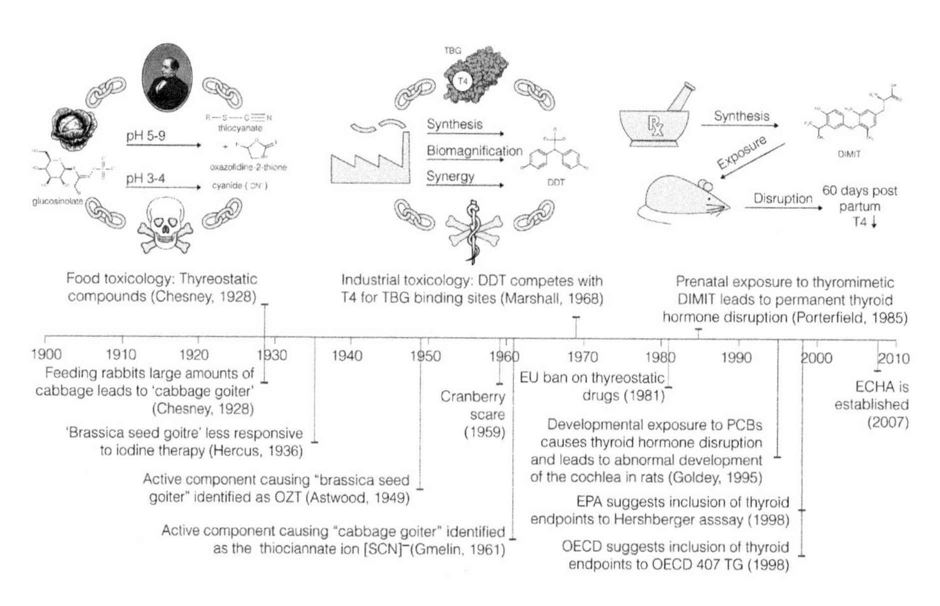

FIGURE 4.3.2 Timeline of events that have shaped our understanding of thyroid hormone disruption in the century preceding the establishment of the European Chemicals Agency (ECHA) in 2007.

TBG stands for thyroxine-binding globulin; thyroxine is another name for T4. DDT, dichlorodiphenyltrichloroethane; PCBs, polychlorinated biphenyls; DIMIT, 3,5-dimethyl-3′-isopropyl-L-thyronine.

—*Barae Jomaa/Wikimedia Commons.*

as the presence of a mixture of active compounds in pesticides, and on the other hand, rodents have been shown to be overly sensitive to thyroid hormone disruption [9]. These uncertainties have led the USEPA to consider aminotriazole as a "probable" human carcinogen (Group B) instead of a definitive human carcinogen (Group A) [10].

The establishment of the European Chemicals Agency (ECHA) in 2007 came on the heels of another public concern, that of endocrine disruption. Again, as one of the three main classes of endocrine disruptors, thyroid disruptors are at the very heart of modern toxicology—the other two classes being estrogen disruptors and androgen disruptors. The main fear today is that a change to thyroid homeostasis could be caused by man-made chemicals and affect the hormone system's well-documented orchestration of sexual and mental development, and this at much smaller doses than predicted from the effects observed at high doses [11].

4.3.2 The thyroid system

The thyroid system is highly responsive to a wide range of external stressors. The hypothalamus responds to conditions requiring increased energy expenditure, such as pregnancy or prolonged periods of cold, by secreting thyrotropin-releasing hormone (TRH), which stimulates the anterior pituitary to produce thyroid-stimulating hormone (TSH), which in turn stimulates the thyroid to secrete thyroid hormones (Figure 4.3.3). The hypothalamus also responds to changes in diet by setting off a similar downstream chain of events. A diet that is high in carbohydrates promotes energy expenditure and leads to increased T3 levels, whereas fasting promotes energy conservation and therefore leads to a drop in T3 levels [12].

In order to keep the rise in T3 production and resulting increase in energy consumption and thermogenesis under control, the body has a negative feedback loop, whereby high levels of T3 downregulate TRH and TSH production. The resultant drop in these tropic hormones leads to decreased stimulation of thyroid hormone production and secretion, and ultimately a drop in serum thyroid hormone levels.

The circulating levels of the hormones T3, T4, and TSH are constantly reacting to environmental factors that include temperature, oxygen levels, light, diet, and physical and emotional stress, in addition to chemicals and drugs [13–18]. TH levels are also influenced by age and sex [19]. Moreover, an individual's range is half the width of the population reference range, which means that TH levels that are normal for one individual can be abnormal for another [20]. This also means that there is a genetically determined set-point, whose heritability was estimated in a study on healthy twins to be around 65%, with the rest being attributed to environmental factors [21].

4.3.3 Mode-of-action-based alternative testing strategies for thyroid activity

The development of in vitro testing strategies for thyroid-active compounds must be directed at a variety of endpoints, taking into account all the physiological targets that are known to be sensitive to interference by exogenous chemicals with various modes of action (MOAs). The next sections provide an overview of the MOAs that are relevant to the thyroid system, a brief overview of the

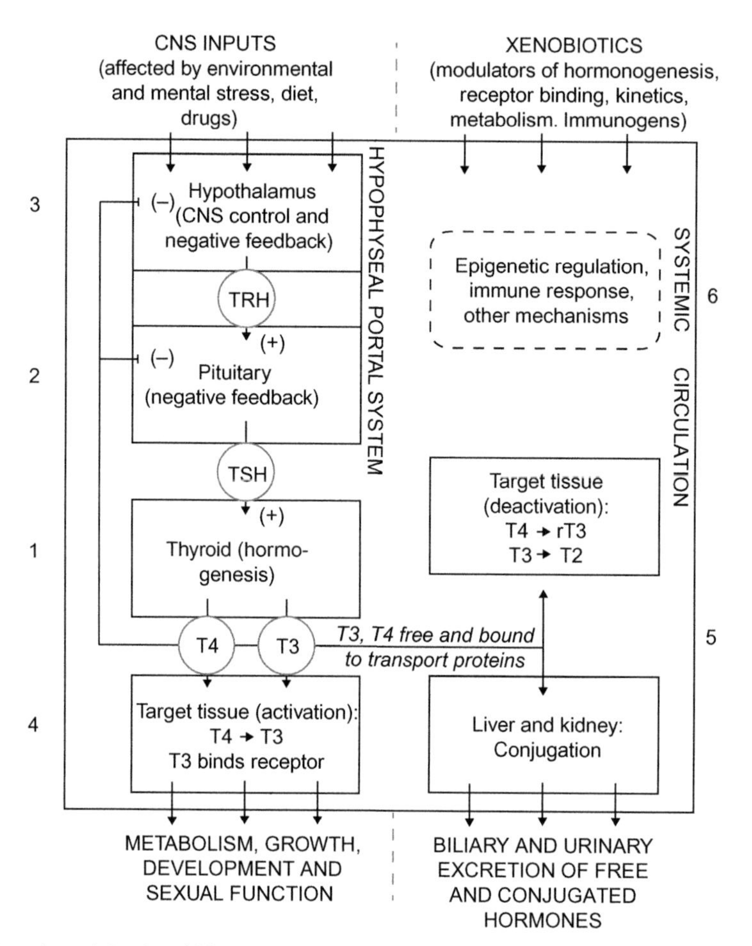

FIGURE 4.3.3 Overview of the thyroid hormone system from synthesis to clearance.

Chemicals can exert their effects through modes of action relating to (1) thyroidal hormonogenesis; (2) pituitary feedback; (3) hypothalamic central nervous system (CNS) control and feedback; (4) thyroid activity at the target tissue; (5) kinetics; and (6) epigenetics, immune response, and other mechanisms.

—Barae Jomaa/Wikimedia Commons.

current state of the art with respect to the development of in vitro testing strategies for these endpoints, as well as possibilities for integrating or developing so-called 'omics-based screens. While classic 'omics studies are based on differential gene expression (transcriptomics), the term now includes proteomics, metabolomics, and interactomics, among others. The number of studies using transcriptomics, let alone other 'omics, when studying thyroid hormone disruption is still very limited. Nonetheless, the aim is still the same—to find signatures and biomarkers of chemical exposures and validate them with unknown compounds, to classify phenotypes of thyroid toxicity, and to delineate models/mechanisms of action [22]. Table 4.3.1 gives an overview of the currently available in vitro assays that could serve as potential candidates for high-throughput screening. The individual

Table 4.3.1 Overview of a Selection of Assays that are Deemed Potential Candidates for High-Throughput Screens

	MOA	Target	Endpoint	Readout	Cellular Model	Origin	Organism	Reference
1	TSHR activity	TSHR	Cell proliferation	FL	FRTL-5 (WT)	Thyroid	Rat	[29]
			[I-125]TSH	GR	Cell-free	Cell-free	Cell-free	[26]
			cAMP levels	FL	rHEK293 (R)	Kidney	Human	[27]
			cAMP levels	LU	CHO (R)	Ovary	Hamster	[28]
	Iodide transport	NIS	FL quenching	FL	FRTL-5 (R)	Thyroid	Rat	[49]
		Na^+/K^+ ATPase	Molybdenum blue	OD	FRTL-5 (WT)	Thyroid	Rat	[185]
			Gene expression	LU	FRTL-5	Thyroid	Rat	–
		Pendrin	^{125}I levels	GR	MDCK (R)	Kidney	Dog	[51]
	Iodination and coupling	TPO	Guaiacol oxidation	OD	Cell-free	Cell-free	Cell-free	[55]
		$DUOX/H_2O_2$	CM-H2DCFDA oxidation	FL	FRTL-5 (WT)	Thyroid	Rat	[61]
	Proteolysis	Cts	Proteolysis	–	–	–	–	–
	Iodide scavenging	IYD	$[^{125}I]DIT$, $[^{125}MIT]$	GR	Cell-free	Cell-free	Cell-free	[75]
	TH synthesis	T4	T4 levels	FL	Whole embryo	Whole embryo	Zebrafish	[186]
2	TRHR activity	TRHR	Calcium flux	FL	Chem-1 (HR)	Bone marrow	Rat	[92]
	TSH synthesis	TSH	TSH levels	OD	TαT1 (WT)	Pituitary	Mouse	[187]
3	TRH synthesis	TRH	TRH levels	GR	U-373-MG (WT)	Brain	Human	[106]
4	THR activity	THR	Cell proliferation	FL	GH3 (WT)	Pituitary	Rat	[116]
			Gene expression	LU	GH3 (R)	Pituitary	Rat	[115]
	Co-regulator activity	THR CoR	TRβ-SRC2-2	FP	Cell-free	Cell-free	Cell-free	[124]

(Continued)

Table 4.3.1 Overview of a Selection of Assays that are Deemed Potential Candidates for High-Throughput Screens (Continued)

	MOA	Target	Endpoint	Readout	Cellular Model	Origin	Organism	Reference
5	TH blood transport	TTR, TBG	Competitive binding	FL	Cell-free	Cell-free	Cell-free	[131]
	TH cellular transport	OATP1C1	$[^{125}I]T4$	GR	HEK293 (R)	Kidney	Human	[142]
		MCT8/10	$[^{125}I]T3$	GR	MDCK1 (R)	Kidney	Dog	[141]
	TH metabolism	DIOs	$[^{125}I]$ levels	GR	Cell-free	Cell-free	Cell free	[188]
		hrSULTs	^{3}H-DHEA	GR	Cell-free	Cell-free	Cell-free	[189]
		hrUGTs	$[^{125}I]T4$-G	GR	Cell-free	Cell-free	Cell-free	[190]
6	ITG $\alpha v \beta 3$ activity	ITG $\alpha v \beta 3$	$[^{125}I]T4$	GR	CV-1 (R)	Kidney	Monkey	[178]

Screens using radioisotopes were avoided when possible. They cover all or part of (1) thyroidal hormonogenesis; (2) pituitary feedback; (3) hypothalamic central nervous system control and feedback; (4) thyroid activity at the target tissue; (5) kinetics; and (6) epigenetics, immune response, and other mechanisms.
cAMP, cyclic AMP; CM-H2DCFDA, a fluorescent dye; CoR, co-regulator; Cts, cathepsin; DHEA, dehyroepiandrosterone; DIT, diiodotyrosine; DIO, deiodinase; DUOX, dual oxidase; FL, fluorescence; FP, fluorescence polarization; GR, gamma radiation; HR, human recombinant; hrSULT, human recombinant sulfotransferase; hrUGT, human recombinant UDP-glucuronosyltransferase; ITG, integrin.; IYD, iodotyrosine deiodinase; LU, luminescence; MIT, monoiodotyrosine; NIS, sodium iodide symporter; OATP, organic-anion-transporting polypeptide; OD, optical density; R, recombinant (stable and transient); TBG, thyroxine-binding globulin; TPO, thyroid peroxidase; TRHR, TRH receptor; TSHR, TSH receptor; TTR, transthyretin; WT, wild-type. A dash (–) indicates when an assay either hasn't been developed or is not suited for the purpose of high-throughput analysis.

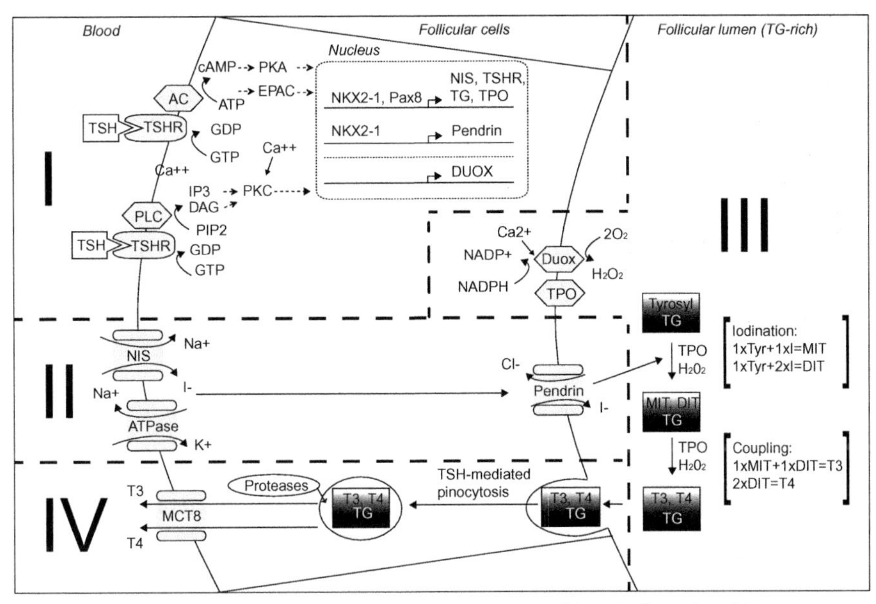

FIGURE 4.3.4 Primary MOAs that take place at the level of the thyroid and affect thyroid hormone synthesis.

(I) TSH secreted from the pituitary reaches the thyroid and binds to its membrane receptor (TSHR), leading to up-regulation of the expression of genes that encode both structural and functional proteins necessary for thyroid hormone production; (II) iodide is transported through the cell and into the lumen; (III) iodination and coupling reactions take place and form iodinated thyroglobulin (TG) molecules; (IV) iodinated TG undergoes lysosomal proteolysis to release thyroid hormones as well as monoiodotyrosine (MIT) and diiodotyrosine (DIT), whose iodide anions are scavenged. AC, adenyl cyclase; DAG, diacylglycerol; DUOX, dual oxidase; EPAC, exchange protein directly activated by cAMP; NIS, sodium iodide symporter; PIP2, phosphoinositol 4,5 biphosphate; PKA, cAMP-dependent protein kinase; PKC, protein kinase C; PLC, phospholipase C; TPO, thyroid peroxidase.

—Barae Jomaa/Wikimedia Commons.

assays and the MOAs that they represent are discussed in the following sections, providing a basis for the future development of thyroid-toxicogenomics-based screens.

Primary effects on the thyroid system

Compounds with a primary effect target hormonogenesis at the level of the thyroid and might lead to primary hypothyroidism/hyperthyroidism. The main MOAs involved include inhibition of TSH receptor (TSHR) activity, iodide transport, iodination and coupling, proteolysis, and iodide scavenging (Figure 4.3.4).

Inhibition of TSHR activity

TSH, also known as thyrotropin, is a glycoprotein hormone that is produced by thyrotrophic cells in the anterior pituitary and secreted into the circulation where it eventually reaches the thyroid. Once at the basolateral side of thyroid follicular cells, it binds the thyrotropin receptor (TSHR), a G-protein coupled receptor (GPCR), and the ligand–receptor pair get internalized, thereby initiating a signaling cascade that leads to thyroid hormone production as well as cell proliferation [23].

There are two pathways that are mostly initiated by TSH-regulated TSHR activation, likely owing to the presence of different active conformations of this receptor [24]. The first involves the **adenyl cyclase (AC) pathway**. TSHR-mediated G-protein alpha s (Gαs) subtype activity leads to AC catalyzed conversion of ATP to cyclic AMP (cAMP), which then activates cAMP-dependent protein kinase (PKA) and exchange proteins directly activated by cAMP (EPACs), both of which are mediators of the mitogenic action of TSH [25]. The second involves the **phospholipase C (PLC) pathway**. TSHR-mediated G-protein alpha q (Gαq) subtype activity leads to PLC-catalyzed hydrolysis of phosphoinositol 4,5 biphosphate (PIP2) and production of diacylglycerol (DAG) and inositol triphosphate (IP3). DAG activates the calcium-dependent protein kinase C (PKC) cascade that results in mitogenic activity, while IP3 signaling leads to the release of the required calcium from the endoplasmic reticulum.

In vitro models for the inhibition of TSHR activity

The inhibition of receptor binding can be tested by measuring the displacement of radiolabeled TSH from TSHR, and a variation of this method using a plate coated with TSHR has also been developed, which eliminates the need for cell culturing facilities [26]. Another approach uses the second messenger cAMP as a reporter of TSH activity and has been developed for high-throughput screening (HTS). Measurements are performed fluorometrically upon addition of a fluorescence-labeled cAMP antibody [27]. Another method involves transfecting TSHR and a cAMP-responsive luciferase gene into Chinese hamster ovary cells [28].

While not many compounds have yet been found to act through the TSH receptor, a new assay called the TSH screen might facilitate the identification of compounds acting through this membrane receptor [29]. This assay measures TSH-dependent cell proliferation without added TSH (to detect agonist activity) as well as in the presence of 1 milli-international unit per milliliter (mIU/ml) TSH (to detect antagonist activity). This simple and inexpensive assay encompasses the binding of TSH to its receptor, second messenger activity, downstream signal transduction, and gene expression, since the end result of this gene expression is TSH-mediated mitogenesis.

Potential for 'omics-based testing

Current in vitro assays only look at the TSH-mediated activation of the AC pathways. While the AC pathway regulates the expression of the sodium iodide symporter (NIS), TSHR, thyroglobulin (TG), and thyroid peroxidase (TPO), the importance of the PLC pathway seems to be in the expression of dual oxidase 2 (DUOX2), a critical oxidase whose mutation or the mutation of its maturation factor (DUOXA2) in humans causes congenital hypothyroidism [30–32]. In mice, TSH regulates the expression of DUOX much less than it regulates the expression of NIS and TPO [33]. Among the PKC isozymes that are expressed in the thyroid, αPKC-ζ has been found to stimulate hormone-independent cell proliferation, in contrast to its effect in other cell types [34]. Hence, the PKA pathway in the thyroid is TSH dependent, whereas the PLC pathway has the potential for TSH-independent activity. In this regard, a microarray might be a necessary addition to a battery of assays in order to uncover the effect of xenobiotics that modulate physiologically relevant thyroid endpoints that are not as tightly regulated by TSH as others. This battery must also include nongenetic parameters such as enzyme inhibition, which might or might not be reflected by microarrays. Other 'omics technologies are increasingly being used, including micro RNA (miRNA) microarrays, which have helped uncover mechanisms by which TG can induce thyroid cell growth [35]. 'Omics-based techniques reflect global patterns of disruption that together might allow more sensitive discrimination of thyroid disruptors and the detection of their early effects [36].

Inhibition of iodide transport

All dietary iodine is reduced to iodide prior to absorption in the small intestine [37]. Once in the circulation, iodide is absorbed by tissues through **NIS**, also known as solute carrier family 5, member 5 (SLC5A5). NIS is a transmembrane glycoprotein that is mainly expressed on the basolateral side of thyroid follicles and which concentrates iodide to a level that is 20 to 50 times that in plasma [38,39]. NIS is also functional in extrathyroidal tissues, including the small intestine where dietary iodide is taken up into the bloodstream, and lactating mammary glands, through which iodide is delivered to the newborn for proper development. Gastric mucosa, salivary glands, and rectal mucosa also express NIS but its function in these tissues is still unknown. In the thyroid, TSH up-regulates the expression of the NIS gene, whereas iodide inhibits it [40]. Anions such as thiocyanate (SCN^-) compete with iodide for transport through NIS.

NIS activity is dependent on the **sodium potassium pump (Na/K-ATPase)** to provide a sodium gradient. It is through this sodium gradient that iodide is transported along with sodium into the cell through NIS. Accordingly, the Na/K-ATPase inhibitor ouabain inhibits NIS-mediated iodide transport [41]. Na/K-ATPase ATP1A1 and ATP1B1 expression levels are more than threefold and tenfold higher, respectively, in the brain, kidney, and thyroid than in other tissues [42].

In 1997, a century after Vaughan Pendred described a genetic condition giving rise to hearing loss and goiter, mutations in the **pendrin gene (*SLC26A4*)** were associated with the disease [43]. As can be seen from Figure 4.3.4, iodide does not only need to be transported in the blood and enter the basolateral side of the thyroid cell, it also needs to cross the apical side in order to reach the follicular lumen where the iodination of TG and the coupling of monoiodotyrosine (MIT) and diiodotyrosine (DIT) take place as penultimate steps in thyroid hormone production. While research on the apical transport of iodide is ongoing, pendrin is at the very least one of the key transporters involved in TSH-mediated iodide efflux at the apical side of follicular cells, thus completing the transport into the follicular lumen that was started by NIS at the basolateral side [44,45].

In vitro models for the inhibition of iodide transport

The epithelial follicular rat thyroid cell line FRTL-5 is the most widely used model for the thyroid owing to its expression of physiologically relevant proteins, including NIS (Table 4.3.1). The cellular transport of monatomic ions was first studied using radioisotopes [46]. The voltage clamp is another technique that can be used to study substrate selectivity of ion transporters by measuring ion currents [47]. This technique was later refined to study the currents of single ion channels in a technique known as the patch clamp [48]. While the use of radiotracers has advantages, such as the possibility to perform direct measurements of ion concentrations, their use for HTS is mired with requirements for waste disposal and radiation safety. The "intracellular anion fluorescence assay" is highly promising, especially for HTS. The technique is based on iodide's ability to quench the fluorescence of YFP–H148Q/I152L, a halide-sensitive variant of yellow fluorescent protein (YFP), which is stably transfected into FRTL-5 cells [49].

Just like NIS studies, pendrin anion transport can be studied using a variety of techniques, including radioisotopic labels and voltage clamp in FRTL-5 cells [50]. The implementation of a bicameral system allows for the study of polarized cells expressing NIS at the basolateral side and pendrin at the apical side, in a way that is closer to the in vivo situation [51].

Potential for 'omics-based testing

NIS expression is stimulated by TSH and inhibited by excess iodide and TG and, therefore, holds promise as a genomic biomarker for thyroid disruption by xenobiotics. In the presence of TG, pendrin gene expression is also stimulated by TSH, while the effect of iodide is unknown [52]. Therefore, both NIS and the pendrin gene hold promise as candidate genomic biomarkers for the disruption of iodide transport by xenobiotics that modulate the expression levels of these genes.

Inhibition of iodination and coupling

At the apical surface of the thyroid follicle, thyroglobulin is iodinated in the presence of H_2O_2 by **TPO** [53]. Iodinated tyrosyl residues on thyroglobulin form MIT and DIT—the building blocks for T3 and T4 (Figure 4.3.4). TPO inhibitors can be divided into three classes: drugs, environmental pollutants, and food products. Antithyroid drugs, which are given in cases of hyperthyroidism, act on TPO, and include methimazole and propylthiouracil. Psychotropic drugs such as tricyclic antidepressants deactivate TPO by binding to its heme [54]. Environmental contaminants shown to inhibit TPO include benzophenone 2 (BP2) [55]. Flavonoids contained in food can also inhibit this enzyme [56].

Since the reactive oxygen species H_2O_2, needed along with TPO for hormonogenesis, can induce oxidative damage, additional safety mechanisms are needed by the thyrocytes. H_2O_2-generating **DUOXs** are localized at the apical membrane bordering the follicular lumen [57]. As will be detailed later in the text, TPO is capable of degrading T4. Hence, TPO and H_2O_2, needed in the production of T4, could both antagonize hormonogenesis and must be constantly kept under control. The extracellular surface of the cell membrane, facing the TG-rich lumen, is covered with a thin film rich in H_2O_2, while the intracellular surface of the cell membrane is lined with regions that are high in TPO [58,59]. The apical surface of the thyroid cell can be thought of as a toxic bioreactor. This functional morphology is critical and might help explain why thyroid hormone production has not yet been achieved in a fully in vitro model. In 2012, a step in the right direction was taken and involved the in vitro overexpression of transcription factors needed for thyroid cell differentiation followed by in vivo grafting onto mice to finally produce hormones. Nonetheless, in vitro results of this breakthrough experiment only indicated the in vitro iodination of TG, suggesting that either the proteolysis of TG to produce thyroid hormones is not taking place or the hormones are being degraded before they can be measured [60]. Follow-up research could soon lead to an in vitro model for thyroid toxicity that encompasses all the complex mechanisms involved in the synthesis of thyroid hormones.

In vitro models for the inhibition of iodination and coupling

Human recombinant TPO (hrTPO) has been produced in vitro by stably transfecting its gene into the human follicular thyroid carcinoma cell line FTC-238. The altered enzymatic activity of the resulting protein upon exposure to benzophenone 2 could then be assessed in the classical guaiacol oxidation method [55]. Intracellular H_2O_2 generation can be quantified by the fluorescent dye CM-H2DCFDA [61].

Potential for 'omics-based testing

Inhibition of TPO by methimazole and propylthiouracil is compensated by increased expression of TPO [62]. Analysis of whole-genome gene expression in FTC/hrTPO recombinant cells using DNA microarrays identified 362 genes that could classify test chemicals as either TPO inhibiting or TPO activating with approximately 70% accuracy [63]. The sensitivity of such a method, compared with directly testing enzymatic activity, and its accuracy in identifying a wider selection of toxicants, are still to be established. In another study using NHEK cells derived from neonatal foreskin, alterations in DUOX1 expression have been assessed using DNA microarrays and increased H_2O_2 was shown to accompany increased expression [64].

Inhibition of proteolysis

Iodinated thyroglobulin is pinocytosed back into the cell from the thyroglobulin-rich follicular lumen and is subjected to proteolysis by **lysosomal proteases**, which releases active hormones [65]. This process is inhibited by excess iodide either from the diet, the environment, or iodine-rich drugs such as amiodarone [66].

In vitro models for the inhibition of proteolysis

Iodinated TG-containing thyroid lysosomes have been isolated from primary pig thyroid cells [65]. While it has never been reported, it is conceivable that these could be used in potential assays that screen for the effects of chemicals on the proteolysis of iodinated thyroglobulin and release of active hormones.

Potential for 'omics-based testing

There are indications that 'omics-based techniques might be indicative of the effect of chemicals on the proteolysis of iodinated thyroglobulin in lysosomes. A systems biology approach examining miRNA-based posttranscriptional regulation of the autophagy-lysosomal pathway has proven the usefulness of such an approach in unraveling complex mechanisms [67]. Dennemärker and colleagues have reported impaired thyroglobulin processing while studying mice deficient for the lysosomal cysteine endopeptidase cathepsin L (Ctsl) [68]. Other studies have found increased expression of lysosomal aspartic protease cathepsin D (Ctsd) in Ctsl knockout mice as well as Ctsl-deficient A549 cells [69]. In cathepsin K (Ctsk) or Ctsb/Ctsk knockout mice, Ctsl is up-regulated, likely as a compensatory mechanism for Ctsk deficiency. These mice also had significantly reduced levels of free thyroxine, which underscores the role lysosomal proteases in thyroid hormone production [70]. The compensatory expression of various cathepsins also raises the possibility of being able to detect such changes, whether they are caused by defective genes or by environmental chemicals that inhibit these enzymes, using 'omics-based techniques.

Inhibition of iodide scavenging

In addition to the release of thyroid hormones, proteolysis of iodinated TG also results in the release of MIT and DIT. These molecules carry the valuable iodide anion, which is scavenged by **iodotyrosine deiodinase (IYD)** and recycled. The importance of this mechanism of intrathyroidal iodide conservation was further elucidated after the cloning and characterization of IYD in 2002, and the implication of its mutation in 2008 in various cases of hypothyroidism [71–73].

In vitro models for the inhibition of iodide scavenging and potential for 'omics-based testing

The crystal structure of IYD was characterized in 2009 [74]. Furthermore, the gene has been successfully transfected in HEK293 cells, and its activity was shown to be NADPH dependent and to be inhibited by 3,5-dinitro-L-tyrosine [75]. Enzymatic activity was assessed in a cell-free method using an enzyme preparation and radioisotope-labeled [^{125}I]DIT and [^{125}MIT] followed by descending paper chromatography [75]. This in vitro assay may prove useful in an integrated testing strategy (ITS). Potential expression biomarkers of the inhibition of this enzyme and their application within 'omics-based testing have yet to be explored.

Secondary effects on the thyroid system

Compounds with a secondary effect on the thyroid system exert their action at the level of the pituitary, affect the gland's ability to regulate thyroid hormone secretion, and thereby have the potential to lead to secondary hypothyroidism/hyperthyroidism. TSH is secreted by the pituitary gland in response to hypothalamic TRH, which arrives directly through the hypophyseal portal system and is quickly degraded (half-life of 5 minutes) in the systemic circulation by a TRH-degrading enzyme [76,77].

Inhibition of TRHR activity and TSH secretion

TRH binds to the **TRH receptor (TRHR)**, a GPCR on the cell membrane, and sets off a signaling cascade whose net effect is the production of more TSH. TRHR has been studied in detail in rodents, where two subtypes were identified and named TRHR1 and TRHR2. As with many GPCRs, the receptors are desensitized by internalization upon agonist binding in order to regulate the length of the signal [78]. Signal transduction that leads to TSH production is mediated by the PLC pathway described earlier. It has been postulated that the anticonvulsant and mood stabilizer carbamazepine inhibits TRH-induced TSH secretion in humans [79]. Further studies revealed decreased binding of TRH to TRHR [80].

The activation of the signal takes place when TRHR is coupled to Gq proteins [81]. Moreover, TRHR is not able to activate Gαs/AC [82]. Even so, the cAMP response-element-binding protein (CREB) is one of the transcription factors involved in gene expression that is mediated by a calcium-dependent Gq/PLC pathway [83]. AP-1 and Elk-1 are two other transcription factors that are induced by this pathway [84]. The three main genes of interest that are being expressed in response to TRH-mediated signaling are *TSH*, *DIO2*, and *THRB*. Secondary mechanisms of disruption affect the pituitary and its ability to regulate thyroid hormone production through the alteration in TSH expression and secretion. Factors that suppress TSH secretion include somatostatin, thyroid hormones, growth hormone, glucocorticoids, melatonin, and opioid peptides.

Calcium channel blockers have been found to significantly decrease serum levels of T3 and T4 and to elevate serum TSH in rabbits, which is thought to result from impaired thyroid hormone synthesis or release at the level of the thyroid [85]. In man, the pituitary is more affected by calcium channel blockers than the thyroid, as studies have revealed that there is a significant decrease in TSH release that is stimulated by TRH but no significant change in levels of either T3 or T4 [86,87].

Thyroid cancer in rats follows continuous **secretion of TSH**, which leads to the overstimulation of thyroid cells and results in hyperplasia that progresses to adenoma and later to carcinoma

[88]. Iodine-deficient and goiterous patients have elevated levels of TSH but not the thyroid cancer that is observed in rats. This implies that rat thyroids are more responsive to TSH than human thyroids and that the human relevance of TSH has more to do with its role in modulating the secretion of thyroid hormones. While this underlines species differences, excessive and prolonged TSH stimulation of the thyroid can lead thyrotoxicosis in humans [89]. Excessive TSH secretion can be the result of increased TRH activity but can also be due to improper negative feedback, which involves the other two genes of interest that are expressed in the thyroid—*DIO2* and *THRB*.

In vitro models for the inhibition of TRHR activity and TSH secretion

The radioisotope-labeled hormone [^{3}H]TRH has been used to detect the inhibitory effects of compounds such as vinblastine on TRHR binding in the rat pituitary somatolactotroph cell line GH3 [90]. Decreased TRHR levels in the human thyroid follicular cell line Nthy-ori 3-1 after PCB153 exposure were detected by Western blotting [91].

Calcium-flux-based assays have been developed to detect TRHR activation, since its downstream target IP3 releases calcium from the ER by binding to IP3R, which is a calcium channel. A TRHR transfected cell line of bone marrow origin and expressing high levels of promiscuous Gα15 to couple the receptor to the calcium signaling cascade is available commercially [92]. However, this requires the addition of a fluorescent dye, which is circumvented in another commercially available cell line that is also transfected with the luminescent protein aequorin [93]. Aequorin undergoes a conformational change upon binding to calcium ions that results in the emission of blue light [94]. Other than having the reporter expressed intracellularly, this technique also overcomes the need to subtract background fluorescence that results from the cell's natural autofluorescence. Whether the readout is fluorescence or luminescence, both cell lines are non-pituitary in origin and the systems they represent are the result of an artificial construct.

In vitro TSH secretion has long been confined to the human T-cell leukemia cell line Molt 4 or the mouse thyrotropic pituitary tumor cell line TtT-97 [95,96]. However, while the former is not pituitary derived, the latter is difficult to grow. The immortalization of the mouse pituitary TαT1 cell line, which retains the expression of both TSH subunits, might be a promising new model cell line for pituitary thyrotropes but its use in HTS still needs to be developed [97].

Potential for 'omics-based testing

Microarrays have been used in in vivo experiments to show the up-regulation of *TSHB* mRNA by glutamine and glutamic acid [98]. The use of microarrays in unraveling the interference of compounds with TSH signaling has great potential, especially when looking at functionally relevant genes such as *TSHB*, *DIO2*, and *THRB*.

Tertiary effects on the thyroid system

Compounds with a tertiary effect exert their action at the level of the hypothalamus and compromise its role as a control center that is able to translate various stimuli into messages to the pituitary where the message is relayed to the thyroid. These effects might lead to tertiary hypothyroidism and tertiary hyperthyroidism; the latter, although theoretically possible, is very rare [99]. Any condition that increases body energy requirements, such as pregnancy or prolonged cold, will stimulate the hypothalamus to increase **TRH secretion**.

TRH is present throughout the central nervous system and is highly expressed in the paraventricular nucleus (PVN) region of the hypothalamus. The median eminence at the base of the brain contains both the hypothalamo-hypophyseal portal vessels and axons of a group of neurons that originate in the PVN. These neuroendocrine cells release TRH directly into the portal vessels that vascularize the pituitary. There, the hormone acts as a stimulant of TSH release by binding the TRH receptor (TRHR) on the membrane of thyrotrophic cells [100].

This function as a stimulant of TSH release seems to have evolved as animals moved to land, since in poikilotherms, including the zebrafish that is increasingly being used in thyroid research, TRH does not stimulate TSH release. The ancestral role of TRH as a neuroregulator is being increasingly recognized and an analog has been successfully used in vitro as a neuroprotective agent in the glutamate-induced toxicity that is associated with neurodegenerative diseases such as Alzheimer's [101].

The secretion of TRH is down-regulated by T3 through local deiodination of T4 by DIO2. This action is thought to be THRB mediated. A rapid fall in serum TSH, T4, and T3 is brought about by fasting and leads to a drop in the appetite suppressant leptin, which has been found to increase DIO2 activity [102]. Other regulators of TRH neurons include neuropeptide Y, melanocortin-stimulating hormone, and agouti-related peptide. What is known about the regulation of the TRH neuron has been reviewed in detail by Eduardo Nillni [103].

In vitro models for the inhibition of TRH secretion

The human A375, TXM18, MeWo, WM35, and WM793 melanoma cell lines and the rat CA77 neoplastic parafollicular cell line all produce TRH [104,105]. However, the human glioblastoma–astrocytoma cell line U-373-MG also produces TRH and, owing to its neuroectodermal origin, is more likely to be representative of hypothalamic neuronal TRH production. Exposure of U-373-MG cells to pilocarpine enhanced TRH production [106]. Pilocarpine is a muscarinic cholinergic agonist and muscarinic control of the central TRH system has been shown in vivo, which underlines the relevance of the physiological response of the U-373-MG cell line and its potential within an ITS [107].

Potential for 'omics-based testing

In theory, genomic studies could help in the discovery of xenobiotics that influence TRH and TRH-related genes; however, most of the research on the current accepted model cell line U-373-MG is focused on cancer [108]. When developing such approaches, attention should be focused on the validation of U-373-MG as an optimal cellular model of TRH-producing neurons. To that end, this cell line has to be exposed to a panel of model compounds and possible modulation of TRH production has to be correlated with TRH level changes in in vivo experiments using the same compounds. In addition, 'omics-based techniques could be useful in identifying mechanisms and biomarkers related to xenobiotic-induced disruption of TRH production.

Peripheral effects

Compounds with a peripheral effect exert their action at the level of the target organ and could result in either hypothyroidism or hyperthyroidism. The effects of thyroid hormones are mediated in a large part by their binding to **thyroid hormone receptors (THRs)**, which act as transcription factors that mediate specific patterns of gene expression. There are four different isoforms: A1, A2, and B1 that are widely expressed, with B1 having its highest levels in the liver, and B2 that is expressed mainly in the hypothalamus and pituitary, where it regulates TH negative feedback, as well as in the inner ear and retina during development [109,110]. THRA2 does not bind thyroid hormones but as an apoprotein can still play a role in the regulation of THR-mediated gene expression, especially in its dephosphorylated state [111].

Inhibition of THR activity

DNA-binding domains on THRs have a high affinity for certain sequences of DNA called thyroid hormone response elements (TREs) and the resulting binding of the receptors to the DNA dictates the expression of linked genes. THRs can be found as monomers, homodimers, and heterodimers with retinoid X receptor (RXR), with the last having the highest affinity for TREs. Positive regulation of gene activity by thyroid hormones involves three general states of expression of genes that are linked to TREs. First, there is basal activity, which can be exemplified in experimental knockout models of THRs. Second, there is repression of gene expression when THRs are bound to DNA in the absence of T3, since this state favors the formation of a co-repressor complex with histone deacetylase (HDA) activity. Third and last, there is activation of gene expression when THRs are bound to DNA in the presence of T3, since this state favors the formation of a co-activator complex with histone acetyltransferase (HAT) activity [112]. This is relevant when considering that acetylated histones pack DNA less tightly and allow for greater gene expression. Negative regulation of gene activity is exemplified by T3-dependent repression of TRH expression in the hypothalamus and repression of TSH expression in the pituitary, a process that involves THRB2 binding to negative TREs (nTREs) and interaction of THRB2 with the co-activator SRC-1 [110,113,114].

In vitro models for thyroid hormone agonism/antagonism

Reporter gene assays are widely used for studying thyroid hormone disruption [115]. These assays make use of an artificially inserted TRE followed by a reporter gene, such as luciferase, that is expressed upon activation of THR by T3. In responsive cells, such as the rat pituitary somatolactotroph GH3 cells, proliferation is another endpoint that can be measured by determining the fluorescence intensity that is observed upon mitochondrial reduction of resazurin to resorufin [116].

Potential for 'omics-based testing

Exposure to methimazole (MMI)/propylthiouracil (PTU) or T3, rendering mice hypothyroid or hyperthyroid, respectively, results in a differential gene expression as determined by DNA microarray analysis, which includes THR-regulated genes [117,118]. These studies are performed in vivo on the liver since it is a major target of thyroid hormone action. However, the costs associated with such studies limit the number of compounds that can be tested. Performing such tests in vitro would allow for the testing of a much larger set of compounds on a variety of cell types, and that at a fraction of the cost. Such an in vitro transcriptomics study has been performed by exposing human thyroid follicular FTC-238 cells to THR antagonists carbaryl and vinclozolin, which led to the identification of an expression profile for such disruption [119].

Co-regulator activity

Resistance to thyroid hormones is a condition in which the majority of tissues are not responsive even when the level of thyroid hormones is high. A mutation in THRB has been implicated and is thought to cause a disruption of the charge clusters (acidic, basic, or a mixture of both) in the tertiary structure of the protein that allow for co-activator binding [191]. It is therefore evident that co-regulators play a physiologically significant role in the regulation of THR-mediated gene expression. As was mentioned above, thyroid-hormone-mediated activity necessitates either **co-activator or co-repressor recruitment**. While canonical steroid hormone receptors are inactive in the unliganded state, THR and retinoic acid receptor (RAR) form heterodimers with RXR that bind to DNA and recruit nuclear receptor co-repressor (NCoR) and silencing mediator for retinoid or thyroid-hormone receptors (SMRT), as well as Sin3 and HDAC that actively repress gene expression [112,120].

In vitro models for co-regulator activity

While yeast two-hybrid screening helped identify SMRT and NCoR as proteins with decreased interaction with THR upon T3 binding, new high-throughput methods are also available [121,122]. These are mostly based on tagged THR ligand-binding domains (LBDs) that are added to a library/array of known co-regulator peptides that contain the LXXLL binding motif (L for leucine, X for any amino acid) [123,124].

Potential for 'omics-based testing

As the above-mentioned co regulator studies eventually reflect effects of compounds on gene expression, transcriptomics can be expected to detect these changes. In fact, not only are these changes reflected but by comparing a THR–T3 inhibitor to a TRH–CoR inhibitor, Sadana et al. found that these affect different subsets of T3 target genes in the human liver cell line HepG2 [125].

Hormone-kinetics-based effects

Thyroid hormone kinetics involves the distribution of T3 and T4 to the body's tissues, their metabolism, and their eventual clearance. Though these are technically peripheral effects, they are classified separately due to their vastly different MOAs and their potential for kinetic modeling.

Circulation

Due to the lipophilic nature of the thyroid hormones T3 and T4, their transport in the body's circulatory system must be assisted by transport proteins. **Thyroxine-binding globulin (TBG), transthyretin (TTR), and albumin (ALB)** are, in order, the three proteins with the highest affinity for T4 and T3. While most of the thyroid hormones in humans are transported by TBG, it is thought that albumin alone is sufficient for carrying these hormones [126]. "Free" T3 and T4 that are unbound to blood transport proteins are called fT3 and fT4 respectively. These free hormones are thought to be good indicators of the amount that is available at the cellular level.

Unlike humans, rodents only express thyroxine-binding globulin (TBG), which acts as a reservoir for thyroid hormones, when they need it the most—during development and again later on during senescence [127]. TTR is the main T4 carrier in the circulation of rodents but TTR null mice did not respond any differently than wild-type mice when challenged with exposure to cold, which is known to result in thyroid-hormone-mediated thermogenesis. Additional experiments including thyroidectomy and gene-expression analysis led the authors to conclude that TTR has no role to play in thyroid hormone homeostasis [128]. In humans, decreased thyroid hormone binding to transport proteins raises the ratio of fT4 to T4, which results in greater clearance of T4 and a drop in total T4 [126]. Patients with undetectable levels of TBG in their serum due to a mutation in the gene had low total T4, which is associated with fatigue [129]. Another study examined patients with sepsis and found a clear association between a drop in TBG, TTR, and ALB, a rise in fT4/T4, and a drop in total

T4 [130]. It can be concluded that using rodents as models in animal experiments does not reflect the human situation and that in vitro experiments should take into consideration all of the main three thyroid-hormone-binding proteins.

In vitro models for the inhibition of binding to transport proteins

Upon binding to transport proteins, e.g. TTR, the fluorescence peak of 8-anilino-1-naphthalenesulfonic acid ammonium (ANSA) shifts from 515 to 465 nm and the fluorescence intensity is substantially increased. By placing an emission filter near 465 nm and a defined ratio of ANSA–TTR, the decrease in fluorescence at that wavelength due to increasing concentrations of T4 is assessed. This gives a dose–response curve from which a half maximal inhibitory concentration (IC_{50}) can be calculated. IC_{50} values for the displacement of ANSA from TTR by test compounds are calculated in a similar manner. The obtained data can then be used to rank the test compounds in terms of T4 equivalency factors (TEQ) by dividing the T4 IC_{50} by the test compound IC_{50}. In other words, the higher the TEQ the more potent the test compound is at displacing T4 from its binding proteins. Such a method has been successfully applied to test the potential of hydroxylated PCBs (polychlorinated biphenyls) and polybrominated diphenyl ethers (PBDEs) for competitive binding to TTR [131].

On the flipside of the inhibition of binding is the inhibition of the expression of the binding proteins. As discussed earlier, sepsis leads to a drop in the level of TBG, TTR, and ALB, which results in a drop in T4. These proteins are produced in the liver and Hep G2 cells can be used as an in vitro cellular model. Metabolic labeling with [^{35}S]methionine is performed in order to identify newly synthesized proteins, which are then identified and quantified by Western blot. Such a method was used to identify a dose-dependent decrease in TBG, TTR, and ALB upon exposure to interleukin-6 [132].

Potential for 'omics-based testing

While a transcriptomics-based assay will not replace binding studies such as ANSA–TTR, it might prove to be a high-throughput way of looking at the changes in TBG, TTR, and ALB gene expression in Hep G2 cells.

Transport into cells

While it was previously thought that the lipophilic nature of thyroid hormones is sufficient for them to passively diffuse across the cell membrane, it is now clear that their transport is facilitated by plasma membrane transporters, including monocarboxylate (MCT) **MCT8**, **MCT10**, and organic anion transporting polypeptide (OATP) **OATP1C1** [133–135]. A mutation in the *MCT8* gene causes Allan–Herndon–Dudley syndrome, which includes high serum T3 levels coupled with severe psychomotor retardation. This transporter has been found to be expressed in a variety of tissues, including the heart, brain, liver, kidney, intestine, and placenta, while MCT10 is expressed in the skeletal muscles, liver, kidney, intestine, and placenta [136]. To date, there have been no clinical cases of thyroid disease linked to either MCT10 or OATP1C1. While MCT8 is a specific thyroid hormone transporter, OATPs are also known to transport xenobiotics [137]. However, OATP1C1 is specific to the brain and has high affinity for T4, and its lack in an OATP1C1 knockout mouse has been shown to lead to central nervous system (CNS)-specific hypothyroidism [138].

In addition to the previously mentioned role of calcium channels in thyroid hormone production, it appears that they are also involved in the transport of these hormones into peripheral organs. Calcium channels have been found to be involved in T4 uptake into cardiomyocytes. In fact, the calcium channel blocker dilitazem reduced T4 uptake by 45%, whereas T3 uptake was unaffected [139]. Moreover, calcium has been found to be a first messenger in thyroid-hormone-mediated increases in sugar uptake, and calcium blockers inhibited this physiologically relevant action of thyroid hormones [140].

In vitro models for the inhibition of transport into cells

Just like anion transport studies of NIS and pendrin, study of MCT8- and OATP-mediated thyroid hormone transport makes use of radioisotopes [^{125}I]T4 and [^{125}I]T3. However, since the target tissues are many and the cells that they contain are diverse, mimicking the physiological setting in an exact manner is inefficient while screening for thyroid hormone disrupters. Hence, an artificial system has been created within controlled settings. Cells are selected based on low endogenous expression of the transporters being studied and then they are transfected with a vector for transporter expression. This way, significant amounts of thyroid hormones can be transported across the cell membrane and by comparing with the wild-type cell line it can be concluded that the effect is specific to the studied transporters. Another important factor that has to be taken into account in order to decrease variability between experiments is the number of transporters that are being expressed at any given point in time. To this end, the amount of thyroid hormone uptake is normalized to the amount of plasma membrane proteins [141,142]. In vitro methods for detecting changes in calcium flux have already been discussed.

Potential for 'omics-based testing

MCT8 expression is induced by retinoic acid and down-regulated by T3 in F9 mouse teratocarcinoma cells [143,144]. In a rabbit model of prolonged critical illness, both MCT8 and MCT10 expression levels were regulated by thyroid hormones. In the same study, but now with patients who were under acute surgical stress, only MCT8 expression levels increased [145]. A rat study has found the expression of OATP1C1 to be regulated by thyroid hormones [146]. Taken together, these results suggest that alterations in circulating levels of thyroid hormones will be compensated to a certain extent by the regulation of thyroid hormone transporters. Such changes in the expression of functional genes could find applications within toxicogenomics-based assays for thyroid hormone disruption.

Metabolism and clearance

The metabolism and clearance of thyroid hormones is an essential process in the regulation of thyroid hormone activity and serum levels (Figure 4.3.5) with deiodination being the main mechanism and accounting for 85% of T4 and around 50% of T3 breakdown [147]. The rest is accounted for by sulfonation and glucuronidation.

Thyroid-hormone-regulated transcriptional activity is largely dependent on **deiodinase type 2 (DIO2)** for the removal of an iodine atom from the outer ring of T4, turning it into T3, in order to boost the hormone's affinity for the THR. **Deiodinase type 3 (DIO3)** removes an iodine from the inner ring of T4, turning it into rT3, an inactive congener (Figure 4.3.5). **Deiodinase type 1 (DIO1)** is capable of both inner and outer ring deiodination. rT3, containing two outer ring iodine atoms, can now be deiodinated into T2 by DIO2, while T3, which has two inner ring iodine atoms, can be deiodinated by DIO3 to form T2.

DIO3 is highly expressed in the placenta and embryonic tissue but has only been detected in a few adult tissues. Nonetheless, DIO3 has been shown to be responsible for the degradation of up to 80% of thyroid hormone production, which might indicate that the anatomical location of the bulk of this

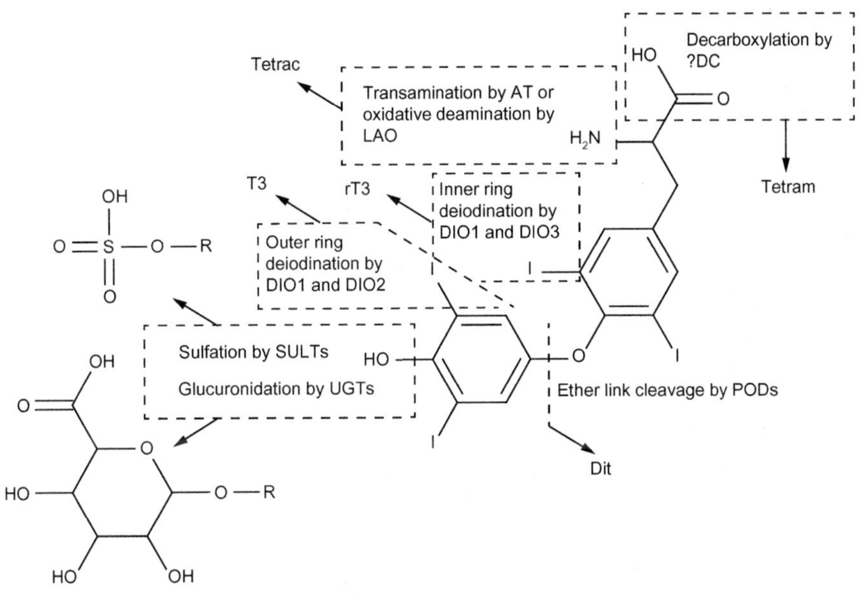

FIGURE 4.3.5 Enzymes involved in the metabolism of T4.

Sulfotransferases (SULTs), UDP-glucuronosyltransferases (UGTs), deiodinases (DIOs), L-amino acid oxidase (LAO), thyroid hormone aminotransferase (AT), peroxidases (PODs), and a yet-to-be-discovered decarboxylase (?DC). Tetrac is the deaminated form of T4, tetram is 3,3,5,5-tetraiodothyronamine, and triam is 3,3,5-triiodothyronamine.

—Barae Jomaa/Wikimedia Commons.

enzyme has yet to be defined. This enzyme has been found to be re-expressed during neoplasia and critical illness, leading to the hypothesis that it is a contributing factor in the low T3 levels that are observed in these situations [148].

In the liver and to a lesser extent the kidney, **sulfotransferases (SULTs)** add a sulfate or, alternatively, **UDP-glucuronosyltransferases (UGTs)** add a glucuronic acid moiety to the phenolic hydroxyl group of T4, T3, rT3, and T2, rendering them more water soluble for facilitated excretion. Among the studied SULTs, 1A1, which is also involved in the inactivation of estrogens, has the lowest Km value and thus the highest affinity for thyroid hormones; in order, T2 > > rT3 > T3 > T4 [149].

Glucuronidation takes place mainly in the liver, with the result being an increase in the water solubility of thyroid hormones, which is followed by biliary excretion. While drug-metabolizing enzymes are not very abundant under normal circumstances, they are up-regulated upon xenobiotic challenge [150]. The importance of glucuronidation for the thyroid system became apparent when compounds such as anti-epileptic drugs (AEDs) were tested on rats and were found to cause thyroid hyperplasia and tumors. These drugs, by acting as microsomal enzyme inducers (MEIs), increase the breakdown of thyroid hormones, which in turn leads to compensatory TSH production and its associated thyroid cell hyperplasia and tumor formation in rats. The only problem with these results is that AEDs are well studied in man and do not result in an increase in TSH production, nor have they ever been correlated with thyroid cancer. This is an important species difference between humans and rodents with regards to thyroid active compounds, both in terms of sensitivity and outcome.

Clearly, there are differences between humans and rodents in the way they metabolize thyroid hormones. Indeed, it was found that in humans, unlike rats, T3 is not significantly glucuronidated. Nonetheless, in both species, T4 and rT3 are conjugated by UGT1A1 and UGT1A9. What is missing in humans is UGT1A9-mediated glucuronidation of T3 [151,152]. This means that T3 is first converted to rT3 by deiodinases before being glucuronidated, and rT3 is by far the preferred iodothyronine substrate for both UGT1A1 and UGT1A9.

Phenobarbital, an anticonvulsant, has been shown to increase thyroid hormone clearance by inducing the expression of OATP and UGT genes [153]. This increased expression of hepatic transport and conjugation genes is mediated by the constitutive androstane receptor (CAR), which in addition to OATP and UGT, has also been shown to regulate SULT, leading to a drop in circulating levels of thyroid hormone. The same is thought to happen with other anticonvulsants, including phenytoin and carbamazepine, as well as certain antibiotics such as rifampicin [154].

Minor degradation pathways for thyroid hormones are much less known and studied but are arousing interest simply because the resulting metabolites, while in small quantities, may have a significant biological role. Tetrac and triac, the deaminated forms of T4 and T3 respectively, are potent metabolites that act on the thyroid receptor in vitro and suppress TSH secretion in vivo [155–157]. The deamination reactions are catalyzed by **L-amino acid oxidase (LAO) and thyroid hormone aminotransferase (AT)**. Triac, which is present in human serum at a picomolar range, represents 14% of T3 degradation, and has a higher clearance rate than T3 owing to its rapid conjugation in the liver [158]. T4, sold as Synthroid, has long been used illicitly for weight loss, and now triac is being sold under various names as a dietary supplement for weight loss, leading the US Food and Drug Administration (FDA) to issue repeated safety alerts and go to court in the case *United States v. Syntrax Innovations, Inc., et al.* [159]. According to the FDA, the health consequences of consuming products containing triac include heart attacks and strokes [160].

Decarboxylation of T4 and T3 by a still unknown decarboxylase can lead to the formation of 3,3,5,5-tetraiodothyronamine (tetram) and 3,3,5-triiodothyronamine (triam), respectively. Though these were not found in vivo, other decarboxylation metabolites of thyroid hormones were. A 2004 *Nature Medicine* article by Scanlan et al. surprised the thyroid community with results suggesting that the observed non-genomic effects of thyroid hormones are in fact mediated by even less recognized metabolites of thyroid hormones. The authors found that 3-iodothyronamine (T_1AM) and thyronamine (T_0AM) are in vitro agonists of G-protein-coupled trace amine receptor TAR1 and in vivo induce hypothermia and bradycardia within minutes [161]. It is increasingly becoming evident that the action of these minor metabolites within new signaling pathways should be of major concern for toxicologists; however, in vitro models of the disruption of thyroid hormone metabolism and clearance still look solely at inhibition of SULTs, UGTs, and DIOs [162].

Ether link cleavage of T4 by **peroxidases** in the presence of H_2O_2 leads to diiodotyrosine (DIT). It is not surprising then that, in the thyroid, TPO and its substrate H_2O_2 can also act on newly synthesized T4 and degrade it. This means that the thyroid follicles have mechanisms in place to safeguard T4 from TPO-mediated degradation. Glutathione and ascorbate, antioxidants that are present in copious amounts intracellularly, have been found to strongly inhibit TPO by disposing of the substrate H_2O_2 [58].

In vitro models for the inhibition of metabolism and clearance

Liver microsomes are currently the gold standard for testing for the inhibition of liver enzymes. This is usually followed by an analytical technique, such as high-performance liquid chromatography (HPLC), in order to quantitatively determine the metabolite profile of TH [163].

Potential for 'omics-based testing

The use of microarrays to study drug metabolism is not new but the sensitivity of microarrays to study the modulation of enzyme activity must be compared with traditional and more direct methods [164].

Other MOAs affecting the thyroid hormone system

Epigenetics

Abnormal thyroid gland development, also called thyroid dysgenesis (TD), accounts for 80% of reported cases of congenital hypothyroidism (CH), with the remaining 20% of patients having a normal thyroid but a dysfunction in the synthesis of thyroid hormones [165]. A study on monozygotic twins led its authors to conclude that a classic genetic hypothesis for thyroid dysgenesis is unlikely [166]. In addition, mutations in genes commonly associated with thyroid development, such as *PAX8*, *FOXE1*, *NKX2.1*, and *NKX2.5*, were identified in only 3% of patients with congenital hypothyroidism from thyroid dysgenesis [167]. The sporadic nature of thyroid dysgenesis suggests that there could be epigenetic factors involved [168,169]. Xenobiotic exposure during critical time-windows of development could affect both biochemical and epigenetic mechanisms and help explain the lack of a clear genetic underpinning for this disease.

While it is a challenge to gather epidemiological data showing the contribution of xenobiotics to thyroid dysgenesis, it is well established that hyperthyroid mothers receiving treatment with anti-thyroid drugs such as methimazole and propylthiouracil or who have been given radioactive iodine (RAI) are at higher risk of having a child with hypothyroidism [170,171]. CH is associated with an increased risk of other congenital malformations and of these, most are cardiac malformations [172]. A recent study correlating pituitary cell proliferation upon exposure to thyroid-active compounds to pituitary gland weight ended up finding high correlation with another organ—the heart [29]. This underlines the strong interplay between the heart and the thyroid system and also emphasizes the potential of in vitro systems to unravel physiological effects.

Autoimmunity

Autoimmune thyroid disease (AITD) can be divided into diseases that lead to hypothyroidism, such as Hashimoto's (chronic autoimmune thyroditis) and postpartum thyroditis, and diseases that lead to hyperthyroidism, such as Grave's disease. Autoimmune hypothyroidism results mainly from autoantibodies against TPO and TG, whereas autoimmune hyperthyroidism results from autoantibodies that bind the TSHR and activate it in a similar way to TSH. Even though the symptoms of autoimmune hypothyroidism and autoimmune hyperthyroidism are different, they both have a high degree of heritability and share susceptibility genes, including those for HLA-DR3, HLA-C*07, PTPN22, and CTLA4—all linked to the deregulation of T-cell activation [173]. Nonetheless, a Danish twin cohort revealed that 21% of Grave's disease could not be explained by heredity alone and must be due to the environment [174]. On the other hand, a steady increase in the incidence of autoimmune thyroditis could be due to increased diagnosis, but increasing evidence is linking iodine to this disease [175,176]. Some have even suggested the involvement of soy isoflavones in the formation of TPO neoantigens during their covalent binding and inactivation of the enzyme [177].

Novel pathways

Integrin alpha v beta 3 (ITG αvβ3) has recently been found to be a cell surface receptor for T4 inducing a downstream signaling cascade that might help elucidate some of the in vivo effects that cannot be accounted for fully by other mechanisms [178].

Multiple effects

The widespread use of lithium that followed its approval in the 1970s by the FDA for the treatment of bipolar disorder quickly led to its association with goiter [179,180]. However, a single MOA cannot explain this drug's antithyroidal effects. The multiple effects of lithium include abnormal iodine kinetics, altered thyroglobulin structure, inhibition of iodotyrosine coupling and thyroid hormone secretion, DIO2 inhibition, immunostimulation, and altered THR gene expression [181]. These effects may be exacerbated by other factors such as iodine exposure, dietary goitrogens, and immunogenic background [182].

4.3.4 Conclusion and future perspectives

In view of the severity of the adverse effects that can be expected from chemical disruption of the thyroid system, cellular models can still be expanded to include minor metabolic routes and newly discovered transporters as part of an integrated testing strategy. The overview of the various MOAs listed in Table 4.3.1 shows that there is still a lot of work to be done in order to develop suitable models for thyroid disruption. There is a general move away from radioisotopes in testing, when possible, in favor of fluorescence [183]. As there's also a move away from animal testing, the zebrafish, from 0–5 days post-fertilization, may prove invaluable as an in vitro model for thyroid disruption [184]. Further research and development into zebrafish-based high-throughput screens might lead to a comprehensive testing program, decreased expenditure for government agencies, and a decreased need to test on rodents and other mammals. The overview presented here illustrates that thyroid activity can be affected at various levels by a wide range of possible modes of action. So far, hardly any of these endpoints have been translated into 'omics-based testing strategies but several of the endpoints may provide opportunities to do so, making thyroid toxicogenomics an up-and-coming field.

References

[1] Terris D, Gourin C. Thyroid and parathyroid diseases: medical and surgical management. Thieme; 2008.

[2] Temple RKG, Needham J. The genius of China: 3,000 years of science, discovery, and invention. Rochester, VT: Inner Traditions; 2007.

[3] Encyclopedia Britannica Online. Paracelsus, <http://www.britannica.com/EBchecked/topic/442424/Paracelsus>; 2013.

[4] Kelly FC. Iodine in medicine and pharmacy since its discovery: 1811–1961. Proc R Soc Med 1961;54:831–6.

[5] Kendall EC. The isolation in crystalline form of the compound containing iodin, which occurs in the thyroid. JAMA 1915;64:2042–3.

[6] Gross J, Pitt-Rivers R. The identification of 3:5:3'-L-triiodothyronine in human plasma. Lancet 1952;1:439–41.

[7] Lieberman AJ. Facts versus fears: a review of the greatest unfounded health scares of recent times. American Council on Science and Health; 2004.

[8] Axelson O, Sundell L, Andersson K, Edling C, Hogstedt C, Kling H. Herbicide exposure and tumor mortality: an updated epidemiologic investigation on Swedish railroad workers. Scand J Work Environ Health 1980;6:73–9.

[9] Alison RH, Capen CC, Prentice DE. Neoplastic lesions of questionable significance to humans. Toxicol Pathol 1994;22:179–86.

[10] USEPA RED facts amitrole. Washington DC: US Environmental Protection Agency, Office of Prevention, Pesticides and Toxic Substances; 1996.

[11] Vandenberg LN, Colborn T, Hayes TB, Heindel JJ, Jacobs Jr DR, Lee D-H, et al. Hormones and endocrine-disrupting chemicals: low-dose effects and nonmonotonic dose responses. Endocr Rev 2012;33:378–455.

[12] Palmblad J, Levi L, Burger A, Melander A, Westgren U, von Schenck H, et al. Effects of total energy withdrawal (fasting) on the levels of growth hormone, thyrotropin, cortisol, adrenaline, noradrenaline, T4, T3, and rT3 in healthy males. Acta Med Scand 1977;201:15–22.

[13] Hackney AC, Feith S, Pozos R, Seale J. Effects of high altitude and cold exposure on resting thyroid hormone concentrations. Aviat Space Environ Med 1995;66:325–9.

[14] Singh DV, Turner CW. Effect of light and darkness upon thyroid secretion rate and on the endocrine glands of female rats. Proc Soc Exp Biol Med 1969;131:1296–9.

[15] Roth GS, Handy AM, Mattison JA, Tilmont EM, Ingram DK, Lane MA. Effects of dietary caloric restriction and aging on thyroid hormones of rhesus monkeys. Horm Metab Res 2002;34:378–82.

[16] Tingley JO, Morris AW, Hill Jr SR, Pittman Jr JA. The acute thyroid response to emotional stress. Ala J Med Sci 1965;2:297–300.

[17] Chan V, Wang C, Yeung RT. Pituitary-thyroid responses to surgical stress. Acta Endocrinol 1978;88:490–8.

[18] Capen CC, Martin SL. The effects of xenobiotics on the structure and function of thyroid follicular and C-cells. Toxicol Pathol 1989;17:266–93.

[19] Franklyn JA, Ramsden DB, Sheppard MC. The influence of age and sex on tests of thyroid function. Ann Clin Biochem 1985;22(5):502–5.

[20] Andersen S, Pedersen KM, Bruun NH, Laurberg P. Narrow individual variations in serum T(4) and T(3) in normal subjects: a clue to the understanding of subclinical thyroid disease. J Clin Endocrinol Metab 2002;87:1068–72.

[21] Hansen PS, Brix TH, Sørensen TIA, Kyvik KO, Hegedüs L. Major genetic influence on the regulation of the pituitary–thyroid axis: a study of healthy Danish twins. J Clin Endocrinol Metab 2004;89:1181–7.

[22] Shirai T, Asamoto M. Application of toxicogenomics to the endocrine disruption issue. Pure Appl Chem 2003;75:2419–22.

[23] Werthmann RC, Volpe S, Lohse MJ, Calebiro D. Persistent cAMP signaling by internalized TSH receptors occurs in thyroid but not in HEK293 cells. FASEB J 2012;26:2043–8.

[24] Wonerow P, Chey S, Führer D, Holzapfel HP, Paschke R. Functional characterization of five constitutively activating thyrotrophin receptor mutations. Clin Endocrinol 2000;53:461–8.

[25] Hochbaum D, Hong K, Barila G, Ribeiro-Neto F, Altschuler DL. Epac, in synergy with cAMP-dependent protein kinase (PKA), is required for cAMP-mediated mitogenesis. J Biol Chem 2008;283:4464–8.

[26] Minich WB, Lenzner C, Bergmann A, Morgenthaler NG. A coated tube assay for the detection of blocking thyrotropin receptor autoantibodies. J Clin Endocrinol Metab 2004;89:352–6.

[27] Titus S, Neumann S, Zheng W, Southall N, Michael S, Klumpp C, et al. Quantitative high throughput screening using a live cell cAMP assay identifies small molecule agonists of the TSH receptor. J Biomol Screen 2008;13:120–7.
[28] Evans C, Morgenthaler NG, Lee S, Llewellyn DH, Clifton-Bligh R, John R, et al. Development of a luminescent bioassay for thyroid stimulating antibodies. J Clin Endocrinol Metab 1999;84:374–7.
[29] Jomaa B, Aarts JMMJG, de Haan LHJ, Peijnenburg AACM, Bovee TFH, Murk AJ, et al. In vitro pituitary and thyroid cell proliferation assays and their relevance as alternatives to animal testing. ALTEX 2013;30:293–307.
[30] Maruo Y, Takahashi H, Soeda I, Nishikura N, Matsui K, Ota Y, et al. Transient congenital hypothyroidism caused by biallelic mutations of the dual oxidase 2 gene in Japanese patients detected by a neonatal screening program, J. Clin Endocrinol Metab 2008;93:4261–7.
[31] Zamproni I, Grasberger H, Cortinovis F, Vigone MC, Chiumello G, Mora S, et al. Biallelic inactivation of the dual oxidase maturation factor 2 (DUOXA2) gene as a novel cause of congenital hypothyroidism. J Clin Endocrinol Metab 2008;93:605–10.
[32] Rigutto S, Hoste C, Grasberger H, Milenkovic M, Communi D, Dumont JE, et al. Activation of dual oxidases Duox1 and Duox2 differential regulation mediated by cAMP-dependent protein kinase and protein kinase C-dependent phosphorylation. J Biol Chem 2009;284:6725–34.
[33] Milenkovic M, De Deken X, Jin L, De Felice M, Di Lauro R, Dumont JE, et al. Duox expression and related H2O2 measurement in mouse thyroid: onset in embryonic development and regulation by TSH in adult. J Endocrinol 2007;192:615–26.
[34] Fernandez N, Caloca MJ, Prendergast GV, Meinkoth JL, Kazanietz MG. Atypical protein kinase C-zeta stimulates thyrotropin-independent proliferation in rat thyroid cells. Endocrinology 2000; 141:146–52.
[35] Akama T, Hara T, Yan D, Kawashima A, Yoshihara A, Sue M, et al. Thyroglobulin induces thyroid cell growth through suppression of miR-16 and miR-195, 2012.
[36] Heijne WHM, Kienhuis AS, van Ommen B, Stierum RH, Groten JP. Systems toxicology: applications of toxicogenomics, transcriptomics, proteomics and metabolomics in toxicology. Expert Rev Proteomics 2005;2:767–80.
[37] Hays MT. Compartmental models for human iodine metabolism. Math Biosci 1984;72:317–35.
[38] Caillou B, Troalen F, Baudin E, Talbot M, Filetti S, Schlumberger M, et al. Na + /I− symporter distribution in human thyroid tissues: an immunohistochemical study. J Clin Endocrinol Metab 1998;83:4102–6.
[39] Eskandari S, Loo DDF, Dai G, Levy O, Wright EM, Carrasco N. Thyroid Na + /I− symporter mechanism, stoichiometry, and specificity. J Biol Chem 1997;272:27230–27238.
[40] Dohán O, la Vieja AD, Paroder V, Riedel C, Artani M, Reed M, et al. The sodium/iodide symporter (NIS): characterization, regulation, and medical significance. Endocr Rev 2003;24:48–77.
[41] Josefsson M, Evilevitch L, Weström B, Grunditz T, Ekblad E. Sodium-iodide symporter mediates iodide secretion in rat gastric mucosa in vitro. Exp Biol Med (Maywood) 2006;231:277–81.
[42] Su AI, Wiltshire T, Batalov S, Lapp H, Ching KA, Block D, et al. A gene atlas of the mouse and human protein-encoding transcriptomes. Proc Natl Acad Sci USA 2004;101:6062–7.
[43] Everett LA, Glaser B, Beck JC, Idol JR, Buchs A, Heyman M, et al. Pendred syndrome is caused by mutations in a putative sulphate transporter gene (PDS). Nat Genet 1997;17:411–22.
[44] Pesce L, Bizhanova A, Caraballo JC, Westphal W, Butti ML, Comellas A, et al. TSH regulates pendrin membrane abundance and enhances iodide efflux in thyroid cells. Endocrinology 2012;153:512–21.
[45] Twyffels L, Massart C, Golstein PE, Raspe E, van Sande J, Dumont JE, et al. Pendrin: the thyrocyte apical membrane iodide transporter? Cell Physiol Biochem 2011;28:491–6.
[46] Cohn WE, Cohn ET. Permeability of red corpuscles of the dog to sodium ion. Proc Soc Exp Biol Med 1939;41:445–9.
[47] Huxley A. From overshoot to voltage clamp. Trends Neurosci 2002;25:553–8.

[48] Hamill OP, Marty A, Neher E, Sakmann B, Sigworth FJ. Improved patch-clamp techniques for high-resolution current recording from cells and cell-free membrane patches. Pflugers Arch 1981;391:85–100.
[49] Di Bernardo J, Iosco C, Rhoden KJ. Intracellular anion fluorescence assay for sodium/iodide symporter substrates. Anal Biochem 2011;415:32–8.
[50] Yoshida A, Taniguchi S, Hisatome I, Royaux IE, Green ED, Kohn LD, et al. Pendrin is an iodide-specific apical porter responsible for iodide efflux from thyroid cells. J Clin Endocrinol Metab 2002;87:3356–61.
[51] Gillam MP, Sidhaye AR, Lee EJ, Rutishauser J, Stephan CW, Kopp P. Functional characterization of pendrin in a polarized cell system: evidence for pendrin-mediated apical iodide efflux. J Biol Chem 2004;279:13004–10.
[52] Suzuki K, Kohn LD. Differential regulation of apical and basal iodide transporters in the thyroid by thyroglobulin. J Endocrinol 2006;189:247–55.
[53] Schweizer U, Chiu J, Köhrle J. Peroxides and peroxide-degrading enzymes in the thyroid. Antioxid Redox Signal 2008;10:1577–92.
[54] Bou Khalil R, Richa S. Thyroid adverse effects of psychotropic drugs: a review. Clin Neuropharmacol 2011;34:248–55.
[55] Schmutzler C, Bacinski A, Gotthardt I, Huhne K, Ambrugger P, Klammer H, et al. The ultraviolet filter benzophenone 2 interferes with the thyroid hormone axis in rats and is a potent in vitro inhibitor of human recombinant thyroid peroxidase. Endocrinology 2007;148:2835–44.
[56] de Souza Dos Santos MC, Gonçalves CFL, Vaisman M, Ferreira ACF, de Carvalho DP. Impact of flavonoids on thyroid function. Food Chem Toxicol 2011;49:2495–502.
[57] Senou M, Khalifa C, Thimmesch M, Jouret F, Devuyst O, Col V, et al. A coherent organization of differentiation proteins is required to maintain an appropriate thyroid function in the pendred thyroid. J Clin Endocrinol Metab 2010;95:4021–30.
[58] Ekholm R, Björkman U. Glutathione peroxidase degrades intracellular hydrogen peroxide and thereby inhibits intracellular protein iodination in thyroid epithelium. Endocrinology 1997;138:2871–8.
[59] Masini-Repiso AM, Bonaterra M, Spitale L, Di Fulvio M, Bonino MI, Coleoni AH, et al. Ultrastructural localization of thyroid peroxidase, hydrogen peroxide-generating sites, and monoamine oxidase in benign and malignant thyroid diseases. Hum Pathol 2004;35:436–46.
[60] Antonica F, Kasprzyk DF, Opitz R, Iacovino M, Liao X-H, Dumitrescu AM, et al. Generation of functional thyroid from embryonic stem cells. Nature 2012;491:66–71.
[61] Yoshihara A, Hara T, Kawashima A, Akama T, Tanigawa K, Wu H, et al. Regulation of dual oxidase (DUOX) expression and H2O2 production by thyroglobulin. Thyroid 2012 doi:10.1089/thy.2012-0003.
[62] Sugawara M, Sugawara Y, Wen K. Methimazole and propylthiouracil increase cellular thyroid peroxidase activity and thyroid peroxidase mRNA in cultured porcine thyroid follicles. Thyroid 1999;9:513–8.
[63] Song M, Kim Y-J, Song M-K, Choi H-S, Park Y-K, Ryu J-C. Identification of classifiers for increase or decrease of thyroid peroxidase activity in the FTC-238/hTPO recombinant cell line. Environ Sci Technol 2011;45:7906–14.
[64] Hirakawa S, Saito R, Ohara H, Okuyama R, Aiba S. Dual oxidase 1 induced by Th2 cytokines promotes STAT6 phosphorylation via oxidative inactivation of protein tyrosine phosphatase 1B in human epidermal keratinocytes. J Immunol 2011;186:4762–70.
[65] Selmi S, Rousset B. Identification of two subpopulations of thyroid lysosomes: relation to the thyroglobulin proteolytic pathway. Biochem J 1988;253:523–32.
[66] Radvila A, Roost R, Bürgi H, Kohler H, Studer H. Inhibition of thyroglobulin biosynthesis and degradation by excess iodide: synergism with lithium. Acta Endocrinol 1976;81:495–506.
[67] Jegga AG, Schneider L, Ouyang X, Zhang J. Systems biology of the autophagy-lysosomal pathway. Autophagy 2011;7:477–89.
[68] Dennemärker J, Lohmüller T, Müller S, Aguilar SV, Tobin DJ, Peters C, et al. Impaired turnover of autophagolysosomes in cathepsin L deficiency. Biol Chem 2010;391:913–22.

[69] Wille A, Gerber A, Heimburg A, Reisenauer A, Peters C, Saftig P, et al. Cathepsin L is involved in cathepsin D processing and regulation of apoptosis in A549 human lung epithelial cells. Biol Chem 2004;385:665–70.
[70] Friedrichs B, Tepel C, Reinheckel T, Deussing J, von Figura K, Herzog V, et al. Thyroid functions of mouse cathepsins B, K, and L. J Clin Invest 2003;111:1733–45.
[71] Moreno JC, Keijser R, Aarraas S, De Vijlder JJM, Ris-Stalpers C. Cloning and characterization of a novel thyroidal gene encoding proteins with a conserved nitroreductase domain. J Endocrinol Invest 2002;25:40.
[72] Afink G, Kulik W, Overmars H, de Randamie J, Veenboer T, van Cruchten A, et al. Molecular characterization of iodotyrosine dehalogenase deficiency in patients with hypothyroidism. J Clin Endocrinol Metab 2008;93:4894–901.
[73] Moreno JC, Klootwijk W, van Toor H, Pinto G, D'Alessandro M, Lèger A, et al. Mutations in the iodotyrosine deiodinase gene and hypothyroidism. N Engl J Med 2008;358:1811–8.
[74] Thomas SR, McTamney PM, Adler JM, LaRonde-LeBlanc N, Rokita SE. Crystal structure of iodotyrosine deiodinase, a novel flavoprotein responsible for iodide salvage in thyroid glands. J Biol Chem 2009;284:19659–19667.
[75] Gnidehou S, Caillou B, Talbot M, Ohayon R, Kaniewski J, Noël-Hudson M-S, et al. Iodotyrosine dehalogenase 1 (DEHAL1) is a transmembrane protein involved in the recycling of iodide close to the thyroglobulin iodination site. FASEB J 2004;18:1574–6.
[76] Marangell LB, George MS, Callahan AM, Ketter TA, Pazzaglia PJ, L'Herrou TA, et al. Effects of intrathecal thyrotropin-releasing hormone (protirelin) in refractory depressed patients. Arch Gen Psychiatry 1997;54:214–22.
[77] Schmitmeier S, Thole H, Bader A, Bauer K. Purification and characterization of the thyrotropin-releasing hormone (TRH)-degrading serum enzyme and its identification as a product of liver origin. Eur J Biochem 2002;269:1278–86.
[78] Jones BW, Hinkle PM. Subcellular trafficking of the TRH receptor: effect of phosphorylation. Mol Endocrinol 2009;23:1466–78.
[79] Joffe RT, Gold PW, Uhde TW, Post RM. The effects of carbamazepine on the thyrotropin response to thyrotropin-releasing hormone. Psychiatry Res 1984;12:161–6.
[80] Rosen JB, Weiss SR, Post RM. Contingent tolerance to carbamazepine: alterations in TRH mRNA and TRH receptor binding in limbic structures. Brain Res 1994;651:252–60.
[81] Sun Y, Lu X, Gershengorn MC. Thyrotropin-releasing hormone receptors: similarities and differences. J Mol Endocrinol 2003;30:87–97.
[82] Engel S, Neumann S, Kaur N, Monga V, Jain R, Northup J, et al. Low affinity analogs of thyrotropin-releasing hormone are super-agonists. J Biol Chem 2006;281:13103–9.
[83] O'Dowd BF, Lee DK, Huang W, Nguyen T, Cheng R, Liu Y, et al. TRH-R2 exhibits similar binding and acute signaling but distinct regulation and anatomic distribution compared with TRH-R1. Mol Endocrinol 2000;14:183–93.
[84] Wang W, Gershengorn MC. Rat TRH receptor type 2 exhibits higher basal signaling activity than TRH receptor type 1. Endocrinology 1999;140:4916–9.
[85] Mittal SR, Mathur AK, Prasad N. Effect of calcium channel blockers on serum levels of thyroid hormones. Int J Cardiol 1993;38:131–2.
[86] Teba L, Smailer S, Taylor HC. Effect of nifedipine on TRH stimulation of TSH and PRL release by the pituitary gland. Metab Clin Exp 1985;34:161–3.
[87] Yamada M, Mori M, Yamaguchi M, Akiyama H, Shiono S, Kobayashi I, et al. Thyrotropin-releasing hormone stimulation of thyrotropin secretion is suppressed by calcium ion antagonists that block transmembrane influx and intracellular mobilization of calcium ion in human subjects. J Endocrinol Invest 1986;9:227–31.
[88] Segev DL, Umbricht C, Zeiger MA. Molecular pathogenesis of thyroid cancer. Surg Oncol 2003;12:69–90.
[89] Ahmed E-L, Steve O. Pituitary thyroid hormone resistance (PTHR). Endocr Abstr 2011;25:P66.

[90] Ravindra R, Forman LJ, Patel SA. Vinblastine and nocodazole inhibit basal and thyrotropin-releasing hormone-stimulated prolactin secretion in GH(3) cells. Endocrine 1995;3:591–6.
[91] Liu C, Ha M, Cui Y, Wang C, Yan M, Fu W, et al. JNK pathway decreases thyroid hormones via TRH receptor: a novel mechanism for disturbance of thyroid hormone homeostasis by PCB153. Toxicology 2012;302:68–76.
[92] Millipore, Millipore TRH receptor assay, 2012.
[93] PerklinElmer, AequoScreen® GPCR Cell Line, 2009.
[94] Shimomura O, Johnson FH, Saiga Y. Extraction, purification and properties of aequorin, a bioluminescent protein from the luminous hydromedusan, *Aequorea*. J Cell Comp Physiol 1962;59:223–39.
[95] Martin A, Platzer M, Davies TF. Retention of cyclic AMP response to TSH in a cloned human thyrocyte/T cell hybridoma (HY2-15). Mol Cell Endocrinol 1988;60:233–8.
[96] James RA, Sarapura VD, Bruns C, Raulf F, Dowding JM, Gordon DF, et al. Thyroid hormone-induced expression of specific somatostatin receptor subtypes correlates with involution of the TtT-97 murine thyrotrope tumor. Endocrinology 1997;138:719–24.
[97] Alarid ET, Windle JJ, Whyte DB, Mellon PL. Immortalization of pituitary cells at discrete stages of development by directed oncogenesis in transgenic mice. Development 1996;122:3319–29.
[98] Aizawa S, Sakai T, Sakata I. Glutamine and glutamic acid enhance thyroid-stimulating hormone β subunit mRNA expression in the rat pars tuberalis. J Endocrinol 2012;212:383–94.
[99] Gavras I, Thomson JA. Late thyrotoxicosis complicating autoimmune thyroiditis. Acta Endocrinol 1972;69:41–6.
[100] Shibusawa N, Hashimoto K, Yamada M. Thyrotropin-releasing hormone (TRH) in the cerebellum. Cerebellum 2008;7:84–95.
[101] Veronesi MC, Yard M, Jackson J, Lahiri DK, Kubek MJ. An analog of thyrotropin-releasing hormone (TRH) is neuroprotective against glutamate-induced toxicity in fetal rat hippocampal neurons in vitro. Brain Res 2007;1128:79–85.
[102] Coppola A, Meli R, Diano S. Inverse shift in circulating corticosterone and leptin levels elevates hypothalamic deiodinase type 2 in fasted rats. Endocrinology 2005;146:2827–33.
[103] Nillni EA. Regulation of the hypothalamic thyrotropin releasing hormone (TRH) neuron by neuronal and peripheral inputs. Front Neuroendocrinol 2010;31:134–56.
[104] Ellerhorst JA, Naderi AA, Johnson MK, Pelletier P, Prieto VG, Diwan AH, et al. Expression of thyrotropin-releasing hormone by human melanoma and nevi. Clin Cancer Res 2004;10:5531–6.
[105] Gkonos PJ, Tavianini MA, Liu CC, Roos BA. Thyrotropin-releasing hormone gene expression in normal thyroid parafollicular cells. Mol Endocrinol 1989;3:2101–9.
[106] García SI, Porto PI, Martinez VN, Alvarez AL, Finkielman S, Pirola CJ. Expression of TRH and TRH-like peptides in a human glioblastoma-astrocytoma cell line (U-373-MG). J Endocrinol 2000;166:697–703.
[107] Garcia SI, Dabsys SM, Santajuliana D, Delorenzi A, Finkielman S, Nahmod VE, et al. Interaction between thyrotrophin-releasing hormone and the muscarinic cholinergic system in rat brain. J Endocrinol 1992;134:215–9.
[108] Tabunoki H, Saito N, Suwanborirux K, Charupant K, Satoh J. Molecular network profiling of U373MG human glioblastoma cells following induction of apoptosis by novel marine-derived anti-cancer 1,2,3,4-tetrahydroisoquinoline alkaloids. Cancer Cell Int 2012;12:14.
[109] Zandieh Doulabi B, Platvoet-ter Schiphorst M, van Beeren HC, Labruyere WT, Lamers WH, Fliers E, et al. TR(beta)1 protein is preferentially expressed in the pericentral zone of rat liver and exhibits marked diurnal variation. Endocrinology 2002;143:979–84.
[110] Abel ED, Ahima RS, Boers ME, Elmquist JK, Wondisford FE. Critical role for thyroid hormone receptor beta2 in the regulation of paraventricular thyrotropin-releasing hormone neurons. J Clin Invest 2001;107:1017–23.
[111] Katz D, Reginato MJ, Lazar MA. Functional regulation of thyroid hormone receptor variant TR alpha 2 by phosphorylation. Mol Cell Biol 1995;15:2341–8.

[112] Yen PM. Physiological and molecular basis of thyroid hormone action. Physiol Rev 2001;81:1097–142.
[113] Carr FE, Wong NC. Characteristics of a negative thyroid hormone response element. J Biol Chem 1994;269:4175–9.
[114] Weiss RE, Xu J, Ning G, Pohlenz J, O'Malley BW, Refetoff S. Mice deficient in the steroid receptor co-activator 1 (SRC-1) are resistant to thyroid hormone. EMBO J 1999;18:1900–4.
[115] Freitas J, Cano P, Craig-Veit C, Goodson ML, Furlow JD, Murk AJ. Detection of thyroid hormone receptor disruptors by a novel stable in vitro reporter gene assay. Toxicol In Vitro 2011;25:257–66.
[116] Gutleb AC, Meerts IATM, Bergsma JH, Schriks M, Murk AJ. T-Screen as a tool to identify thyroid hormone receptor active compounds. Environ Toxicol Pharmacol 2005;19:231–8.
[117] Paquette MA, Dong H, Gagné R, Williams A, Malowany M, Wade MG, et al. Thyroid hormone-regulated gene expression in juvenile mouse liver: identification of thyroid response elements using microarray profiling and in silico analyses. BMC Genomics 2011;12:634.
[118] Feng X, Jiang Y, Meltzer P, Yen PM. Thyroid hormone regulation of hepatic genes in vivo detected by complementary DNA microarray. Mol Endocrinol 2000;14:947–55.
[119] Song M, Kim YJ, Lee J-N, Ryu JC. Genome-wide expression profiling of carbaryl and vinclozolin in human thyroid follicular carcinoma (FTC-238) cells. BioChip J 2010;4:89–98.
[120] Safi R, Muramoto GG, Salter AB, Meadows S, Himburg H, Russell L, et al. Pharmacological manipulation of the RAR/RXR signaling pathway maintains the repopulating capacity of hematopoietic stem cells in culture. Mol Endocrinol 2009;23:188–201.
[121] Hörlein AJ, Näär AM, Heinzel T, Torchia J, Gloss B, Kurokawa R, et al. Ligand-independent repression by the thyroid hormone receptor mediated by a nuclear receptor co-repressor. Nature 1995;377:397–404.
[122] Chen JD, Evans RM. A transcriptional co-repressor that interacts with nuclear hormone receptors. Nature 1995;377:454–7.
[123] Moore JMR, Galicia SJ, McReynolds AC, Nguyen N-H, Scanlan TS, Guy RK. Quantitative proteomics of the thyroid hormone receptor-coregulator interactions. J Biol Chem 2004;279:27584–27590.
[124] Johnson RL, Hwang JY, Arnold LA, Huang R, Wichterman J, Augustinaite I, et al. A quantitative high throughput screen identifies novel inhibitors of the interaction of thyroid receptor beta with a peptide of steroid receptor coactivator 2. J Biomol Screen 2011;16:618–27.
[125] Sadana P, Hwang JY, Attia RR, Arnold LA, Neale G, Guy RK. Similarities and differences between two modes of antagonism of the thyroid hormone receptor. ACS Chem Biol 2011;6:1096–106.
[126] Schussler GC. The thyroxine-binding proteins. Thyroid 2000;10:141–9.
[127] Savu L, Vranckx R, Marielle R-R, Maya M, Nunez EA, Tréton J, et al. A senescence up-regulated protein: the rat thyroxine-binding globulin (TBG). Biochim Biophys Acta 1991;1097:19–22.
[128] Sousa JC, de Escobar GM, Oliveira P, Saraiva MJ, Palha JA. Transthyretin is not necessary for thyroid hormone metabolism in conditions of increased hormone demand. J Endocrinol 2005;187:257–66.
[129] Carvalho GA, Weiss RE, Refetoff S. Complete thyroxine-binding globulin (TBG) deficiency produced by a mutation in acceptor splice site causing frameshift and early termination of translation (TBG-Kankakee). J Clin Endocrinol Metab 1998;83:3604–8.
[130] Afandi B, Vera R, Schussler GC, Yap MG. Concordant decreases of thyroxine and thyroxine binding protein concentrations during sepsis. Metabolism 2000;49:753–4.
[131] Montaño M, Cocco E, Guignard C, Marsh G, Hoffmann L, Bergman A, et al. New approaches to assess the transthyretin binding capacity of bioactivated thyroid hormone disruptors. Toxicol Sci 2012;130:94–105.
[132] Bartalena L, Farsetti A, Flink IL, Robbins J. Effects of interleukin-6 on the expression of thyroid hormone-binding protein genes in cultured human hepatoblastoma-derived (Hep G2) cells. Mol Endocrinol 1992;6:935–42.
[133] Friesema ECH, Jansen J, Jachtenberg J-W, Visser WE, Kester MHA, Visser TJ. Effective cellular uptake and efflux of thyroid hormone by human monocarboxylate transporter 10. Mol Endocrinol 2008;22:1357–69.

[134] Friesema ECH, Ganguly S, Abdalla A, Manning Fox JE, Halestrap AP, Visser TJ. Identification of monocarboxylate transporter 8 as a specific thyroid hormone transporter. J Biol Chem 2003;278:40128–40135.
[135] Pizzagalli F, Hagenbuch B, Stieger B, Klenk U, Folkers G, Meier PJ. Identification of a novel human organic anion transporting polypeptide as a high affinity thyroxine transporter. Mol Endocrinol 2002;16:2283–96.
[136] Nishimura M, Naito S. Tissue-specific mRNA expression profiles of human solute carrier transporter superfamilies. Drug Metab Pharmacokinet 2008;23:22–44.
[137] van der Deure WM, Peeters RP, Visser TJ. Molecular aspects of thyroid hormone transporters, including MCT8, MCT10, and OATPs, and the effects of genetic variation in these transporters. J Mol Endocrinol 2010;44:1–11.
[138] Mayerl S, Visser TJ, Darras VM, Horn S, Heuer H. Impact of Oatp1c1 deficiency on thyroid hormone metabolism and action in the mouse brain. Endocrinology 2012;153:1528–37.
[139] Verhoeven FA, Moerings EP, Lamers JM, Hennemann G, Visser TJ, Everts ME. Inhibitory effects of calcium channel blockers on thyroid hormone uptake in neonatal rat cardiomyocytes. Am J Physiol 2001;281:H1985–H1991.
[140] Segal J. Calcium is the first messenger for the action of thyroid hormone at the level of the plasma membrane: first evidence for an acute effect of thyroid hormone on calcium uptake in the heart. Endocrinology 1990;126:2693–702.
[141] Kinne A, Roth S, Biebermann H, Köhrle J, Grüters A, Schweizer U. Surface translocation and tri-iodothyronine uptake of mutant MCT8 proteins are cell type-dependent. J Mol Endocrinol 2009;43:263–71.
[142] Westholm DE, Salo DR, Viken KJ, Rumbley JN, Anderson GW. The blood-brain barrier thyroxine transporter organic anion-transporting polypeptide 1c1 displays atypical transport kinetics. Endocrinology 2009;150:5153–62.
[143] Kogai T, Liu Y-Y, Richter LL, Mody K, Kagechika H, Brent GA. Retinoic acid induces expression of the thyroid hormone transporter, monocarboxylate transporter 8 (Mct8). J Biol Chem 2010;285:27279–88.
[144] Capelo LP, Beber EH, Fonseca TL, Gouveia CHA. The monocarboxylate transporter 8 and L-type amino acid transporters 1 and 2 are expressed in mouse skeletons and in osteoblastic MC3T3-E1 cells. Thyroid 2009;19:171–80.
[145] Mebis L, Paletta D, Debaveye Y, Ellger B, Langouche L, D'Hoore A, et al. Expression of thyroid hormone transporters during critical illness. Eur J Endocrinol 2009;161:243–50.
[146] Sugiyama D, Kusuhara H, Taniguchi H, Ishikawa S, Nozaki Y, Aburatani H, et al. Functional characterization of rat brain-specific organic anion transporter (Oatp14) at the blood–brain barrier: high affinity transporter for thyroxine. J Biol Chem 2003;278:43489–95.
[147] Benedetti MS, Whomsley R, Baltes E, Tonner F. Alteration of thyroid hormone homeostasis by antiepileptic drugs in humans: involvement of glucuronosyltransferase induction. Eur J Clin Pharmacol 2005;61:863–72.
[148] Huang SA, Bianco AC. Reawakened interest in type III iodothyronine deiodinase in critical illness and injury. Nat Clin Pract Endocrinol Metab 2008;4:148–55.
[149] Wu S-Y, Green WL, Huang W-S, Hays MT, Chopra IJ. Alternate pathways of thyroid hormone metabolism. Thyroid 2005;15:943–58.
[150] Xu C, Li CY-T, Kong A-NT. Induction of phase I, II and III drug metabolism/transport by xenobiotics. Arch Pharm Res 2005;28:249–68.
[151] Visser TJ, Kaptein E, Gijzel AL, de Herder WW, Ebner T, Burchell B. Glucuronidation of thyroid hormone by human bilirubin and phenol UDP-glucuronyltransferase isoenzymes. FEBS Lett 1993;324:358–60.
[152] Findlay KA, Kaptein E, Visser TJ, Burchell B. Characterization of the uridine diphosphate-glucuronosyltransferase-catalyzing thyroid hormone glucuronidation in man. J Clin Endocrinol Metab 2000;85:2879–83.

[153] Wieneke N, Neuschäfer-Rube F, Bode LM, Kuna M, Andres J, Carnevali Jr LC, et al. Synergistic acceleration of thyroid hormone degradation by phenobarbital and the PPAR alpha agonist WY14643 in rat hepatocytes. Toxicol Appl Pharmacol 2009;240:99–107.
[154] Zavacki AM, Larsen PR. CARs and drugs: a risky combination. Endocrinology 2005;146:992–4.
[155] Everts ME, Visser TJ, Moerings EP, Docter R, van Toor H, Tempelaars AM, et al. Uptake of triiodothyroacetic acid and its effect on thyrotropin secretion in cultured anterior pituitary cells. Endocrinology 1994;135:2700–7.
[156] Juge-Aubry CE, Morin O, Pernin AT, Liang H, Philippe J, Burger AG. Long-lasting effects of Triac and thyroxine on the control of thyrotropin and hepatic deiodinase type I. Eur J Endocrinol 1995;132:751–8.
[157] Lameloise N, Siegrist-Kaiser C, O'Connell M, Burger A. Differences between the effects of thyroxine and tetraiodothyroacetic acid on TSH suppression and cardiac hypertrophy. Eur J Endocrinol 2001;144:145–54.
[158] Gavin LA, Livermore BM, Cavalieri RR, Hammond ME, Castle JN. Serum concentration, metabolic clearance, and production rates of 3,5,3′triiodothyroacetic acid in normal and athyreotic man. J Clin Endocrinol Metab 1980;51:529–34.
[159] FDA, FDA and the U.S. Attorney for the Western District of Texas Announce Guilty Plea in Drug Counterfeiting Case, 2004 Press release accessed on http://www.fda.gov/NewsEvents/Newsroom/PressAnnouncements/2004/ucm108266.htm.
[160] FDA, Tiratricol (triiodothyroacetic acid). Safety Alerts for Human Medical Products 2009.
[161] Scanlan TS, Suchland KL, Hart ME, Chiellini G, Huang Y, Kruzich PJ, et al. 3-iodothyronamine is an endogenous and rapid-acting derivative of thyroid hormone. Nat Med 2004;10:638–42.
[162] Murk AJ, Rijntjes E, Blaauboer BJ, Clewell R, Crofton KM, Dingemans MML, et al. Mechanism-based testing strategy using in vitro approaches for identification of thyroid hormone disrupting chemicals. Toxicol In Vitro 2013 doi:10.1016/j.tiv.2013.02.012.
[163] Shah RB, Bryant A, Collier J, Habib MJ, Khan MA. Stability indicating validated HPLC method for quantification of levothyroxine with eight degradation peaks in the presence of excipients. Int J Pharm 2008;360:77–82.
[164] Režen T, Juvan P, Tacer KF, Kuzman D, Roth A, Pompon D, et al. The Sterolgene v0 cDNA microarray: a systemic approach to studies of cholesterol homeostasis and drug metabolism. BMC Genomics 2008;9:76.
[165] Gruters A. Molecular genetic defects in congenital hypothyroidism. Eur J Endocrinol 2004;151:U39–44.
[166] Kuehnen P, Grueters A, Krude H. Two puzzling cases of thyroid dysgenesis. Horm Res 2009;71:93–7.
[167] Castanet M, Sura-Trueba S, Chauty A, Carré A, de Roux N, Heath S, et al. Linkage and mutational analysis of familial thyroid dysgenesis demonstrate genetic heterogeneity implicating novel genes. Eur J Hum Genet 2005;13:232–9.
[168] Felice MD, Lauro RD. Thyroid development and its disorders: genetics and molecular mechanisms. Endocr Rev 2004;25:722–46.
[169] Vassart G, Dumont JE. Thyroid dysgenesis: multigenic or epigenetic … or both? Endocrinology 2005;146:5035–7.
[170] Atkins P, Cohen SB, Phillips BJ. Drug therapy for hyperthyroidism in pregnancy: safety issues for mother and fetus. Drug Saf 2000;23:229–44.
[171] Gorman CA. Radioiodine and pregnancy. Thyroid 1999;9:721–6.
[172] Olivieri A, Stazi MA, Mastroiacovo P, Fazzini C, Medda E, Spagnolo A, et al. A population-based study on the frequency of additional congenital malformations in infants with congenital hypothyroidism: data from the Italian Registry for Congenital Hypothyroidism (1991–1998). J Clin Endocrinol Metab 2002;87:557–62.
[173] Panicker V. Genetics of thyroid function and disease. Clin Biochem Rev 2011;32:165–75.
[174] Brix TH, Kyvik KO, Christensen K, Hegedüs L. Evidence for a major role of heredity in Graves' disease: a population-based study of two Danish twin cohorts. J Clin Endocrinol Metab 2001;86:930–4.

[175] McLeod DSA, Cooper DS. The incidence and prevalence of thyroid autoimmunity. Endocrine 2012;42:252–65.

[176] Rose NR, Rasooly L, Saboori AM, Burek CL. Linking iodine with autoimmune thyroiditis. Environ Health Perspect 1999;107:749–52.

[177] Doerge DR, Sheehan DM. Goitrogenic and estrogenic activity of soy isoflavones. Environ Health Perspect 2002;110:349–53.

[178] Bergh JJ, Lin H-Y, Lansing L, Mohamed SN, Davis FB, Mousa S, et al. Integrin αVβ3 contains a cell surface receptor site for thyroid hormone that is linked to activation of mitogen-activated protein kinase and induction of angiogenesis. Endocrinology 2005;146:2864–71.

[179] Maletzky BM, Shore JH. Lithium treatment for psychiatric disorders. West J Med 1978;128:488–98.

[180] Bocchetta A, Loviselli A. Lithium treatment and thyroid abnormalities. Clin Pract Epidemiol Ment Health 2006;2:23.

[181] Chakrabarti S. Thyroid functions and bipolar affective disorder. J Thyroid Res 2011;2011:1–13.

[182] Lazarus JH. The effects of lithium therapy on thyroid and thyrotropin-releasing hormone. Thyroid 1998;8:909–13.

[183] Sundberg SA. High-throughput and ultra-high-throughput screening: solution- and cell-based approaches. Curr Opin Biotechnol 2000;11:47–53.

[184] van der Ven L. Lower Organisms as Alternatives for Toxicity Testing in Rodents; with a Focus on *Caenorhabditis elegans* and the Zebrafish (*Danio rerio*), RIVM Report 340720003, National Institute for Public Health and the Environment (RIVM), Bilthoven, Netherlands 2009.

[185] Pekary AE, Levin SR, Johnson DG, Berg L, Hershman JM. Tumor necrosis factor-alpha (TNF-alpha) and transforming growth factor-beta 1 (TGF-beta 1) inhibit the expression and activity of Na+/K(+)-ATPase in FRTL-5 rat thyroid cells. J Interferon Cytokine Res 1997;17:185–95.

[186] Raldúa D, Babin PJ. Simple, rapid zebrafish larva bioassay for assessing the potential of chemical pollutants and drugs to disrupt thyroid gland function. Environ Sci Technol 2009;43:6844–50.

[187] Zatelli MC, Gentilin E, Daffara F, Tagliati F, Reimondo G, Carandina G, et al. Therapeutic concentrations of mitotane (o,p-DDD) inhibit thyrotroph cell viability and TSH expression and secretion in a mouse cell line model. Endocrinology 2010;151:2453–61.

[188] Kuiper GGJM, Klootwijk W, Visser TJ. Expression of recombinant membrane-bound type I iodothyronine deiodinase in yeast. J Mol Endocrinol 2005;34:865–78.

[189] Ekuase EJ, Liu Y, Lehmler H-J, Robertson LW, Duffel MW. Structure-activity relationships for hydroxylated polychlorinated biphenyls as inhibitors of the sulfation of dehydroepiandrosterone catalyzed by human hydroxysteroid sulfotransferase SULT2A1. Chem Res Toxicol 2011;24:1720–8.

[190] Martin LA, Wilson DT, Reuhl KR, Gallo MA, Klaassen CD. Polychlorinated biphenyl congeners that increase the glucuronidation and biliary excretion of thyroxine are distinct from the congeners that enhance the serum disappearance of thyroxine. Drug Metab Dispos 2012;40:588–95.

SECTION

Organ Toxicity

5

CHAPTER

5.1 Hepatotoxicity Screening on In Vitro Models and the Role of 'Omics

Joost van Delft*, Karen Mathijs*, Jan Polman*, Maarten Coonen*, Ewa Szalowska, Geert R. Verheyen†, Freddy van Goethem†, Marja Driessen††, Leo van de Ven††, Sreenivasa Ramaiahgari‡ and Leo S. Price‡**

**Department of Toxicogenomics, Maastricht University, Maastricht, The Netherlands, **RIKILT—Institute of Food Safety, Wageningen University and Research Centre, Wageningen, The Netherlands, †Drug Safety Sciences, Janssen Research and Development, Beerse, Belgium, ††RIVM—National Institute for Public Health and the Environment, Bilthoven, The Netherlands, ‡Division of Toxicology, Leiden Amsterdam Center for Drug Research, Leiden University, Leiden, The Netherlands*

5.1.1 General introduction to hepatotoxicity and its main pathologies

The liver is one of the five most common target organs of toxicity, both during acute and chronic (repeated dose) toxicity, not only for drugs but also for cosmetic ingredients [1,2]. Chemical entities can trigger liver damage in humans, and hepatotoxicity is the leading cause of withdrawal of drugs from the market, accounting for 40% of withdrawals worldwide [3]. About 20% of the chemicals listed in Haz-Map—an occupational toxicology database for >3300 compounds—are known or posses the potency to cause hepatic injury in the occupational setting (http://hazmap.nlm.nih.gov) [4,5].

The liver is the major site for metabolism of compounds and therefore toxicities can be expected to occur frequently. Olson et al. [6] estimated that only 50% of human liver toxicities could be predicted using animal models. The other 50% could not be predicted owing to unavailability of suitable models and idiosyncratic toxicity. Given the enormous loss of time and resources spent before these attritions occur and given the risks to human patients, alternative strategies to develop new and safe compounds are needed.

In contrast to drugs, for which a wealth of toxicological information is available, the (hepato)toxic potential of many industrial chemicals and cosmetic ingredients is poorly documented and human data are mostly lacking. The ultimate goal of the safety evaluation of many of these chemicals, as opposed to drugs, is to define a margin of safety, which is based upon "no observed adverse effect" (NOAEL) levels. These are typically derived from either 28-day or 90-day repeated-dose toxicity studies, with measurement of biochemical (blood) parameters and postmortem histopathological findings. Hepatotoxicity-related effects are frequently observed in long-term rodent assays. The high attrition rate of compounds in in vivo toxicity testing following a positive evaluation in in vitro tests demonstrates

Toxicogenomics-Based Cellular Models.

the need for improved model systems. Furthermore, unexpected hepatotoxicity is regularly observed in humans in clinical studies or once a product has been put on the market. These idiosyncratic effects are a major reason to withdraw drugs from the market, having huge financial consequences.

Hepatocellular injury can manifest in a number of ways, including hepatitis, steatosis, cirrhosis, inflammation, phospholipidosis, and cholestasis. Cholestasis and steatosis are among the most prominent and well-documented types of liver injury [7]. Therefore, in The Netherlands Toxicogenomics Center (NTC) organ toxicity studies, focus is on these specific liver pathologies. Furthermore, attention is also paid to necrosis, a mode of cell death that is observed in most of the aforementioned types of hepatocellular injury.

Cholestasis

Cholestasis is defined as the retention of cholephilic compounds in the liver and/or reduction of bile flow [7]. Depending on the site of interference with bile secretion, cholestasis is referred to as extrahepatic (i.e. obstruction in biliary passages outside the liver) or intrahepatic (i.e. obstruction in the liver). Cholestatic injury clinically manifests as a marked increase in serum alkaline phosphatase (ALP) and/or γ-glutamyltransferase (GGT), and a concomitant rise in total bilirubin levels. Occasionally, a subsequent increase in serum levels of alanine aminotransferase (ALT; glutamic pyruvic transaminase) and aspartate aminotransferase (AST; glutamic oxaloacetic transaminase) may occur, reflecting secondary hepatocellular damage, and thus leakage of these cytosolic enzymes into the blood [8]. In part, cholestasis results from inhibiting the expression and/or functionality of xenobiotic export pumps located on the apical membrane domain of hepatocytes (e.g. MDRs, MRPs, BSEP). The use of synthetic conjugates of cholic acid (ChA) for in vitro assessment combined with flow cytometry can detect compounds that affect bile acid uptake [9].

Steatosis

Steatosis is increased lipid accumulation, mainly as triglycerides, in the liver [10]. Although relatively benign, simple steatosis can eventually lead to inflammation and subsequent tissue injury and eventual liver failure, if the underlying cause is not eliminated [11]. Because some xenobiotics target liver fatty acid metabolism, especially mitochondrial beta-oxidation, it is important to avoid potential drug candidates that can contribute to either the initiation of liver steatosis or progression to the more injurious steatohepatitis.

Necrosis

Cell death is the ultimate endpoint of cellular injury induced by xenobiotics. Historically, two modes of cell death have been distinguished, namely apoptosis (programmed cell death) and necrosis (accidental cell death). Necrosis, as opposed to apoptosis, is characterized by cell swelling and subsequent cell plasma membrane rupture, followed by an inflammatory response. Increased necrotic activity is seen in the end stages of most types of liver injury. In its more acute form, necrosis arises as a result of damage to cellular constituents [7]. Hepatocellular necrotic injury can be approached by measuring, for example, extracellular release of cytosolic enzymes, such as lactate dehydrogenase, by in situ staining with propidium iodide, and by viability assays based on cell proliferation.

5.1.2 'Omics-based in vitro approaches for hepatotoxicity screening: the NTC strategy

The major aim of NTC organ toxicity studies is to develop, by using a systems toxicogenomics approach, in vitro methods or short-term in vivo methods for testing the organ-toxic properties of compounds, as an alternative to the chronic rodent toxicity assays for long-term toxicity. Numerous in vitro studies are designed to assess potential liver toxicity, but none of these provide the complete picture [12,13]. In order to obtain more relevant mechanistic information on perturbed pathways, many toxicogenomics studies have been conducted on hepatotoxicity in rodents (mainly in the rat) [14–19]. Several demonstrate that specific liver pathologies can be predicted by these so-called 'omics approaches [20–24]. Furthermore, an increasing number of in vitro toxicogenomics studies derived from in vitro models of hepatotoxicity (e.g. hepatocytes, HepG2, HepaRG cells) are appearing and data are becoming available [25–30].

To improve the predictability of in vitro tests, three main aims were defined:

- To develop novel mechanism-based in vitro assays based on the application of genomics (i.e. genome-wide transcriptomics, including mRNA and microRNA (miRNA) profiling), proteomics, and metabolomic technologies.
- To evaluate the improved predictability of several three-dimensional (3D) cell models over existing models in toxicity testing with panels of characterized compounds representing different classes, and compare these to responses in mice, rats, and humans. The in vitro models include a cell line, primary hepatocytes, liver slices, and zebrafish embryos.
- Finally, to apply the obtained knowledge to build an iterative in silico model of hepatotoxicity. Valorization of this research would include the provision of improved in vitro assays for high-throughput screening of compounds for hepatotoxicity in fast and cost-effective ways, thereby reducing in vivo testing.

Selection of compounds

The selection of prototypical compounds with good knowledge of mechanisms and effects is important to reach the goals set above. The research focuses on steatosis, cholestasis, and cytotoxicity caused by oxidative stress. Three compounds per category, including compounds used in translational studies and negative controls, were selected (Table 5.1.1); three of these have been prioritized for the studies within NTC, namely cyclosporin A, amiodarone, and acetaminophen (Figure 5.1.1). These compounds have not only been abundantly documented in terms of their hepatotoxic potential with in-depth mechanistic toxicological information, they have also been the subject of a number of published in vivo and in vitro toxicogenomics studies, in which their effects on global hepatic gene expression patterns have already been profiled using DNA microarrays (among others, in HepG2 cells and human primary hepatocytes at the University of Maastricht (UM) [31]).

Toxicogenomics, metabolomics, and systems toxicology

Many toxicogenomics studies have been conducted on hepatotoxicity in rodents, mainly rats, thereby providing a wealth of information [14–20,23,24]. Unfortunately, the data are scattered over the public

Table 5.1.1 Prototypical Hepatotoxic Compounds and Negative Controls Used Within NTC

Compound	Abbreviation	CAS Number	Liver Phenotype
Cyclosporin A	CSA	59865-13-3	Cholestasis
Chlorpromazine	CPZ	69-09-0	Cholestasis
Ethinyl estradiol	EE2	57-63-6	Cholestasis
Paraquat	PQ	1910-42-5	Necrosis
Isoniazid	INAH	54-85-3	Necrosis, nonzonal
Acetaminophen	APAP	103-90-2	Necrosis, zone 3
Amiodarone	AM	19774-82-4	Steatosis, microvesicular
Valproic acid	VPA	1069-66-5	Steatosis, microvesicular
Tetracycline	TET	64-75-5	Steatosis, microvesicular, zone 3
Adefovir	ADV	113852-37-2	Non-hepatotoxic
Lithium carbonate	Li_2CO_3	554-13-2	Non-hepatotoxic
D-mannitol	D-mann	69-65-8	Non-hepatotoxic

CAS, Chemical Abstracts Service.

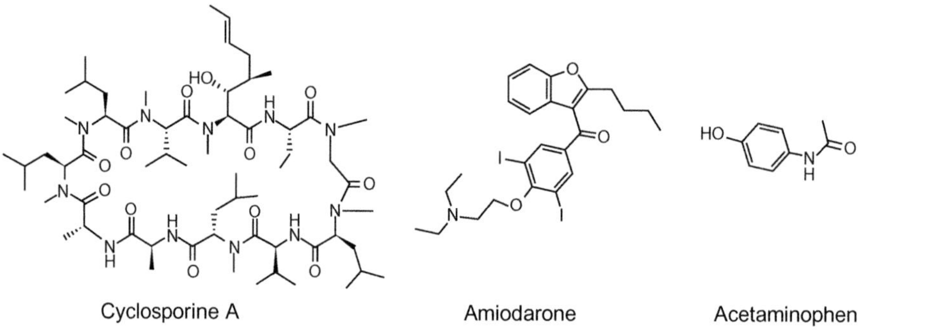

Cyclosporine A

Amiodarone

Acetaminophen

FIGURE 5.1.1

Chemical Structures of Cyclosporin A, Amiodarone, and Acetaminophen.

and private domain. The largest public transcriptomics repository is currently at the Chemical Effects in Biological Systems (CEBS) database from the US National Institute of Environmental Health Sciences (NIEHS), which includes data from Iconix's DrugMatrix database and NTP studies [17]. Recently, the data from the Toxicogenomics Project in Japan have also become available (OPEN TG-GATEs). Unfortunately, a myriad of study designs, array platforms, and analysis methods has been used, which hampers integration of these data. For in vitro liver models, the toxicogenomics data are far more scarce, fragmented over various models and species, and rarely validated against in vivo data [31–34].

The number of methods for mining and analyzing the high-content data from these transcriptomics studies for mechanistic and prediction purposes is continuously growing. Tools like connectivity and molecular concept mapping are applied for association and grouping, whereas supervised machine learning techniques are used for class prediction [35–38].

5.1.3 In vitro liver models used within NTC

Over recent decades, many in vitro liver models have been developed, ranging from cell lines like HepG2 and HepaRG to cultures of primary hepatocytes in 2D or 3D, co-cultures of hepatocytes with other cell types, and precision-cut liver slices [39,40]. Stem cell technologies to generate hepatocyte-like cells are emerging [41], and promising applications in toxicology have been shown [42]. One of the latest new models for hepatotoxicity screening is zebrafish embryos [43]. Within NTC, we focus on a selection of these well-established in vitro models (Table 5.1.2), which will be further discussed below.

Immortalized cell lines

Immortalized cell lines are cells that continue to grow and divide indefinitely in vitro under optimal culture conditions [44]. HepaRG and HepG2 cell lines are the most frequently used for toxicity studies among the currently available human hepatic cell lines. However, these cell lines are usually derived from tumors and have adapted to growth in culture: they lack liver tissue architecture, and cell–cell interactions and liver-specific functions tend to vanish as culture time increases [45,46]. They often acquire a molecular phenotype quite different from liver cells in vivo. Their main limitation is a relatively low expression of drug-metabolizing enzymes, although HepaRG clearly outperforms HepG2 in this respect.

HepG2 are very popular in high-throughput (HTP) platforms, such as for high-content screening (HCS) of cytotoxicity and other molecular endpoints [47–52], and are the core of the US Environmental Protection Agency (EPA)'s ToxCast™ and Tox21™ programs [52–54]. HepG2 cells are well characterized and frequently used in toxicological and pharmacological studies [55–57]. A 2012 toxicogenomics study has demonstrated that accurate genotoxicity assessment can be obtained using HepG2 cells; also, chemicals that require metabolic activation can be detected by means of several assays [58]. One of the earliest results from NTC, based on a mechanistic proteome analysis, showed that with HepG2 cells modes of action of the cholestatic compound cyclosporin A can be distinguished from the other hepatotoxic compounds amiodarone and acetaminophen. Therefore, the HepG2 in vitro cell system probably has distinctive characteristics that allow detection of cholestasis at an early stage of drug discovery [59].

Table 5.1.2 In Vitro Models for Hepatotoxicity Studies Used within NTC

Model	Complexity	Liver-Specific Functions (Biotransformation)	Potential Exposure Period
Immortalized cell lines (2D)	Some cell–cell*	Poor to good	Indefinite
Immortalized cell lines (3D)	Cell–cell	Good	Days to weeks
Primary hepatocyte sandwich culture	Cell–cell	Fair to excellent**	Days to weeks
Liver slices	Cell–cell (all cell types)	Good to excellent	One to a few days
Zebrafish embryo	Cell–cell (all cell types)	Good to excellent	One to a few days

Varies with cell line.
**Varies with culture conditions.*

In 2004, another human-liver-carcinoma-derived cell line was introduced, namely HepaRG. These cells, when cultured until quiescence and with specific medium conditions, possess features corresponding to well-differentiated hepatocytes [60]. In particular, they demonstrate expression of various cytochrome P450 (CYP) enzyme activities at levels that are more comparable to primary human hepatocytes (PHH) [60–64]. HepaRG cultures contain a mixture of hepatocyte-like and biliary-like epithelial cells. The fraction of hepatocyte-like cells in HepaRG cultures varies from 45 to 90% between batches and passages [65].

In a few studies, full-genome basal expression profiles generated from HepG2 and HepaRG cells have been compared with those from PHH [25,26,66]. These show that the basal gene-expression profiles of HepaRG and HepG2 are distinct from the expression profile for PHH, although the HepaRG profile seems slightly more PHH-like.

In light of the comparability and applicability of these cell types for toxicogenomics studies, compound-induced gene-expression profiles are at least as important as basal gene-expression profiles. A 2010 toxicogenomics study demonstrated that HepaRG is a more suited in vitro liver model for biological interpretation of the effects of exposure to chemicals, whereas HepG2 is more promising for classification studies using the toxicogenomics approach [26].

3D cell models

In vivo, hepatocytes are highly polarized with distinct basal–lateral sinusoids and apical canalicular domains [67] that are essential for proper functioning of the liver. This highly polarized morphology is lost when cells are cultured under non-physiological conditions [68].

Hepatocytes cultured in a 3D environment using bioreactors [69,70], hanging drop methods [71], collagen sandwich cultures [72], micro-space cultures [73], collagen and matrigel cultures [74,75], and other synthetic biomaterials [76,77] have been shown to reacquire tissue-specific properties and possess many hallmarks of in vivo epithelial cells.

As discussed in earlier sections, there are several advantages in using immortalized cell lines for toxicity assessment, with a common first choice being HepG2 cells [78]. One of the concerns with HepG2 cells and other hepatocyte cell lines is the lack of metabolic competence compared with primary hepatocytes [78–80]. This may be partly due to their hepatocarcinoma origin and oncogenic transformation, but also because they have been passaged extensively, resulting in drift from the original hepatocyte genetic profile. Furthermore, the absence of an in vivo-like environment in a tissue culture dish results in a dedifferentiated phenotype with inevitable functional changes.

HepG2 cells show a spheroid morphology when cultured in micro-space cultures [73], bioreactors [69], and on extra-cellular matrix (ECM) protein gels (Figure 5.1.2), with distinctive characteristics of polarized epithelial hepatocytes. The gene expression of metabolic enzymes is also higher in HepG2 spheroids cultured on micro-space cultures [73] or ECM gels [18], in contrast to low levels with conventional 2D cultures. Drug metabolic enzymes consist of phase 1 enzymes, that modify or introduce a functional group to the drug to make it a polar metabolite [80], and phase 2 enzymes whose activity involves conjugation steps that make metabolites more water soluble for excretion via drug transporters [80]. The low levels of metabolic enzymes observed in cells grown on tissue culture plastic will inevitably undermine the quality of safety assessment studies. mRNA expression levels of CYP450 enzymes like CYP1A2 (Figure 5.1.3) and various other CYP450 enzymes (data not shown) are strongly up-regulated in HepG2 cells cultured under 3D conditions. Increased expression of CYP450

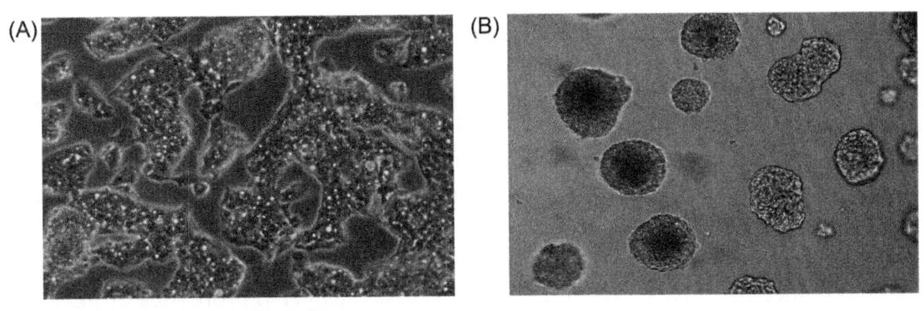

FIGURE 5.1.2 Morphology of HepG2 Cells.

Cells cultured (A) in a 2D tissue culture plate and (B) on a 3D ECM gel.

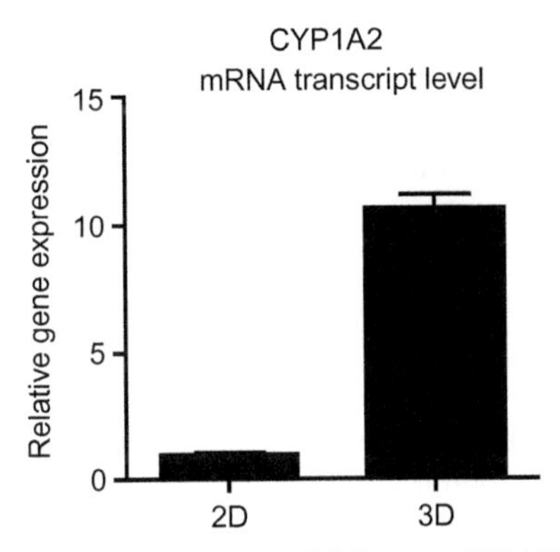

FIGURE 5.1.3 mRNA Expression Levels of Cytochrome P450 Enzyme CYP1A2 in 2D and 3D Cultured HepG2 Cells.

Fold change compared to 2D gene expression is shown. Data normalized to glyceraldehyde 3-phosphate dehydrogenase (GAPDH).

—Unpublished results [Sreenivasa and Price, Leiden University).

enzymes might offer a great improvement to the safety assessment studies and studies of drug–drug interactions where the activation of a xenobiotic response by one (not necessarily toxic) compound may increase the metabolism of a second compound into toxic intermediates [81]. The absence or impairment of CYP450 enzymes may account for the failure to identify some hepatotoxic compounds in vitro.

Some functional activities of polarized hepatocytes were also recapitulated in 3D cell culture models. The formation of bile canaliculi is an important feature that was shown in sandwich cultures of primary rat hepatocytes [75] and HepG2 cells cultured on peptide hydrogels [77]. The restoration of excretory function is likely due to the improved morphological differentiation of the hepatocytes—in particular, the establishment of apical–basal polarity—but also the restored expression of transporters and other components of the excretory machinery. Many drugs disrupt excretion pathways; for example, rifampicin inhibits activity of OATP1B3, a transporter essential for bile acid flow [82]. Three-dimensional culture models may therefore allow the evaluation of the effects of new chemical entities on transporter function at the in vitro screening stage.

HepG2 cells were shown to be reasonably effective in classifying genotoxic and non-genotoxic compounds using a toxicogenomics approach [26]. High-content screening of HepG2 cells can also identify hepatotoxic compounds, detecting mitochondrial membrane potential, intracellular calcium levels, nuclear area, and plasma membrane permeability with an overall sensitivity of 93% (percentage of toxic drugs testing positive) and specificity of 98% (percentage of nontoxic drugs testing negative) [48]. Despite this, in vitro toxicity assays have yet to have a significant impact on efforts to replace animal tests. The application of toxicogenomics and high-content screening to 3D spheroid models may increase the predictive power of in vitro toxicity screening assays and provide sufficient functional data to reduce the reliance on animal models. The near-vivo properties of 3D cultures, their ease of use, low cost, and availability suggest that these models offer great promise and are likely to play a significant part in animal-free toxicity testing in the future.

Primary mammalian hepatocytes

Primary cells such as hepatocytes can be isolated directly from healthy animals (e.g. biopsy material or whole organs) and maintained in culture in vitro. Primary mammalian hepatocytes largely retain their liver-specific functions when freshly derived from the animal, when compared with immortalized cell lines. Long-term cultures of functional hepatocytes with stable hepatocyte characteristics, however, are difficult to establish. To increase the longevity and maintain differentiated functions of hepatocytes in primary cultures, cells are cultured in a sandwich configuration of collagen–collagen and in serum-free culture medium [83]. Hepatocytes cultured in a sandwich configuration reorganize to form an architecture similar to that found in the liver and are able to form functional bile canalicular networks and gap junctions. In such a sandwich configuration, hepatocytes can be cultured for a longer time period compared with cultures on single layers of collagen [84–87].

The use of primary human hepatocytes is preferential in order to predict in vivo toxicity in humans but is hampered by the limited availability of donor material and the large variability between the donors [88]. Therefore, primary hepatocytes isolated from other mammals are used as an alternative. Human hepatocytes also tend to show a decrease in most CYP450 enzymes, although some level of CYP450 gene expression can be restored after a few days [31,89,90].

In rat hepatocytes, the rapid decline in liver-specific functions, in particular CYP450 enzyme activity, limits their use in studies testing chemicals for which metabolism depends on the CYP450 enzymes [84,91]. Improving their metabolic competence appeared possible by adding a mixture of CYP450 inducers, which resulted in the expression of phase I and the majority of phase II genes more closely resembling liver in vivo [85].

A mouse in vitro hepatocyte system might be an alternative to the rat and human systems, especially if the metabolic competence is preserved. Indeed, a genomics study indicated that the sandwich-cultured primary mouse hepatocyte system is robust and seems to maintain its metabolic competence better than does the rat hepatocyte system [92]. Another important advantage of using mouse instead of rat hepatocytes is that the murine genome is better characterized, its complete sequence is known, and transgenic mouse models are widely available that are usable for performing mechanistic investigations of liver toxicity [93,94]. Gene-expression profiling in mice primary hepatocytes has shown that the system is promising for discriminating various classes of carcinogenic compounds, and that the discrimination improves with increasing treatment period [93,94].

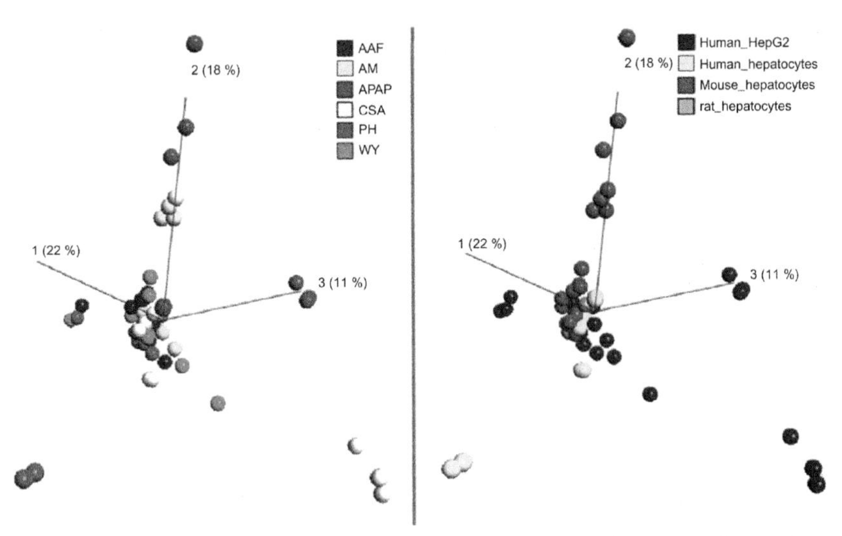

FIGURE 5.1.4 PCA of Toxicogenomics Responses for Several Compounds Following a 24-h Treatment in HepG2 cells, and Primary Hepatocytes from Mice, Rats, and Humans.

Left panel is colored for compound, and right panel for model. The principal component analysis (PCA) is based on 1161 genes with a variance >0.2.

—Data from [58,95] and TG-GATES.

A comparison of toxicogenomics responses for several compounds following a 24-h treatment in HepG2 cells, and in primary hepatocytes from mice, rats, and humans, shows that responses are strongly driven by both the in vitro model used and the test compound (Figure 5.1.4). The biological repeats for a compound within a model are always close together, but most compounds induce profoundly different effects between the models, as can be seen by the wide scattering of the data points. Not one compound induced similar responses in all in vitro models. Based on these results, if the human primary were considered the "gold standard," not one of the other in vitro systems would be a good mimic for that.

Precision-cut liver slices

The precision-cut liver slices (PCLS) technique originates from the organ culture method developed in early twentieth century by Warburg and Krebs, with the aim of studying organ functions in vitro. Their experiments were renewed by Krumdieck in 1980, who introduced equipment for generation of precision-cut tissue slices (the Krumdieck tissue slicer). Krumdieck's efforts overcame a problem of reproducibility and gave a solid base for further development of this technique [96]. Although, in principle, the PCLS technique can be applied to diverse tissues, the most common organ used for this purpose is the liver [97]. Depending on the purpose of the study and properties of the PCLS, the PCLS culture can be maintained up to 5 days [98].

Besides parenchymal cells (hepatocytes), PCLS contain all the other cell types characteristic of the liver, such as Kupffer cells, endothelial cells, stellate cells, immune cells, and stem cells. Additionally, PCLS maintain liver native architecture where original cell–cell interactions are intact and the three-dimensional structure is preserved, allowing the study of liver functions in its unique configuration, which is absent in, for example, cultures of hepatocytes [99].

Despite many similarities between liver slices and the liver, it has to be noted that gene expression alters substantially from its original state in cultured PCLS. During culturing of PCLS, genes involved in inflammation are up-regulated and genes involved in drug and lipid metabolism are significantly down-regulated. The differences in gene expression between the in vitro and in vivo situations can be explained by activation of several adaptation and stress responses in PCLS that need to adjust their new environmental conditions [100]. Regardless of these changes, PCLS preserve sufficient biochemical and molecular characteristics to allow study of diverse liver functions. Consistent with this notion, PCLS express both phase I (hydroxylation) and phase II (sulfation or glucuronidation) genes and other enzymes that are necessary for biotransformation of xenobiotics and drugs and maintenance of other liver functions [101]. Therefore, PCLS are often used as a model to study biotransformation of diverse compounds such as steroids, drugs and other xenobiotics, or bile acids [102].

The emerging toxicogenomics field (the use of gene-expression profiling in toxicology) represents an attractive approach to predict toxicity and to understand the mechanistic mode of action of the compounds under study [103]. It was reported that studies employing toxicogenomics in PCLS can correctly predict the toxicity and pathology observed in vivo [104]. PCLS were also used to study the carcinogenic potency of polycyclic aromatic hydrocarbons (PAHs). Those studies showed that at a specific pathway level, namely oxidative stress response, PAHs with high and low carcinogenic potency could be discriminated [105], and suggest that PAH mixtures have a lower carcinogenic potency than an estimation based on the additivity of the individual compounds [106].

PCLS are mostly prepared from rat livers; however, the technique can easily be applied to different species, such as the cow, mouse, monkey, pig, dog, and, importantly, human. This allows interspecies comparison and extrapolation to the human situation [97]. PCLS obtained from one animal can be used in many exposure experiments and, therefore, contribute to at least partial replacement, refinement, and reduction (the "3Rs") of animal experiments [107].

A possible disadvantage of the use of PCLS in toxicology and for high-throughput purposes is the laborious preparation and culture procedure. Also, the limited culture time can be too short to study chronic effects of drug exposure. In addition, poor penetration of compounds into the inner cell layers of slices and inter-assay variability due to different preservation of cells in different slices and a very short life span have been reported [108,109].

Zebrafish embryos

The zebrafish (*Danio rerio*) is a diploid minnow of the teleost family Cyprinidae and was introduced in the early 1990s to study development and neurobiology. From there, the zebrafish emerged as a powerful model to study human disease and the effects of chemical exposure [110–114].

Several features make the zebrafish assay attractive, among which are the potential to use it in a high-throughput screening setting and the small size of both the adult zebrafish and the zebrafish embryo. Furthermore, maintenance in large stocks is easy owing to their high fecundity. In the embryo, direct observation and experimental manipulation of tissue and organs is relatively simple, owing to the transparency and the rapid ex utero development. As a result of this rapid development, most organs become fully functional between 3 and 5 days post-fertilization [110]. In addition, during the earliest developmental stages (until independent free-feeding), the zebrafish embryo is not a protected animal under recent European legislation (Directive 2010/63/EU), and can therefore be considered as an alternative to animal experimentation [115].

A fully functional liver is present in the zebrafish embryo by 72 h post-fertilization, at which time it is perfused with blood. Regarding biotransformation, the zebrafish embryo expresses 94 Cyps, and most of them are human orthologs. Furthermore, the Cyp1–4 families are mainly involved in metabolizing xenobiotics, which is similar to the human situation. In addition, the presence of Cyp3a65, the ortholog of the human CYP3A4, is important because the majority of xenobiotic metabolic reactions are catalyzed through this Cyp [116–118].

A few studies have shown that zebrafish embryos are suitable to detect human hepatotoxicants. In a study by Jones et al., the whole-zebrafish-embryo model was evaluated through morphological endpoints after exposure to a set of compounds, including drugs falsely characterized by HepG2 cells. In this study, the whole-zebrafish-embryo model successfully detected hepatotoxicants with higher specificity than the HepG2 cells [119]. In another study, Amali et al. carried out histopathological, molecular, and biochemical analysis in zebrafish embryos exposed to a single dose of thioacetamide, and showed that the model is suitable to detect steatohepatitis [120]. These studies indicate the potential of the zebrafish embryo as a model for hepatotoxicity testing at the level of the phenotype, but they provide limited insights into the underlying toxicity mechanisms [121]. However, for drug development, understanding the mechanisms of toxicity is imperative, not only to compare between reference toxicants and test compounds, but also to differentiate between toxic and targeted effects, and to develop relevant markers of toxicity. Toxicogenomics enables detailed analysis of the mechanisms of cellular responses upon chemical exposure. To apply toxicogenomics in the whole zebrafish embryo, detection of hepatotoxicity-associated signals in the noise of other tissues in whole-zebrafish-embryo extracts has to be confirmed. This is needed to avoid dissection of the liver from the embryo, which would limit high- or even medium-throughput analysis. An adequate overlap in induced transcripts after exposure to a set of reference hepatotoxicants was observed with next-generation sequencing (NGS) between whole zebrafish embryo and adult zebrafish liver, including *fabp10a*, *pparγ*, and *apoA2*. This comparison confirmed that hepatotoxicity-specific expression can be detected in whole zebrafish embryos [122]. Subsequent gene-expression analysis of single-compound exposure experiments with the same set of hepatotoxicants showed confirmative overlap of biomarkers with the NGS results, and this analysis suggests that a selection of these expression markers would be sufficient to characterize hepatotoxic properties of a test compound [123].

5.1.4 Non-'omics-based in vitro approaches for hepatotoxicity screening

Prioritization of compounds based on human hepatotoxicity potential is currently a key unmet need in drug discovery and early development, as it can become a major problem for lead compounds in later stages of the drug discovery pipeline. Ideally, highly reliable predictive assays should be available that allow the screening of many compounds before preclinical development is initiated. As described earlier (Section 5.1.3), different in vitro test models can be used to identify the hazard of compounds early in development.

In vitro cytotoxicity assays have been used for decades as a tool to understand hypothesis-driven questions regarding mechanisms of toxicity. However, when used in a prospective manner, they have not been highly predictive of in vivo toxicity. Although conventional cytotoxicity assays frequently have more than 80% specificity, they have less than 25% sensitivity for detecting human hepatotoxic compounds.

High-content screening (HCS) technology in general represents a major step towards improving the drug discovery process [124]. The platform enables the simultaneous evaluation of multiple morphological, biochemical, and functional parameters at the single-cell level by combining the automation of quantitative epifluorescence microscopy and image analysis with the application of microfluorescent multiprobe technology. The implementation of HCS in the drug development process and the potential it offers in compound prioritization and toxicity prediction has been described by several authors [48,50,125].

When using human HepG2 cells, multiparametric high-content cytotoxicity screening (HCCS) showed high concordance with drug-induced human hepatotoxicity. Human toxicity potential was detected with 80% sensitivity and 90% specificity at a concentration of 30 × the maximal efficacious concentration [48]. In addition, this approach appears to identify early and late events in the hepatotoxic process and suggests the mechanism(s) involved in the toxicity of compounds. This allows the classification of new drug candidates according to their degree and type of injury [126].

Multiparametric HCCS in HepG2 cells, as implemented at Janssen Research and Development, consists of automated fluorescent microscopy and image analysis of approximately 500 cells per well following a 72-h treatment. Six cytotoxicity endpoints are determined, which are associated with nuclear morphology (Hoechst 33342 staining: nuclear count, nuclear size), plasma membrane integrity (SytoxGreen™), mitochondrial function (TMRE: hyper-/depolarization, number of affected cells, mitochondrial area), and cell proliferation (Hoechst 33342: nuclear count). In-house validation of the HCCS based on a classifier (cutoff) value of 30 μM, demonstrated 67% sensitivity (33% false negatives) and 100% specificity (no false positives) for the identification of severely hepatotoxic compounds. When moderate hepatotoxicants were included, sensitivity was 53% (47% false negatives), while specificity remained unchanged.

As mentioned earlier (Section 5.1.2), the mechanism-based selection of reference compounds is of crucial importance. Within the NTC context, 12 tool compounds with well-documented in vivo toxicity were selected according to different hepatotoxic phenotypes observed (see Table 5.1.1). This set of prototypical compounds was tested in the HCCS with HepG2 (Table 5.1.3). The ratio of the lowest cytotoxic concentration in vitro to the in vivo concentration associated with efficacy (Cmax) was assumed to provide an indication of the safety margin for the drug. In addition, drug-induced liver injury (DILI) labels were included based on the classification reported by the US Food and Drug Administration (FDA) [127].

Based on the lowest toxic concentration (LTC), the HCCS assay identified four of the nine liver toxicants. However, when taking a 30-fold safety margin cutoff, seven were correctly classified. Importantly, no false positives were generated. This is key in a screening strategy in order to avoid promising candidate drugs being falsely flagged as potentially hepatotoxic and therefore rejected for further development. As illustrated by this small-scale analysis, HCCS was not able to identify all hepatotoxic compounds. This is to be expected and is related to the complex nature of drug-induced liver injury, where metabolic and immunologic aspects are often involved. It is anticipated that the

Table 5.1.3 HCCS Data for the Prototypical Hepatotoxic and Control Compounds Described in Table 5.1.1

Compound	Liver Phenotype	Cmax (μM)	HCS LTC	HCCS LTEP	HCCS Class	Safety Margin	FDA Class
Cyclosporin A	Cholestasis	0.2	5.3	MMP	Pos	26.5	Less
Chlorpromazine	Cholestasis	0.5	7.7	NC	Pos	15.4	Less
Ethinyl estradiol	Cholestasis	0.0003	4.8	NC	Pos	16,000	–
Diclofenac	Necrosis	4.2	66.4	NC	Neg	15.8	Most
Isoniazid	Necrosis	40	3812	NC	Neg	95.3	Most
Acetaminophen	Necrosis	130	486	NS	Neg	3.7	–
Amiodarone	Steatosis	0.81	7.9	NC	Pos	9.8	Most
Valproic acid	Steatosis	540	137	MA	Neg	0.3	Most
Tetracycline	Steatosis	9.0	39.7	MMP	Neg	4.4	Less
Adefovir	NH	0.061	56	NC	Neg	918	–
Lithium carbonate	NH	–	>200	–	Neg	–	–
D-mannitol	n.c.	–	>10,000	–	Neg	–	–

Cmax, efficacious concentration at therapeutic dose; LTC, lowest toxic concentration (μM, IC_{20} values); LTEP, lowest toxic endpoint; MA, mitochondrial area; MMP, mitochondrial membrane potential; n.c., negative control; NC, nuclear count; NH, not hepatotoxic (kidney toxicant); NS, nuclear size; Safety Margin, ratio LTC/Cmax; FDA class, see Chen et al.; IC_{20}, inhibiting concentration of 20% inhibition; Pos, positive; Neg, negative; More or Less, see the paper by Chen et al..

phenotypic output from the HCCS assay in combination with other (more complex) test systems, such as the use of metabolically competent cells [128], co-cultures [129], 3D liver tissues [69], and zebrafish models [43], and with the inclusion of 'omics technologies for improved mechanistic elucidation, will contribute to a better understanding and predictivity of human drug-induced liver injury.

References

[1] Worth AP, Balls M. Alternative (non-animal) methods for chemicals testing: current status and future prospects. A report prepared by ECVAM and the ECVAM working group on chemicals. ATLA 2002;30:1–125.
[2] Chang CY, Schiano TD. Review article: drug hepatotoxicity. Aliment Pharmacol Ther 2007;25:1135–51.
[3] Russmann S, Kullak-Ublick GA, Grattagliano I. Current concepts of mechanisms in drug-induced hepatotoxicity. Curr Med Chem 2009;16:3041–53.
[4] Brown JA. An internet database for the classification and dissemination of information about hazardous chemicals and occupational diseases. Am J Ind Med 2008;51:428–35.
[5] Fitzpatrick RB. Haz-Map: information on hazardous chemicals and occupational diseases. Med Ref Serv Q 2004;23:49–56.
[6] Olson H, Betton G, Robinson D, Thomas K, Monro A, Kolaja G, et al. Concordance of the toxicity of pharmaceuticals in humans and in animals. Regul Toxicol Pharmacol 2000;32:56–65.
[7] Boelsterli UA, editor. Mechanistic toxicology: the molecular basis of how chemicals disrupt biological targets. London: Taylor and Francis; 2002.
[8] Antoine DJ, Mercer AE, Williams DP, Park BK. Mechanism-based bioanalysis and biomarkers for hepatic chemical stress. Xenobiotica 2009;39:565–77.
[9] Rohacova J, Marin ML, Martinez-Romero A, Diaz L, O'Connor JE, Gomez-Lechon MJ, et al. Fluorescent benzofurazan-cholic acid conjugates for in vitro assessment of bile acid uptake and its modulation by drugs. Chem Med Chem 2009;4:466–72.
[10] Donato MT, Gomez-Lechon MJ. Drug-induced liver steatosis and phospholipidosis: cell-based assays for early screening of drug candidates. Curr Drug Metab 2012
[11] Amacher DE. Strategies for the early detection of drug-induced hepatic steatosis in preclinical drug safety evaluation studies. Toxicology 2011;279:10–18.
[12] Adler S, Basketter D, Creton S, Pelkonen O, van Benthem J, Zuang V, et al. Alternative (non-animal) methods for cosmetics testing: current status and future prospects—2010. Arch Toxicol 2011;85:367–485.
[13] Food and Drug Administration. Guidance for industry drug-induced liver injury: premarketing clinical evaluation 2009.
[14] Ruepp S, Boess F, Suter L, de Vera MC, Steiner G, Steele T, et al. Assessment of hepatotoxic liabilities by transcript profiling. Toxicol Appl Pharmacol 2005;207:161–70.
[15] Natsoulis G, Pearson CI, Gollub J, Eynon P, Ferng B, Nair J, et al. The liver pharmacological and xenobiotic gene response repertoire. Mol Syst Biol 2008;4:175.
[16] Hirode M, Ono A, Miyagishima T, Nagao T, Ohno Y, Urushidani T. Gene expression profiling in rat liver treated with compounds inducing phospholipidosis. Toxicol Appl Pharmacol 2008;229:290–9.
[17] Waters M, Stasiewicz S, Merrick BA, Tomer K, Bushel P, Paules R, et al. CEBS—Chemical Effects in Biological Systems: a public data repository integrating study design and toxicity data with microarray and proteomics data. Nucleic Acids Res 2008;36:D892–900.
[18] McMillian M, Nie AY, Parker JB, Leone A, Bryant S, Kemmerer M, et al. A gene expression signature for oxidant stress/reactive metabolites in rat liver. Biochem Pharmacol 2004;68:2249–61.
[19] McMillian M, Nie AY, Parker JB, Leone A, Kemmerer M, Bryant S, et al. Inverse gene expression patterns for macrophage-activating hepatotoxicants and peroxisome proliferators in rat liver. Biochem Pharmacol 2004;67:2141–65.

[20] Ellinger-Ziegelbauer H, Adler M, Amberg A, Brandenburg A, Callanan JJ, Connor S, et al. The enhanced value of combining conventional and "omics" analyses in early assessment of drug-induced hepatobiliary injury. Toxicol Appl Pharmacol 2011;252:97–111.

[21] Steiner G, Suter L, Boess F, Gasser R, de Vera MC, Albertini S, et al. Discriminating different classes of toxicants by transcript profiling. Environ Health Perspect 2004;112:1236–48.

[22] van Ravenzwaay B, Cunha GC, Leibold E, Looser R, Mellert W, Prokoudine A, et al. The use of metabolomics for the discovery of new biomarkers of effect. Toxicol Lett 2007;172:21–8.

[23] Boitier E, Amberg A, Barbie V, Blichenberg A, Brandenburg A, Gmuender H, et al. A comparative integrated transcript analysis and functional characterization of differential mechanisms for induction of liver hypertrophy in the rat. Toxicol Appl Pharmacol 2011;252:85–96.

[24] Suter L, Schroeder S, Meyer K, Gautier JC, Amberg A, Wendt M, et al. EU framework 6 project: predictive toxicology (PredTox)—overview and outcome. Toxicol Appl Pharmacol 2011;252:73–84.

[25] Hart SN, Li Y, Nakamoto K, Subileau EA, Steen D, Zhong XB. A comparison of whole genome gene expression profiles of HepaRG cells and HepG2 cells to primary human hepatocytes and human liver tissues. Drug Metab Dispos 2010;38:988–94.

[26] Jennen DG, Magkoufopoulou C, Ketelslegers HB, van Herwijnen MH, Kleinjans JC, van Delft JH. Comparison of HepG2 and HepaRG by whole-genome gene expression analysis for the purpose of chemical hazard identification. Toxicol Sci 2010;115:66–79.

[27] Hockley SL, Mathijs K, Staal YC, Brewer D, Giddings I, van Delft JH, et al. Interlaboratory and interplatform comparison of microarray gene expression analysis of HepG2 cells exposed to benzo(a)pyrene. OMICS 2009;13:115–25.

[28] Lambert CB, Spire C, Claude N, Guillouzo A. Dose- and time-dependent effects of phenobarbital on gene expression profiling in human hepatoma HepaRG cells. Toxicol Appl Pharmacol 2009;234:345–60.

[29] Dere E, Lee AW, Burgoon LD, Zacharewski TR. Differences in TCDD-elicited gene expression profiles in human HepG2, mouse Hepa1c1c7 and rat H4IIE hepatoma cells. BMC Genomics 2011;12:193.

[30] Horiuchi S, Ishida S, Hongo T, Ishikawa Y, Miyajima A, Sawada J, et al. Global gene expression changes including drug metabolism and disposition induced by three-dimensional culture of HepG2 cells: involvement of microtubules. Biochem Biophys Res Commun 2009;378:558–62.

[31] Kienhuis AS, van de Poll MC, Wortelboer H, van Herwijnen M, Gottschalk R, Dejong CH, et al. Parallelogram approach using rat–human in vitro and rat in vivo toxicogenomics predicts acetaminophen-induced hepatotoxicity in humans. Toxicol Sci 2009;107:544–52.

[32] Kienhuis AS, van de Poll MC, Dejong CH, Gottschalk R, van Herwijnen M, Boorsma A, et al. A toxicogenomics-based parallelogram approach to evaluate the relevance of coumarin-induced responses in primary human hepatocytes in vitro for humans in vivo. Toxicol In Vitro 2009;23:1163–9.

[33] Elferink MG, Olinga P, Draaisma AL, Merema MT, Bauerschmidt S, Polman J, et al. Microarray analysis in rat liver slices correctly predicts in vivo hepatotoxicity. Toxicol Appl Pharmacol 2008;229:300–9.

[34] Uehara T, Hirode M, Ono A, Kiyosawa N, Omura K, Shimizu T, et al. A toxicogenomics approach for early assessment of potential non-genotoxic hepatocarcinogenicity of chemicals in rats. Toxicology 2008;250:15–26.

[35] Smalley JL, Gant TW, Zhang SD. Application of connectivity mapping in predictive toxicology based on gene-expression similarity. Toxicology 2010;268:143–6.

[36] Lamb J, Crawford ED, Peck D, Modell JW, Blat IC, Wrobel MJ, et al. The connectivity map: using gene-expression signatures to connect small molecules, genes, and disease. Science 2006;313:1929–35.

[37] Fan X, Lobenhofer EK, Chen M, Shi W, Huang J, Luo J, et al. Consistency of predictive signature genes and classifiers generated using different microarray platforms. Pharmacogenomics 2010;10:247–57.

[38] Rhodes DR, Kalyana-Sundaram S, Tomlins SA, Mahavisno V, Kasper N, Varambally R, et al. Molecular concepts analysis links tumors, pathways, mechanisms, and drugs. Neoplasia 2007;9:443–54.

[39] Walsky RL, Boldt SE. In vitro cytochrome P450 inhibition and induction. Curr Drug Metab 2008;9:928–39.

[40] Guguen-Guillouzo C, Guillouzo A. General review on in vitro hepatocyte models and their applications. Methods Mol Biol 2010;640:1–40.
[41] Snykers S, De Kock J, Rogiers V, Vanhaecke T. In vitro differentiation of embryonic and adult stem cells into hepatocytes: state of the art. Stem Cells 2009;27:577–605.
[42] Yildirimman R, Brolen G, Vilardell M, Eriksson G, Synnergren J, Gmuender H, et al. Human embryonic stem cell derived hepatocyte-like cells as a tool for in vitro hazard assessment of chemical carcinogenicity. Toxicol Sci 2011;124:278–90.
[43] Hill A, Mesens N, Steemans M, Xu JJ, Aleo MD. Comparisons between in vitro whole cell imaging and in vivo zebrafish-based approaches for identifying potential human hepatotoxicants earlier in pharmaceutical development. Drug Metabolism Reviews 2012;44:127–40.
[44] Lundberg AS, Randell SH, Stewart SA, Elenbaas B, Hartwell KA, Brooks MW, et al. Immortalization and transformation of primary human airway epithelial cells by gene transfer. Oncogene 2002;21:4577–86.
[45] Jover R, Bort R, Gomez-Lechon MJ, Castell JV. Re-expression of C/EBP alpha induces CYP2B6, CYP2C9 and CYP2D6 genes in HepG2 cells. FEBS Lett 1998;431:227–30.
[46] Rodriguez-Antona C, Donato MT, Boobis A, Edwards RJ, Watts PS, Castell JV, et al. Cytochrome P450 expression in human hepatocytes and hepatoma cell lines: molecular mechanisms that determine lower expression in cultured cells. Xenobiotica 2002;32:505–20.
[47] O'Brien P, Haskins JR. In vitro cytotoxicity assessment. Methods Mol Biol 2007;356:415–25.
[48] O'Brien PJ, Irwin W, Diaz D, Howard-Cofield E, Krejsa CM, Slaughter MR, et al. High concordance of drug-induced human hepatotoxicity with in vitro cytotoxicity measured in a novel cell-based model using high content screening. Arch Toxicol 2006;80:580–604.
[49] Schoonen WG, Westerink WM, Horbach GJ. High-throughput screening for analysis of in vitro toxicity. EXS 2009;99:401–52.
[50] Abraham VC, Towne DL, Waring JF, Warrior U, Burns DJ. Application of a high-content multiparameter cytotoxicity assay to prioritize compounds based on toxicity potential in humans. J Biomol Screen 2008;13:527–37.
[51] Westerink WM, Stevenson JC, Horbach GJ, Schoonen WG. The development of RAD51C, cystatin A, p53 and Nrf2 luciferase-reporter assays in metabolically competent HepG2 cells for the assessment of mechanism-based genotoxicity and of oxidative stress in the early research phase of drug development. Mutat Res 2010;696:21–40.
[52] Judson RS, Houck KA, Kavlock RJ, Knudsen TB, Martin MT, Mortensen HM, et al. In vitro screening of environmental chemicals for targeted testing prioritization: the ToxCast Project. Environ Health Perspect 2009;11:8.
[53] Martin MT, Judson RS, Reif DM, Kavlock RJ, Dix DJ. Profiling chemicals based on chronic toxicity results from the U.S. EPA ToxRef Database. Environ Health Perspect 2009;117:392–9.
[54] Shukla SJ, Huang R, Austin CP, Xia M. The future of toxicity testing: a focus on in vitro methods using a quantitative high-throughput screening platform. Drug Discov Today 2010
[55] van Delft JH, van Agen E, van Breda SG, Herwijnen MH, Staal YC, Kleinjans JC. Discrimination of genotoxic from non-genotoxic carcinogens by gene expression profiling. Carcinogenesis 2004;25:1265–76.
[56] Olsavsky KM, Page JL, Johnson MC, Zarbl H, Strom SC, Omiecinski CJ. Gene expression profiling and differentiation assessment in primary human hepatocyte cultures, established hepatoma cell lines, and human liver tissues. Toxicol Appl Pharmacol 2007;222:42–56.
[57] Hockley SL, Arlt VM, Brewer D, Giddings I, Phillips DH. Time- and concentration-dependent changes in gene expression induced by benzo(a)pyrene in two human cell lines, MCF-7 and HepG2. BMC Genomics 2006;7:260.
[58] Magkoufopoulou C, Claessen SMH, Tsamou M, Jennen DGJ, Kleinjans JCS, van Delft JHM. A transcriptomics-based in vitro assay for predicting chemical genotoxicity in vivo. Carcinogenesis 2012;33:1421–9.

[59] van Summeren A, Renes J, Bouwman FG, Noben JP, van Delft JH, Kleinjans JC, et al. Proteomics investigations of drug-induced hepatotoxicity in HepG2 cells. Toxicol Sci 2011;120:109–22.
[60] Aninat C, Piton A, Glaise D, Le Charpentier T, Langouet S, Morel F, et al. Expression of cytochromes P450, conjugating enzymes and nuclear receptors in human hepatoma HepaRG cells. Drug Metab Dispos 2006;34:75–83.
[61] Turpeinen M, Tolonen A, Chesne C, Guillouzo A, Uusitalo J, Pelkonen O. Functional expression, inhibition and induction of CYP enzymes in HepaRG cells. Toxicol In Vitro 2009;23:748–53.
[62] Lambert CB, Spire C, Renaud MP, Claude N, Guillouzo A. Reproducible chemical-induced changes in gene expression profiles in human hepatoma HepaRG cells under various experimental conditions. Toxicol In Vitro 2009;23:466–75.
[63] Kanebratt KP, Andersson TB. HepaRG cells as an in vitro model for evaluation of cytochrome P450 induction in humans. Drug Metab Dispos 2008;36:137–45.
[64] Guillouzo A, Corlu A, Aninat C, Glaise D, Morel F, Guguen-Guillouzo C. The human hepatoma HepaRG cells: a highly differentiated model for studies of liver metabolism and toxicity of xenobiotics. Chem Biol Interact 2007;168:66–73.
[65] Schulze A, Mills K, Weiss TS, Urban S. Hepatocyte polarization is essential for the productive entry of the hepatitis B virus. Hepatology 2012;55:373–83.
[66] Josse R, Dumont J, Fautrel A, Robin MA, Guillouzo A. Identification of early target genes of aflatoxin B1 in human hepatocytes, inter-individual variability and comparison with other genotoxic compounds. Toxicol Appl Pharmacol 2012;258:176–87.
[67] Wang L, Boyer JL. The maintenance and generation of membrane polarity in hepatocytes. Hepatology 2004;39:892–9.
[68] Maurice M, Rogier E, Cassio D, Feldmann G. Formation of plasma membrane domains in rat hepatocytes and hepatoma cell lines in culture. J Cell Sci 1988;90(1):79–92.
[69] Fey SJ, Wrzesinski K. Determination of drug toxicity using 3D spheroids constructed from an immortal human hepatocyte cell line. Toxicol Sci 2012;127:403–11.
[70] Leite SB, Wilk-Zasadna I, Zaldivar Comenges JM, Airola E, Reis-Fernandes MA, Mennecozzi M, et al. 3D HepaRG model as an attractive tool for toxicity testing. Toxicol Sci 2012
[71] Tung Y-C, Hsiao AY, Allen SG, Torisawa Y-s, Ho M, Takayama S. High-throughput 3D spheroid culture and drug testing using a 384 hanging drop array. Analyst 2011;136:473–8.
[72] Dunn JC, Tompkins RG, Yarmush ML. Hepatocytes in collagen sandwich: evidence for transcriptional and translational regulation. J Cell Biol 1992;116:1043–53.
[73] Nakamura K, Mizutani R, Sanbe A, Enosawa S, Kasahara M, Nakagawa A, et al. Evaluation of drug toxicity with hepatocytes cultured in a micro-space cell culture system. J Biosci Bioeng 2011;111:78–84.
[74] Zhang F, Xu R, Zhao M-J. QSG-7701 human hepatocytes form polarized acini in three-dimensional culture. J Cell Biochem 2010;110:1175–86.
[75] Matsui H, Takeuchi S, Osada T, Fujii T, Sakai Y. Enhanced bile canaliculi formation enabling direct recovery of biliary metabolites of hepatocytes in 3D collagen gel microcavities. Lab Chip 2012;12:1857–64.
[76] Kim Y, Rajagopalan P. 3D hepatic cultures simultaneously maintain primary hepatocyte and liver sinusoidal endothelial cell phenotypes. PLoS ONE 2010;5:e15456.
[77] Malinen MM, Palokangas H, Yliperttula M, Urtti A. Peptide nanofiber hydrogel induces formation of bile canaliculi structures in three-dimensional hepatic cell culture. Tissue Eng Part A 2012
[78] Gerets HHJ, Tilmant K, Gerin B, Chanteux H, Depelchin BO, Dhalluin S, et al. Characterization of primary human hepatocytes, HepG2 cells, and HepaRG cells at the mRNA level and CYP activity in response to inducers and their predictivity for the detection of human hepatotoxins. Cell Biol Toxicol 2012
[79] Xu JJ, Diaz D, O'Brien PJ. Applications of cytotoxicity assays and pre-lethal mechanistic assays for assessment of human hepatotoxicity potential. Chem Biol Interact 2004;150:115–28.

[80] Westerink WMA, Schoonen WGEJ. Cytochrome P450 enzyme levels in HepG2 cells and cryopreserved primary human hepatocytes and their induction in HepG2 cells. Toxicol In Vitro 2007;21:1581–91.
[81] Lin JH. CYP induction-mediated drug interactions: in vitro assessment and clinical implications. Pharm Res 2006;23:1089–116.
[82] Hirano M, Maeda K, Shitara Y, Sugiyama Y. Drug–drug interaction between pitavastatin and various drugs via OATP1B1. Drug Metab Dispos 2006;34:1229–36.
[83] Tuschl G, Mueller SO. Effects of cell culture conditions on primary rat hepatocytes: cell morphology and differential gene expression. Toxicology 2006;218:205–15.
[84] Gandolfi AJ, Wijeweera J, Brendel K. Use of precision-cut liver slices as an in vitro tool for evaluating liver function. Toxicol Pathol 1996;24:58–61.
[85] Kienhuis AS, Wortelboer HM, Maas WJ, van Herwijnen M, Kleinjans JC, van Delft JH, et al. A sandwich-cultured rat hepatocyte system with increased metabolic competence evaluated by gene expression profiling. Toxicol In Vitro 2007;21:892–901.
[86] LeCluyse EL, Ahlgren-Beckendorf JA, Carroll K, Parkinson A, Johnson J. Regulation of glutathione S-transferase enzymes in primary cultures of rat hepatocytes maintained under various matrix configurations. Toxicol In Vitro 2000;14:101–15.
[87] LeCluyse EL, Fix JA, Audus KL, Hochman JH. Regeneration and maintenance of bile canalicular networks in collagen-sandwiched hepatocytes. Toxicol In Vitro 2000;14:117–32.
[88] Schaeffner I, Petters J, Aurich H, Frohberg P, Christ B. A microtiterplate-based screening assay to assess diverse effects on cytochrome P450 enzyme activities in primary rat hepatocytes by various compounds. Assay Drug Dev Technol 2005;3:27–38.
[89] LeCluyse EL. Human hepatocyte culture systems for the in vitro evaluation of cytochrome P450 expression and regulation. Eur J Pharm Sci 2001;13:343–68.
[90] Morel F, Beaune PH, Ratanasavanh D, Flinois JP, Yang CS, Guengerich FP, et al. Expression of cytochrome P-450 enzymes in cultured human hepatocytes. Eur J Biochem 1990;191:437–44.
[91] Hoen PA, Commandeur JN, Vermeulen NP, van Berkel TJ, Bijsterbosch MK. Selective induction of cytochrome P450 3A1 by dexamethasone in cultured rat hepatocytes: analysis with a novel reverse transcriptase-polymerase chain reaction assay section sign. Biochem Pharmacol 2000;60:1509–18.
[92] Mathijs K, Kienhuis AS, Brauers KJ, Jennen DG, Lahoz A, Kleinjans JC, et al. Assessing the metabolic competence of sandwich-cultured mouse primary hepatocytes. Drug Metab Dispos 2009;37:1305–11.
[93] Mathijs K, Brauers KJ, Jennen DG, Boorsma A, van Herwijnen MH, Gottschalk RW, et al. Discrimination for genotoxic and nongenotoxic carcinogens by gene expression profiling in primary mouse hepatocytes improves with exposure time. Toxicol Sci 2009;112:374–84.
[94] Mathijs K, Brauers KJ, Jennen DG, Lizarraga D, Kleinjans JC, van Delft JH. Gene expression profiling in primary mouse hepatocytes discriminates true from false-positive genotoxic compounds. Mutagenesis 2010;25:561–8.
[95] Mathijs K, Brauers KJJ, Jennen DGJ, Lizarraga D, Kleinjans JCS, van Delft JHM. Gene expression profiling in primary mouse hepatocytes discriminates true from false-positive genotoxic compounds. Mutagenesis 2010;25:561–8.
[96] Parrish AR, Gandolfi AJ, Brendel K. Precision-cut tissue slices: applications in pharmacology and toxicology. Life Sci 1995;57:1887–901.
[97] de Graaf I, Olinga P, de Jager MH, Merema MT, de KR, van de Kerkhof EG, et al. Preparation and incubation of precision-cut liver and intestinal slices for application in drug metabolism and toxicity studies. Nat Protoc 2010;5:1540–51.
[98] Vickers AE, Fisher RL. Organ slices for the evaluation of human drug toxicity. Chem Biol Interact 2004;150:87–96.
[99] de Kanter R, Monshouwer M, Meijer DK, Groothuis GM. Precision-cut organ slices as a tool to study toxicity and metabolism of xenobiotics with special reference to non-hepatic tissues. Curr Drug Metab 2002;3:39–59.

[100] Boess F, Kamber M, Romer S, Gasser R, Muller D, Albertini S, et al. Gene expression in two hepatic cell lines, cultured primary hepatocytes, and liver slices compared to the in vivo liver gene expression in rats: possible implications for toxicogenomics use of in vitro systems. Toxicol Sci 2003;73:386–402.

[101] De Kanter R, De Jager MH, Draaisma AL, Jurva JU, Olinga P, Meijer DK, et al. Drug-metabolizing activity of human and rat liver, lung, kidney and intestine slices. Xenobiotica 2002;32:349–62.

[102] van Midwoud PM, Groothuis GM, Merema MT, Verpoorte E. Microfluidic biochip for the perifusion of precision-cut rat liver slices for metabolism and toxicology studies. Biotechnol Bioeng 2010;105:184–94.

[103] Blomme EA, Yang Y, Waring JF. Use of toxicogenomics to understand mechanisms of drug-induced hepatotoxicity during drug discovery and development. Toxicol Lett 2009;186:22–31.

[104] Elferink MG, Olinga P, van Leeuwen EM, Bauerschmidt S, Polman J, Schoonen WG, et al. Gene expression analysis of precision-cut human liver slices indicates stable expression of ADME-Tox related genes. Toxicol Appl Pharmacol 2011;253:57–69.

[105] Staal YCM, van Herwijnen MHM, Pushparajah DS, Umachandran M, Ioannides C, van Schooten FJ, et al. Modulation of gene expression and DNA-adduct formation in precision-cut liver slices exposed to polycyclic aromatic hydrocarbons of different carcinogenic potency. Mutagenesis 2007;22:55–62.

[106] Staal YCM, Pushparajah DS, van Herwijnen MHM, Gottschalk RWH, Maas LM, Ioannides C, et al. Interactions between polycyclic aromatic hydrocarbons in binary mixtures: effects on gene expression and DNA adduct formation in precision-cut rat liver slices. Mutagenesis 2008;23:491–9.

[107] Balls M. Replacement of animal procedures: alternatives in research, education and testing. Lab Anim 1994;28:193–211.

[108] de Graaf IA, de Kanter R, de Jager MH, Camacho R, Langenkamp E, van de Kerkhof EG, et al. Empirical validation of a rat in vitro organ slice model as a tool for in vivo clearance prediction. Drug Metab Dispos 2006;34:591–9.

[109] Guillouzo A. Liver cell models in in vitro toxicology. Environ Health Perspect 1998;106(Suppl. 2):511–32.

[110] Barros TP, Alderton WK, Reynolds HM, Roach AG, Berghmans S. Zebrafish: an emerging technology for in vivo pharmacological assessment to identify potential safety liabilities in early drug discovery. Br J Pharmacol 2008;154:1400–13.

[111] Crawford AD, Esguerra CV, de Witte PA. Fishing for drugs from nature: zebrafish as a technology platform for natural product discovery. Planta Med 2008;74:624–32.

[112] Streisinger G, Walker C, Dower N, Knauber D, Singer F. Production of clones of homozygous diploid zebra fish (*Brachydanio rerio*). Nature 1981;291:293–6.

[113] Yang L, Kemadjou JR, Zinsmeister C, Bauer M, Legradi J, Muller F, et al. Transcriptional profiling reveals barcode-like toxicogenomic responses in the zebrafish embryo. Genome Biol 2007;8:R227.

[114] Kimmel CB, Ballard WW, Kimmel SR, Ullmann B, Schilling TF. Stages of embryonic development of the zebrafish. Dev Dyn 1995;203:253–310.

[115] Sukardi H, Chng HT, Chan EC, Gong Z, Lam SH. Zebrafish for drug toxicity screening: bridging the in vitro cell-based models and in vivo mammalian models. Expert Opin Drug Metab Toxicol 2011;7:579–89.

[116] Hill A, Mesens N, Steemans M, Xu JJ, Aleo MD. Comparisons between in vitro whole cell imaging and in vivo zebrafish-based approaches for identifying potential human hepatotoxicants earlier in pharmaceutical development. Drug Metab Rev 2012;44:127–40.

[117] Goldstone JV, McArthur AG, Kubota A, Zanette J, Parente T, Jonsson ME, et al. Identification and developmental expression of the full complement of cytochrome P450 genes in zebrafish. BMC Genomics 2010;11:643.

[118] McGrath P, Li CQ. Zebrafish: a predictive model for assessing drug-induced toxicity. Drug Discov Today 2008;13:394–401.

[119] Jones M, Ball JS, Dodd A, Hill AJ. Comparison between zebrafish and Hep G2 assays for the predictive identification of hepatotoxins. Toxicology 2009;262:13–14.

[120] Amali AA, Rekha RD, Lin CJ, Wang WL, Gong HY, Her GM, et al. Thioacetamide-induced liver damage in zebrafish embryo as a disease model for steatohepatitis. J Biomed Sci 2006;13:225–32.
[121] Sawle AD, Wit E, Whale G, Cossins AR. An information-rich alternative chemicals testing strategy using a high definition toxicogenomics and zebrafish (*Danio rerio*) embryos. Toxicol Sci 2010;118:128–39.
[122] Driessen M, Kienhuis AS, Pennings JL, Pronk TE, van der Brandhof E, Roodbergen M, et al. Exploring the zebrafish embryo as an alternative model for the evaluation of liver toxicity by histopathology and expression profiling. Arch Toxicol 2013;87:807–23.
[123] Driessen M, Kienhuis AS, Pennings JL, Pronk TE, van de Water B, van der Ven LTM. (In preparation). Hepatoxicity-specific regulation of gene expression in zebrafish embryos.
[124] Abraham VC, Taylor DL, Haskins JR. High content screening applied to large-scale cell biology. Trends Biotechnol 2004;22:15–22.
[125] Xu JJ, Henstock PV, Dunn MC, Smith AR, Chabot JR, de Graaf D. Cellular imaging predictions of clinical drug-induced liver injury. Toxicol Sci 2008;105:97–105.
[126] Tolosa L, Pinto S, Donato MT, Lahoz A, Castell JV, O'Connor JE, et al. Development of a multiparametric cell-based protocol to screen and classify the hepatotoxicity potential of drugs. Toxicol Sci 2012;127:187–98.
[127] Chen M, Vijay V, Shi Q, Liu Z, Fang H, Tong W. FDA-approved drug labeling for the study of drug-induced liver injury. Drug Discov Today 2011;16:697–703.
[128] Antherieu S, Chesne C, Li R, Camus S, Lahoz A, Picazo L, et al. Stable expression, activity, and inducibility of cytochromes P450 in differentiated HepaRG cells. Drug Metab Dispos 2010;38:516–25.
[129] Khetani SR, Kanchagar C, Ukairo O, Krzyzewski S, Moore A, Shi J, et al. The use of micropatterned co-cultures to detect compounds that cause drug-induced liver injury in humans. Toxicol Sci Nov 2012;14 [Epub ahead of print].

CHAPTER

An Overview of Toxicogenomics Approaches to Mechanistically Understand and Predict Kidney Toxicity

5.2

Giulia Benedetti, Bob van de Water and Marjo de Graauw
Division of Toxicology, LACDR, Leiden University, Leiden, The Netherlands

5.2.1 Brief introduction to toxicant-induced renal injury

Renal morphology and physiology

Kidneys are responsible for preserving the body's internal environment. They maintain the total body salt, water, potassium, and acid–base balance, while excreting toxins and other waste products. There are three mechanisms by which the kidneys accomplish homeostasis in the internal environment, namely glomerular filtration, tubular reabsorption, and tubular excretion. Each human kidney consists of about a million nephrons and one single nephron consists of several subunits, including the glomerulus, proximal tubule, distal tubule, and the collecting duct, which all have their own specific transport properties. The renal tubule and collecting duct are composed of a single layer of cells surrounding a tubular lumen. The cell structure and function varies considerably from one segment to another, but each contributes to the transport function of the kidney. As a result of this transport function, cells within the different segments of the nephron may be extensively exposed to various toxins.

Pathophysiology of acute renal failure

Acute renal failure (ARF) refers to loss of kidney function occurring over a short period of time (up to days) and is clinically characterized by an abrupt and sustained drop in glomerular filtration rate (GFR). High incidence of hospitalization and mortality makes it a frequent and significant clinical problem, particularly in the intensive care unit (ICU). It has been reported that up to 25% of hospitalized patients suffer from ARF and the mortality rate of these is 15–60% even after dialysis [1]. The regenerative capacity of the kidney is well documented and acute renal failure may be reversible with complete recovery of renal function. However, half of patients do not respond to current therapies [2,3] and develop chronic renal failure and/or end-stage renal disease, a prognosis that has not changed over recent years [4].

Ischemia and toxicant-induced tubular necrosis accounts for more than 80% of cases of ARF [5]. The proximal tubule is considered the major target with both insults, due to its high level of reabsorption capacity and non-homogeneous renal parenchymal oxygenation. Renal proximal tubular epithelial (RPTE) cell injury is characterized by mitochondrial dysfunction, adenosine triphosphate (ATP)

Toxicogenomics-Based Cellular Models.

depletion, activation of stress signaling pathways, impaired solute and ion transport, loss of brush border morphology, loss of cell polarity, and cytoskeletal disruption [6,7]. Proximal tubular cells may lose their interaction with the basement membrane, leaving a denuded proximal tubule and causing cast formation in the tubular lumen (Figure 5.2.1). Denudation of the basement membrane causes an increased back-leak of glomerular filtrate [8,9]. Together with tubular obstruction caused by detached cells, this will lead to impaired renal function. The finding that up to 100% of exfoliated tubular cells found in the urine of ARF patients were viable shows that exfoliation does not necessarily result in cell death [10]. However, when injury is too severe, irreversible cell injury occurs, which does result in cell death. Two different types of cell death have been distinguished in ARF—necrosis and apoptosis [11,12]—but accumulating evidence indicates that pathways generally associated with apoptosis are important in renal tubular injury [13]. During toxicant exposure, the initiation of apoptosis may occur at exposure levels that are less severe than those needed to induce necrosis. RPTE cells that do not die or detach from the basement membrane are thought to contribute to renal regeneration [14]. These surviving cells migrate to the denuded areas, proliferate, re-polarize and/or dedifferentiate, and restore nephron structure and function [14,15].

Nephrotoxic acute renal failure

Chemicals with very diverse structures and sources, including drugs, natural products, industrial chemicals, and environmental pollutants, account for a large number of cases of ARF [16]. Aminoglycoside antibiotics, radiographic contrast media, anticancer drugs, conventional nonselective nonsteroidal anti-inflammatory drugs (NSAIDs), and angiotensin-converting enzyme (ACE) inhibitors are most frequently implicated in drug-induced nephrotoxicity [17]. Nephrotoxicants act at different nephron segments and thereby cause various alterations of renal function [16,18]. Thus, the mechanisms causing ARF may vary with different toxicants, and are generally categorized based on the target sites that are injured [5,17] (see Table 5.2.1).

The disruption of hemodynamics and direct tubular epithelial cell damage are the most common mechanisms of toxicant-induced ARF. Blood flow to the glomerulus is tightly regulated by vasoconstriction and vasodilation of the afferent and efferent blood vessels, to maintain glomerular pressures. Several chemicals can disrupt this delicate homeostasis and result in compromised renal perfusion, such as NSAIDS [19], ACE inhibitors [20], radiocontrast agents [21], and cyclosporin [22]. In addition to their ischemic effect, many nephrotoxicants damage tubules directly, such as amphotericin B [23], platinum-containing compounds [24], and radiocontrast agents [21]. In general, nephrotoxicants mediate cell injury through either covalent or non-covalent binding to critical macromolecules that results in altered activity of these molecules, or through a direct increase in cellular reactive oxygen species (ROS) that results in oxidative damage to critical macromolecules [16].

Role of inflammation in nephrotoxicity

Inflammation plays a major role in the pathophysiology of ARF. Several nephrotoxicants have been shown to induce an inflammatory response, and attenuation of the inflammation demonstrated renal-protective effects [25–29]. It is hypothesized that during nephrotoxicity, the initial insult results in changes in vascular endothelial cells and/or in tubular epithelial cells, leading to the generation of

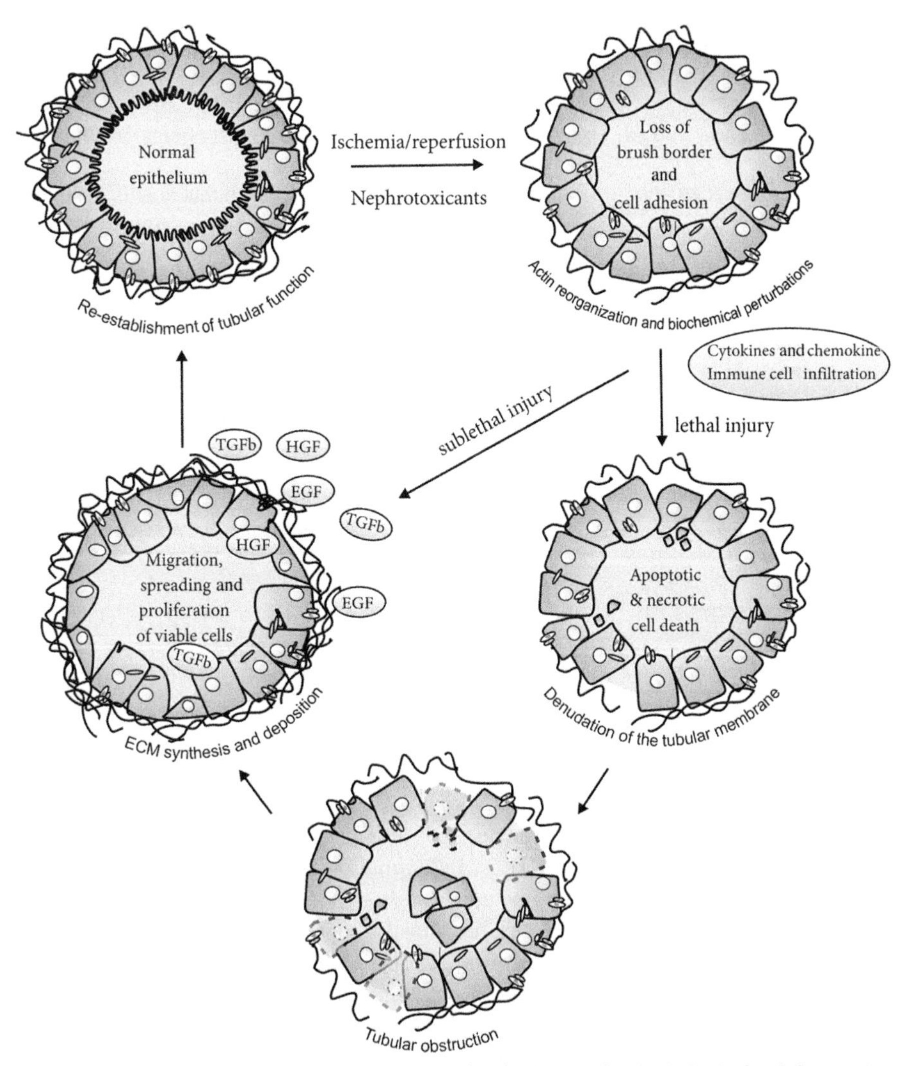

FIGURE 5.2.1 Diagram of morphological changes occurring in the proximal tubule during injury and regeneration.

Proximal tubular cells exposed to ischemia/reperfusion or nephrotoxicants undergo biochemical perturbations as well as disruption of the cytoskeletal network, resulting in loss of brush border morphology and loss of cell polarity. Proximal tubular cells detach from the basement membrane, leaving a denuded proximal tubule. Irreversible cell injury results in the onset of cell death via necrosis or apoptosis, and the subsequent further denudation of the basement membrane causes tubular obstruction and back-leak glomerular filtrate. Inflammation via secretion of cytokines and chemokines and subsequent infiltration of immune cells leads to further damage of the kidneys. During the regeneration phase, cells migrate, spread, and proliferate to fill the denuded areas and regain normal kidney function. When severely damaged, such as after triggering inflammation, the kidney will not be able to regenerate and chronic renal injury results.

Table 5.2.1 Mechanisms and Associated Toxicants that Cause Acute Renal Failure

Mechanisms	Toxicants
Reduction in renal blood flow through alterations of intrarenal hemodynamics	NSAIDs, ACE inhibitors, radiocontrast agents, amphotericin B, cyclosporin, FK506
Direct tubular epithelial cell damage	Aminoglycosides, cisplatin, radiocontrast agents, amphotericin B, pentamidine, cyclosporin, FK506, foscarnet, heavy metals, organic solvents, DCVC, adefovir, cidofovir, tenofovir, zoledronate
Acute and chronic interstitial nephritis	NSAIDS, cyclooxygenase-2 inhibitors, cephalosporins, cadmium chloride, penicillins, ciprofloxacin, proton pump inhibitors, loop diuretics, vancomycin, acyclovir, rifampin, sulfonamides, indinavir, acetaminophen, aspirin, carmustine, semustine, cisplatin, Chinese herbals with aristolochic acid, allopurinol, phenytoin, ranitidine
Intratubular obstruction by precipitation of tissue degradation products or drugs and/or their metabolites	Acyclovir, indinavir, sulfonamides, foscarnet, methotrexate, triamterene
Rhabdomyolysis	Cocaine, heroin, ketamine, methadone, methamphetamine, ethanol, lovastatin (in combination with cyclosporin), amitriptyline, doxepin, fluoxetine, diphenhydramine, doxylamine, benzodiazepines, statins, haloperidol
Renal vasculitis and thrombosis (hemolytic–uremic syndrome)	Mitomycin, cyclosporin, FK506, cocaine, gemcitabine, penicillamine, quinine, clopidogrel, ticlopidine
Glomerular dysfunction with proteinuria	Gold, lithium, interferon-α, penicillamine, NSAIDs, pamidronate

ACE, angiotensin-converting enzyme; DCVC, dichlorovinyl-L-cysteine; NSAIDs, nonsteroidal anti-inflammatory drugs.

inflammatory mediators like cytokines and chemokines by these cells. These inflammatory mediators induce the migration and infiltration of leukocytes into the injured kidneys and aggravate the primary injury induced by the nephrotoxicant [30]. Thus, inflammation plays an important role in the initiation and extension phases of ARF.

The role of inflammation in nephrotoxicity has mostly been studied using the chemotherapeutic drug cisplatin as nephrotoxicant. Cisplatin exposure induces a myriad of proinflammatory cytokines and chemokines [31,32] and the proinflammatory cytokine tumor necrosis factor α (TNF-α) appears to be a key upstream regulator. Several studies have shown that pharmacological inhibitors of and antibodies against TNF-α markedly suppressed the induction of other cytokines and ameliorated cisplatin-induced nephrotoxicity [31–36]. Our laboratory has shown that exposure of immortalized proximal tubular cells (IM-PTECs) to TNF-α prior to cisplatin significantly increased cisplatin-induced apoptosis via altering the JNK–c-Jun–NF-κB (c-Jun N-terminal kinase–c-Jun–nuclear factor kappa B) balance of the cells (Benedetti et al. submitted).

Despite our knowledge of the mechanisms of nephrotoxicity of several drugs and toxins, the fact that the rate of ARF has remained unchanged over recent years indicates that current diagnostic biomarkers of nephrotoxicity (creatinine and urea) are not sensitive enough and inadequate. Fortunately, the use of innovative technologies such as toxicogenomics in several human and animal models of nephrotoxicant-induced ARF has led to the identification of several promising new biomarkers and therapeutic targets.

5.2.2 Use of toxicogenomics in kidney toxicity studies

Toxicogenomics allows detailed analysis of potential mechanisms of toxicity without the need for a priori knowledge of the mode of toxic action. Toxicogenomics enables the study of the functions of the genome in relation to its composition and measures gene expression via transcriptomics, protein levels via proteomics, and metabolite contents via metabolomics. The toxicogenomics studies that have been performed on kidney toxicity and are presented below are summarized in Table 5.2.2 and Table 5.2.3.

Transcriptomic strategies

Determination of changes in gene expression, by measuring complementary messenger RNA (mRNA) levels, is used to predict changes in protein levels and activity. Microarray platforms are useful tools to rapidly identify changes in the expression level of genes following treatment with test compounds. This transcriptomic information, coupled with other toxicological observations, is useful to detect toxic changes in the target organ and formulate hypotheses related to the mechanisms of toxicity.

Transcriptomics to identify molecular mechanisms underlying kidney toxicity

There are numerous nephrotoxicants, acting by a variety of mechanisms. Using toxicogenomics, researchers have been able to elucidate new molecular mechanisms for several different nephrotoxicants, belonging to the groups of chemotherapeutics, antibiotics, immunosuppressants, and environmental toxicants (see Table 5.2.2). Below, we summarize the key findings.

Transcriptomic analysis of cisplatin-induced nephrotoxicity in rats suggested an apoptotic response and perturbations of intracellular calcium homeostasis in combination with increased expression of multidrug resistance genes, such as those for P-glycoproteins, and tissue-remodeling proteins, such as clusterin [37]. Thompson et al. demonstrated a conserved profile of 48 statistically significant cisplatin-induced gene changes on multiple array platforms and multiple sites, suggesting that gene-expression profiles linked to specific types of tissue injuries and mechanisms of toxicity may be extrapolated across platform technologies, laboratories, and in-life studies. These genes included some associated with damage to proximal tubules, tissue remodeling and regeneration, and immune cell invasion [38].

For antibiotic-induced nephrotoxicity, several studies have indicated that oxidative stress was involved, along with other deregulations of cellular processes. Gentamicin exposure in rats deregulated 211 genes involved in lysosomal phospholipidosis, glucose transport and metabolism, apoptosis, and oxidative stress [39]. Cephaloridine-induced nephrotoxicity in rats revealed several characteristic expression patterns of genes associated with specific cellular processes, including oxidative stress and proliferative response [40]. The deregulated genes associated with vancomycin-induced nephrotoxicity in mice suggested the possibility of oxidative stress and mitochondrial damage, as well as involvement of the complement and inflammatory pathways [41].

Gene-expression profiling of human proximal tubular cells exposed to the immunosuppressant cyclosporin A (CsA) suggested that CsA may induce an endoplasmic reticulum (ER) stress in tubular cells in vitro and up-regulation of the ER stress marker binding immunoglobulin protein (BIP)

Table 5.2.2 Overview of Recent Mechanistic Toxicogenomics Studies Performed on Kidney Toxicity

Aim	Toxicant	Model System	Toxicogenomic Tool	Toxicity Pathways/Genes	Publication
Mechanism	Cisplatin	Rats	Transcriptomics	Apoptotic response and tissue injury, perturbations of intracellular calcium homeostasis, increased expression of multidrug resistance genes, tissue remodeling and regeneration, immune cell invasion	[37,38]
Mechanism	Cisplatin ± TNF-α	Immortalized mouse PTECs	Transcriptomics	Cell death and proinflammatory pathways and genes transcribed by NF-κB and c-Jun enhanced with cisplatin + TNF-α	Benedetti, et al., data submitted
Mechanism	Cisplatin	Rats	Proteomics	Identification of excsomal fetuin-A as a potential biomarker	[103]
Mechanism	Cisplatin	Rats	Metabolomics	Altered urinary levels of glucose, amino acids, and Krebs cycle intermediates	[78]
Mechanism	Cisplatin and gentamicin	Rats	Transcriptomics and metabolomics	Decline in mRNA transcripts for several luminal membrane transporters handling each of the respective elevated urinary metabolites	[89]
Mechanism	Cisplatin and methotrexate	Yeast	Functional toxicogenomics	Several genes involved in DNA repair	[90]
Evaluation	Cisplatin	Human and rat kidney slices in comparison to rat kidneys	Transcriptomics	Transcription, DNA damage, cell cycle, proliferation, apoptosis, protein damage, the disruption of transport and calcium homeostasis, cellular metabolism, and oxidative stress	[55]
Evaluation	Cisplatin	Rat PTECs in comparison to rats	Transcriptomics	Genes correlating: clusterin, *c-Myc*, and *Waf-1*	[37]
Mechanism	Gentamicin	Rats	Transcriptomics	Lysosomal phospholipidosis, glucose transport and metabolism, apoptosis, and oxidative stress	[39]
Mechanism	Gentamicin	Rats	Proteomics	Stress responses; glucose metabolism and lipid biosynthesis; mitochondrial dysfunction; and AGAT, PRBP4 and transthyretin as potential biomarkers	[57–60]
Mechanism	Gentamicin	Rats	Metabolomics	Increased urinary levels of glucose and decreased levels of TMAO, xanthurenic kynurenic acids, changes in sulfation patterns, and increase in urinary levels of 6-hydroxymelatonin	[75,77]

Mechanism	Gentamicin, BI-3 and IMM125		Transcriptomics and proteomics	Detection of 12 genes deregulated at both the transcriptomic and proteomic levels involved in oxidative stress, energy metabolism, fatty acid and amino acid metabolism, vesicle transport, and membrane trafficking	[88]
Mechanism	Cephaloridine	Rats	Transcriptomics	Oxidative stress and proliferative response	[40]
Mechanism	Vancomycin	Mice	Transcriptomics	Oxidative stress, mitochondrial damage, and involvement of the complement and inflammatory pathways	[41]
Mechanism	Cyclosporin A	Human PTECs	Transcriptomics	Endoplasmic reticulum stress, epithelial-to-mesenchymal transition	[42,43]
Mechanism	Cyclosporin A	Rats	Proteomics	Calbindin-D as a potential biomarker	[62,63]
Mechanism	Cyclosporin A	Zebrafish	Proteomics	Cytoskeletal/structural assembly, lipid binding, stress response, and metabolism	[65]
Mechanism	Cyclosporin A	PTECs	Proteomics	Protein metabolism, response to damage, cell organization and cytoskeleton, energy metabolism, cell cycle, and nucleobase/nucleoside/nucleotide metabolism	[66]
Mechanism	Cyclosporin A	Rats	Metabolomics	Elevated levels of urinary glucose and acetate along with reduced levels of urinary TMAO	[74–76]
Mechanism	Cyclosporin A and tacrolimus	Yeast	Functional toxicogenomics	The chemical–genetic profiles of the compounds clustered with the genetic interaction profile of *CNB1*, their known target	[104]
Mechanism	Ochratoxin A	Rats and rat PTECs	Transcriptomics	DNA damage response and apoptosis, oxidative stress, and inflammatory reactions	[44,45]
Mechanism	Lead acetate	Rats	Proteomics	Oxidative stress and mitochondrial function	[56]
Mechanism	Puromycin aminonucleoside	Rats	Proteomics	Detailed understanding of the nature and progression of the proteinuria associated with glomerular toxicity	[61]
Mechanism	DCVC	LLC-PK1 cells	Phosphoproteomics	Metabolism, stress responses, and cytoskeletal reorganization	[71,72]

DCVC, dichlorovinyl-L-cysteine; NF-κB, nuclear factor kappa B; PTECs, proximal tubular epithelial cells; TMAO, trimethylamine-N-oxide; TNF-α, tumor necrosis factor alpha.

Table 5.2.3 Overview of Recent Toxicogenomics Studies Performed for the Discovery of New Predictive Biomarkers of Nephrotoxicity

Aim	Toxicant	Model System	Toxicogenomic Tool	Toxicity Pathways/Genes	Publication
Prediction	41 nephrotoxic and non-nephrotoxic compounds	Rats	Transcriptomics	19 genes involved in DNA replication, cell cycle control, apoptosis and responses to oxidative stress and chemical stimuli	[50,51]
Prediction	7 nephrotoxic and non-nephrotoxic compounds	Rat primary renal cortical tubular cells	Transcriptomics	32 genes including *Map3k12* and *Hmox1*	[52]
Prediction	4 nephrotoxicants and 3 hepatotoxicants	Rats	Transcriptomics	48 genes including *Kim1*, lipocalin 2, and osteopontin	[105]
Prediction	10 different nephrotoxicants	Rats	Transcriptomics	Top 100 up-regulated genes and 25 down-regulated genes used but not specified	[49]
Prediction	6 different nephrotoxicants	Rats	Transcriptomics	Several transporters, *Kim1*, *IGFbp1*, osteopontin, α-fibrinogen and Gstα	[53]
Prediction	64 nephrotoxic and non-nephrotoxic compounds	Rats	Transcriptomics	35 genes including glutamate receptor *GRIK4*, heat shock protein *HSPB7*, and cathepsin L	[48]
Prediction	Cisplatin, gentamicin, and puromycin	Rats	Transcriptomics	Genes associated with injury in specific portions of the nephron and reflecting the mechanism of action of these various nephrotoxicants	[47]
Prediction	Cisplatin	Rats	Proteomics	Several biomarkers including fragments of collagen	[68]
Prediction	Cisplatin and gentamicin	Rats	Proteomics	101 peptides similarly regulated by gentamicin or cisplatin with mostly collagen type I and type III fragments	[69]
Prediction	Cisplatin, 4-aminophenol and D-serine	Rats	Proteomics	Changes in T-kininogen protein and a set of proteins found to be a component of the complement cascade and other blood clotting factors	[70]
Prediction	Gentamicin, cisplatin, and tobramycin	Rats	Metabolomics	Increased urinary levels of polyamines and amino acids	[79]
Prediction	50 different compounds including known nephrotoxicants	Rats	Metabolomics	Not specified	[80]
Prediction	80 different compounds including nephrotoxicants and hepatotoxicants	Rats and mice	Metabolomics	Not specified	[81–83]

was associated with CsA exposure in kidney transplant biopsies [42]. This study also indicated that the immunosuppressant sirolimus induced a different transcriptional response in the cells, modifying more biological processes within the nucleus related to transcriptional activity, in comparison to CsA altering biological processes located at the cell membrane. In addition to this, Slattery et al. identified 128 genes that were differentially regulated in renal tubular cells after CsA treatment, including known pro-fibrotic factors, oncogenes, and transcriptional regulators. Morphological changes of the cells and up-regulation of several epithelial-to-mesenchymal transition (EMT) markers were associated with CsA-induced toxicity [43].

Gene-expression profiling has also been used to delineate the mechanisms of toxicity of environmental toxicants. Luhe et al. used transcription profiling to study nephrotoxicity induced by ochratoxin A (OTA), a mycotoxin often found as a contaminant in grains, red wine, coffee beans, nuts, and several spices. DNA microarray analysis of in vivo samples identified treatment-specific transcriptional changes implicating DNA damage response and apoptosis, oxidative stress, and inflammatory reactions [44]. Comparison of these findings with the rat proximal tubular epithelial cell line showed that, under the right conditions, in vitro gene-expression changes were comparable to those identified in vivo [44]. These results were confirmed by another in vivo transcriptomic study [45].

These different transcriptomic studies have revealed the variety of mechanisms that reflect the biochemical and cellular perturbations observed for these various nephrotoxic compounds, and demonstrate the necessity to compare several groups of toxicants in a combined strategy using one cellular model and one series of array in order to find common biomarkers of nephrotoxicity.

Transcriptomics to identify biomarkers for kidney toxicity

Beyond its use for interrogating mechanisms of toxicity, transcriptomics represents an approach to identify novel biomarkers that can be used as (early) sensitive indicators of nephrotoxicity. Because genomic biomarkers are frequently more sensitive than the traditional functional and morphological markers, they represent a useful supplement to more reliably detect toxicity before the full phenotypic manifestation of toxicity may occur. Predictive toxicogenomics assumes that compounds inducing toxicity via similar mechanisms will induce characteristic gene-expression patterns, and by grouping the gene-expression profiles of well-known model compounds, a gene-expression signature related to a specific mechanism of toxicity can be generated. The development of such genomic biomarkers requires an appropriate repository of gene-expression profiles—from multiple structurally diverse compounds sharing similar modes of toxicity and compounds unrelated to the toxicity class—as well as the use of a variety of sophisticated statistical methodologies, such as linear discriminant analysis, artificial neural networks, and support vector machines. The performance of the resulting biomarkers is then estimated with gene-expression profiles independent from those used to derive the biomarkers and performance characteristics (accuracy, sensitivity, and specificity) are derived [46].

Several studies have been performed in order to generate genomic biomarkers specific for nephrotoxicity (Table 5.2.3). All of them had sensitivities and selectivities above 70% with a gene signature comprising at least 10 genes [47–54]. Most of these genomic biomarker signatures were derived from experiments performed in rats, except the study of Suzuki et al. which used rat primary renal cortical tubular cells. Yet, the overall results were consistent with previously reported in vivo data [52]. To increase the sensitivity and selectivity of nephrotoxic biomarkers, one should compare transcriptomic profiles of a large number of nephrotoxicants, divided into different subgroups. A 2012 study of Minowa et al. achieved 93% sensitivity and 90% selectivity with only 19 genomic biomarkers,

Table 5.2.4 List of Biomarkers of Nephrotoxicity Currently Being Clinically Evaluated

Kidney Segment Specificity	Biomarkers
Glomerulus	Total protein, cystatin C, β2-microglobulin, α1-microglobulin, albumin
Proximal tubules	Kim1, clusterin, NGAL, GSTα, β2-microglobulin, α1-microglobulin, NAG, osteopontin, cystatin C, netrin-1, RBP, interleukin 18, HGF, Cyr61, NHE-3, exosomal fetuin-A, L-FABP, albumin
Loop of Henle	Osteopontin and NHE-3
Distal tubules	Osteopontin, clusterin, GST-μ/π, NGAL, H-FABP, calbindin D28
Collecting duct	Calbindin D28

GSTα, glutathione-S-transferase class α; HGF, hepatocyte growth factor; Kim1, kidney injury molecule 1; L-FABP, liver-type fatty-acid-binding protein; NAG, N-acetyl-β-glucosaminidase; NGAL, neutrophil gelatinase associated lipocalin; NHE-3, sodium/hydrogen exchanger isoform 3; RBP, retinol-binding protein.

including genes involved in DNA replication, cell cycle control, apoptosis, and responses to oxidative stress and chemical stimuli. This model was derived from rats exposed to 41 nephrotoxic and non-nephrotoxic compounds with three different doses and four different time-points after single or repeated dosing [51]. The genomic biomarkers kidney injury molecule 1 (KIM1), lipocalin 2 (LCN2) and osteopontin (SPP1) were identified in the majority of these studies.

A few of these potential biomarkers, including KIM1, are already being evaluated in the clinical setting (Table 5.2.4), but for most of them translation to the human situation is still needed. Furthermore, given the heterogeneity of ARF, it is likely that detection of ARF will not be possible using only one biomarker but that a panel of biomarkers will be required.

Transcriptomics to characterize non-clinical models of nephrotoxicity

Transcriptomics can also be useful to evaluate new in vivo and in vitro models that could potentially be used in new toxicity screening systems (Table 5.2.2). Vickers et al. assessed cisplatin-induced nephrotoxicity in human and rat kidney slices in comparison to rat kidneys from an in vivo study [55]. The acute nephrosis of the tubular epithelium induced by cisplatin in vivo was reproduced in both human and rat kidney slices. Kidney gene-expression changes of in vivo and in vitro samples were indicative of transcription, DNA damage, cell cycle, proliferation, and apoptosis, as well as protein damage, the disruption of transport and calcium homeostasis, cellular metabolism, and oxidative stress, in agreement with the mechanism of toxicity of cisplatin [55]. However, not all studies showed resemblance between the in vivo and in vitro situations. Huang et al. assessed the reproducibility of the fingerprint pattern obtained from the cisplatin in vivo study in a rat proximal tubular epithelial cell line and showed overall significant discrepancies between the two models in gene-expression changes and morphologically. However, certain target genes correlated, such as clusterin, *c-Myc*, and *Waf1*[37], which could be key regulators of nephrotoxicity.

This type of transcriptomic comparison study could be applied not only to compare in vivo versus in vitro, but also for interspecies comparisons with the ultimate goal to extrapolate to the human situation.

Proteomics strategies

Not all changes in cellular mechanisms can be measured at the gene-expression level. Protein modification and subcellular redistribution of proteins provide an organism with rapid mechanisms to respond to stimuli, without the need for gene-expression changes. Proteomics technologies measure the proteins in a cell or body fluid and allow visualization of posttranslational modifications and protein complexes.

Proteomics and nephrotoxicity

Several kidney proteome studies investigating drug-induced kidney injury have been reported and shed light on the mechanisms of toxicity of nephrotoxic drugs (Tables 5.2.2 and 5.2.3). Witzmann et al. identified a number of proteins associated with oxidative stress and mitochondrial function that were regulated in the rat kidney in response to treatment with lead acetate [56]. Studies using the antibiotic gentamicin reported changes in a variety of proteins responsible for stress response, glucose metabolism, and lipid biosynthesis, supporting the notion that energy production is impaired and mitochondrial dysfunction is involved in gentamicin-induced nephrotoxicity [57,58]. Two more recent proteomic studies with gentamicin exposure in rats confirmed alterations in these processes and identified AGAT, PRBP4, and transthyretin, strongly modulated in the kidney, as potential nephrotoxicity biomarkers [59,60]. Cutler et al. showed that proteomic analysis after a single dose of puromycin aminonucleoside (PAN) to rats allowed a more detailed understanding of the nature and progression of the proteinuria associated with glomerular toxicity [61].

CsA-induced nephrotoxicity was also investigated, and proteomics analysis from rat kidney homogenates led to the identification of calbindin-D, showing a close correlation between its decrease and CsA-induced nephrotoxicity in several species, including humans [62,63]. Other studies of CsA nephrotoxicity performed in several other models, including in kidney cell lines and zebrafish, identified changes in proteins involved in cytoskeletal/structural assembly, lipid binding, stress response, and metabolism [64–66].

Zhou et al. used a proteomic approach to identify urinary protein biomarkers of cisplatin-induced acute kidney injury in rat exosomes, and identified exosomal fetuin-A as a potential biomarker of acute kidney injury [67]. Two other studies using cisplatin as well as gentamicin led to the identification of new potential biomarkers, which were further validated in independent studies [68,69]. Bandara et al. compared the proteomics profiles of rats after exposure to the three different nephrotoxins cisplatin, 4-aminophenol, and D-serine, and identified changes in several proteins, including T-kininogen protein and a set of proteins found to be a component of the complement cascade and other blood clotting factors, indicating a contribution of the immune system to the observed toxicity [70].

Most of these studies focused only on a single compound and none of them compared over 10 different nephrotoxicants, as described for toxicogenomic studies (see Section Transcriptomics to identify biomarkers for kidney toxicity of this chapter). However, the proteomics-based studies were able to validate some of the genomic biomarkers previously identified, such as KIM1 and cystatine C. Moreover, proteomic analysis of nephrotoxicity has led to the identification of completely new potential biomarkers, like transthyretin [59,60] and exosomal fetuin-A [67].

Phosphoproteomic investigations of nephrotoxicity

The activity of most cellular proteins is regulated by posttranslational modifications. Reversible protein phosphorylation is one of the most important posttranslational modifications and may result in alteration in protein–protein interactions, protein intracellular localization, and protein activity. Therefore, it is anticipated that protein-phosphorylation-mediated signal transduction pathways should be altered in response to toxic effects of chemicals.

Several phosphoproteomic methods have been developed to identify phosphorylated proteins in the kidney and liver from rat, mouse, and human tissue as well as in cell lines. However, only very few studies have examined the differential phosphorylation of proteins after toxicant exposure (Table 5.2.2). Two studies were performed by our laboratory in order to determine the differential phosphorylation of proteins after exposure of renal epithelial cells to the model nephrotoxicant 1,2-(dichlorovinyl)-L-cysteine (DCVC). A total of 14 and 9 proteins were up- and down-regulated, respectively, with DCVC treatment before the onset of apoptosis, including proteins involved in metabolism, stress responses, and cytoskeletal reorganization. The most prominent changes were found for Hsp27, which was shown to control DCVC-induced focal adhesion organization and apoptosis in further validation experiments [71]. Further phosphoproteomic studies in DCVC-exposed LLC-PK1 cells revealed changes in the tyrosine phosphorylation status of a subset of proteins playing a role in the regulation of the actin cytoskeleton reorganization that precedes renal cell apoptosis [72].

Therefore, phosphoproteomics has helped in the identification of new mechanisms of nephrotoxicity, but, thus far, has not resulted in the discovery of new biomarkers. Future studies should include far more sensitive and quantitative approaches based on SILAC proteomics (see also Chapter 2.2).

Metabolomics and metabonomics strategies

Metabolomic methods measure the levels of metabolites in the cell and in extracellular fluids, which may reflect changes in cellular processes. This method enables the study of processes such as metabolism and biotransformation of xenobiotics in cells or plasma as well as excreted metabolites in the urine. Metabonomics extends metabolic profiling to include information about perturbations of metabolism caused by environmental factors, disease processes, and the involvement of extragenomic influences [73] (Tables 5.2.2 and 5.2.3). Metabolic profiling methods have been used in preclinical studies of compounds that cause nephrotoxicity and provided a more complete understanding of the metabolic changes occurring following drug-induced nephrotoxicity [73]. Several metabolomic studies of CsA-induced nephrotoxicity showed elevated levels of urinary glucose and acetate as well as other metabolites, along with reduced levels of urinary trimethylamine-N-oxide (TMAO), that correlated with the onset of nephrotoxicity [74–76]. Gentamicin-induced nephrotoxicity metabolomic studies revealed increased urinary levels of glucose and decreased levels of TMAO and xanthurenic kynurenic acids, as well as changes in sulfation patterns [75] and an increase in urinary levels of 6-hydroxymelatonin, an oxidized byproduct of melatonin, directly correlated with histopathology findings [77]. Portilla et al. showed altered urinary levels of glucose, amino acids, and Krebs cycle intermediates preceding changes in serum creatinine after cisplatin exposure [78]. Studies including several nephrotoxicants have also been performed. A study with gentamicin, cisplatin, and tobramycin in rats detected increased urinary levels of polyamines and amino acids after a single dose observed prior to histological kidney damage, and after 28 days of chronic dosing, an increase in branched amino acids in urine with a decrease in amino acids and nucleosides in kidney tissue were observed [79]. Another study with rats treated with different compounds including cisplatin and gentamicin showed that proximal tubule kidney toxicity could be predicted with 85% accuracy, 88% specificity, and 78% sensitivity using nuclear magnetic resonance (NMR)-based urine analysis [80].

One of the largest metabonomic studies was conducted in collaboration with six pharmaceutical companies in a study called the Consortium for Metabolomic Toxicology (COMET). This study built predictive models of target organ toxicity based upon proteomics analyses of rodent urine and serum from multiple toxicity studies including renal and hepatic toxins [81,82], and using a subset of 80 treatments could predict renal and liver toxicities with sensitivities of 67 and 41% respectively [83].

Overall, these metabolomic studies indicate that metabolic profiling provides a way to identify nephrotoxicity noninvasively and could be used in the preclinical studies of new potential drugs as well as for the monitoring of patients during drug treatment.

Investigation of the role of the immune system in drug-induced organ toxicity by toxicogenomics

The role of immune system activation in drug-induced kidney toxicity is well established and is described in the first section of this chapter. Yet, the signaling pathways activated synergistically by the toxicant and the immune system components leading to cellular death are still not very clear. However, since only one study performed by our laboratory used a toxicogenomics approach to unravel these pathways, and since the immune system also plays a major role in liver toxicity, toxicogenomics studies performed on liver toxicity are also discussed here. We have shown that the nuclear factor kappa B (NF-κB) pathway is important for both kidney and liver immune-mediated toxicity, suggesting that common mechanisms may exist. To date, only a few toxicogenomics studies have been performed in order to identify the mechanisms underlying immune-mediated kidney and liver toxicity.

Our laboratory studied the synergistic mechanisms activated by the nephrotoxic drug cisplatin and the proinflammatory cytokine TNF-α inducing enhanced apoptosis in IM-PTECs. Transcriptomic analysis revealed that cisplatin enhanced the cell death and proinflammatory pathways elicited by cisplatin or TNF-α alone, and the number of genes transcribed by the two transcription factors NF-κB and c-Jun increased with the combined cisplatin/TNF-α treatment. Further mechanistic studies enabled us to build a model of the synergistic response in the cells in which the combination of cisplatin and TNF-α enhanced and prolonged the activation of c-Jun N-terminal kinase (JNK), leading to inhibition of NF-κB activity and to a subsequent decrease in the transcription of NF-κB-induced anti-apoptotic target genes (Benedetti et al., submitted) (Table 5.2.2). Similarly, the TNF-α-mediated enhancement of diclofenac-induced cell death observed in HepG2 cells was due to an alteration of the JNK–c-Jun–NF-κB balance [84]. Further investigation revealed that hepatotoxic drugs other than diclofenac, including carbamazepine and ketoconazole, gave a synergistic apoptotic response with TNF-α. Genome-wide transcriptomics analysis revealed activation of the NRF2-related oxidative stress response and the ER stress response as well as the death receptor-signaling pathway as critical cell toxicity pathways independent of and preceding TNF-α-mediated cell killing. A systematic short interfering RNA (siRNA)-mediated knockdown approach of genes related to these stress-induced pathways allowed us to build a model in which enhanced drug-induced translation initiates PERK-mediated CHOP signaling, thereby sensitizing towards casaspe-8-dependent TNF-α-induced apoptosis (Fredriksson et al., submitted).

Other toxicogenomic studies have pointed at different mechanisms involved in immune-mediated hepatotoxicity. Usnic acid (UA), a natural botanical product constituent of some dietary supplements used for weight loss, has been associated with clinical hepatotoxicity leading to liver failure in humans. Sahu et al. studied the mechanisms responsible for UA-induced toxicity in combination with LPS by gene-expression profiling in HepG2 cells. Compared with the controls, low, nontoxic concentrations of UA and LPS separately showed no effect on the cells but the simultaneous mixed exposure of the cells resulted in increased cytotoxicity, oxidative stress, and mitochondrial injury. Most of the altered gene expression induced by the binary mixture of UA and LPS was not modulated by either UA or LPS alone[85]. The two phosphoproteomic studies performed by Cosgrove et al., in which cultured hepatocytes from multiple human donors were treated with combinations of hepatotoxic drugs and cytokines, suggested that hepatocytes specify their cell death responses to toxic drug/cytokine conditions by integrating signals from four key pathways—Akt, p70-S6K, ERK, and p38. These findings highlight the critical role of kinase signaling in drug/cytokine hepatic cytotoxicity synergies and reveal that hepatic cytotoxicity responses are governed by a multi-pathway signaling network balance [86,87].

Overall, these studies demonstrate that a combination of drug and immune system components activates different signaling pathways than activated by the drug alone, and leads to enhanced toxicity. Therefore, the implementation of immune system components in future toxicogenomics studies and toxicity screening systems is essential.

Integration of toxicogenomics approaches

The different toxicogenomics methods allow the measurement of genes, gene expression, proteins, and metabolites separately. However, integrating and linking these measurements could provide the insights required to understand the processes within a cell after a toxic insult. Below, we highlight several studies that integrated different toxicogenomics methods to better understand drug-induced nephrotoxicity (Tables 5.2.2 and 5.2.3).

In the context of the European InnoMed PredTox project, transcriptomic and proteomic studies were performed to provide new insights into the molecular mechanisms of three nephrotoxic compounds—gentamicin, BI-3 (a platelet fibrinogen receptor antagonist for the treatment of thromboembolic diseases), and IMM125 (a derivative of CsA almost equipotent with regard to its immunosuppressive activity). Transcriptomic and proteomic data turned out to be complementary and their integration gave a more comprehensive insight into the putative mode of nephrotoxicity of the compounds, which was in accordance with histopathological observations in the kidney [59,88]. The study from Xu et al. combined rat urine metabolic profiling and kidney transcriptomic profiling to gain more insight into the mechanisms of toxicity of cisplatin and gentamicin. MetaCore pathway analysis software was employed to identify nephrotoxicant-associated biochemical changes via an integrated quantitative analysis of both urine metabolomic and kidney transcriptomic profiles. Correlation analysis was applied to establish quantitative linkages between the metabolites and the gene-transcript profiles in both cisplatin and gentamicin studies, and revealed that cisplatin and gentamicin treatments were strongly linked to declines in mRNA transcripts for several luminal membrane transporters that handle each of the respective elevated urinary metabolites. The integrated pathway analysis performed on these studies indicated that cisplatin- or gentamicin-induced renal Fanconi-like syndromes might be better explained by the reduction of functional proximal tubule transporters rather than by the perturbation of metabolic pathways inside kidney cells [89].

All these innovative toxicogenomics technologies offer promise of new in vitro screening systems for the detection of potential adverse health effects of new drugs, and have facilitated the detection of several promising biomarkers of nephrotoxicity. However, these so-called 'omics technologies are correlative and do not determine the functional requirements of these biomarkers for the cellular response to a toxicant. Functional genomics, in contrast to other 'omics approaches, can provide a direct link between gene and toxicant.

5.2.3 Functional genomics: a new tool to study target organ toxicity

Functional genomics provides a link between a specific gene and the requirement for that gene product in the cellular response to drug treatment. This functional information is obtained by screening collections of cells/organisms that lack either genes (through deletion) or proteins (through blocking translation by using technologies such as RNA interference (RNAi)). While RNAi will be an important future tool to understand mechanisms of toxicity, so far relatively little work has been done in this context. Therefore, we discuss below both screening strategies in relation to nephrotoxicity as well as in relation to liver toxicity.

Functional toxicogenomics in yeast

Owing to the in-depth genetic knowledge of yeast, yeast functional genomics has been enormously fruitful since the 1990s. Chemical genomic tools developed in yeast have contributed to the understanding of compound and drug mechanisms of action. However, very few studies have used yeast chemogenomic approaches to identify mechanisms of toxicity of compounds inducing nephrotoxicity and hepatotoxicity (Table 5.2.2). Giaever et al. screened 10 diverse compounds, including cisplatin and methotrexate—known inducers of nephrotoxicity—and fluconazole and miconazole—known inducers of hepatotoxicity—in 80 genome-wide experiments against the complete collection of heterozygous yeast deletion strains. In several cases, previously known interactions were identified, and in each case, the analysis revealed novel cellular interactions, even when the relationship between a compound and its cellular target had been well established [90]. For example, profiling the homozygous pool of strains with cisplatin revealed that the majority of the sensitive strains have homozygous deletions of genes involved in DNA repair. Parsons et al. screened 12 inhibitory compounds against the *S. cerevisiae* viable deletion set to generate chemical–genetic interaction profiles. These compounds included the nephrotoxicants cyclosporin A and tacrolimus, and the hepatotoxicant fluconazole. A chemical–genetic profile was generated for each compound, and target pathways or proteins for each drug were identified [91]. For example, the chemical–genetic profiles of FK506 and CsA clustered with the genetic interaction profile of *CNB1*, their known target. The two uncharacterized genes *VID21* and *YBR094W* were identified as highly sensitive to both camptothecin and hydroxyurea. Domain analysis of the sequence of these genes indicated a possible role for Vid21 in chromatin remodeling after DNA damage and suggested that *YBR094W* may encode an enzyme responsible for posttranslational modification of proteins involved in the DNA damage response.

Although functional screening in yeast is a powerful tool to understand toxicity of compounds, it does not provide information to understand human biology. To understand human biology by direct observation in the relevant cellular context, loss-of-function assays have been developed for mammalian cells, including mouse- and human-derived.

RNAi screens in mammalian cells

Functional genomics in mammalian systems has been made possible by the discovery of the mechanism of RNAi [92]. RNAi is a cellular regulatory process in eukaryotic cells in which small RNA molecules are used to inhibit expression of target proteins by degradation of mRNA. The application of genome-wide RNAi has become an integral functional genomic tool for drug target identification and validation, pathway analysis, and drug discovery and is starting to be applied to toxicology [93]. Loss-of-function genetic screens using RNAi in cell lines can be carried out in a high-throughput manner and can be conducted using two different methods, either using short hairpin RNA (shRNA) vector libraries, or using chemically or enzymatically generated small interfering RNAs (siRNAs). So far, functional screening has predominantly been used for direct medical applications, such as cancer studies, and for some drug development, but none has aimed to obtain more insight into the mechanisms of toxicity of nephro- and hepatotoxicants.

As described above, TNF-α-induced NF-κB signaling is implicated in both drug-induced nephrotoxicity and drug-induced hepatotoxicity. Although not focusing on organ toxicity, several studies have been performed to increase our understanding of the NF-κB signaling pathway. These gain- and loss-of-function studies were based on NF-κB luciferase reporter constructs [94–99] using complementary DNA (cDNA)[95] or siRNA screens [94,96–100]. Most of the RNAi screens used targeted siRNA libraries, but two of them used a genome-wide siRNA screen strategy. Gewurz et al. used an Epstein Barr virus latent membrane protein (LMP1) mutant, canonically NF-κB dependent, to identify 155 proteins significantly and substantially important for NF-κB activation in HEK293 cells. These proteins included many kinases, phosphatases, ubiquitin ligases, and deubiquitinating enzymes not previously known to be important for NF-κB activation. Relevance to other canonical NF-κB pathways was extended by finding that 118 of the 155 LMP1 NF-κB activation pathway components were similarly important for interleukin 1 beta (IL-1β)-, and 79 for TNF-α-mediated NF-κB activation in the same cells [99]. Nickles et al. conducted a genome-wide siRNA screen in human cells and identified several new candidate modulators of TNF-α signaling, which were confirmed in independent experiments. Specifically, they showed that caspase 4 was required for the induction of NF-κB activity, while it appeared to be dispensable for the activation of the JNK signaling branch [96]. One RNAi screen took a different approach and probed the entire complement of human kinases and phosphatases in order to find gene products that tilt the balance of TNF signal transduction in favor of cell death or cell viability. This screen found that loss of hexokinase 1 resulted in the greatest elevation in TNF-dependent death [101].

Given the importance of the NF-κB signaling in nephro- and hepatotoxicity, our laboratory set up a live-cell imaging-based siRNA screen to understand the mechanisms involved in the perturbation of the TNF-α-induced NF-κB oscillation in response to a drug. For this, the model hepatotoxicant diclofenac was used, as we had previously shown that exposure to TNF-α enhanced diclofenac-induced apoptosis [84]. NF-κB oscillation in silenced HepG2 cells exposed to the cytokine TNF-α in combination or not with diclofenac was measured by high-content confocal laser scan microscopy in combination with multi-parametric image analysis. Out of the~1500 genes screened, we identified 115 that significantly affected the NF-κB oscillatory response, and 46 of those genes were further confirmed. Knockdown of these genes affected the amplitude or duration of nuclear oscillations; or the time between oscillations, leading to an increase or decrease in the number of nuclear translocations; or an inhibition of the response altogether. In this last category, we identified five genes, three novel, whose reduced expression protected against the diclofenac/TNF-α-induced apoptosis, indicating their role in the synergistic apoptosis (Benedetti et al., data not published). Identical screens could be used to unravel mechanisms of immune-mediated nephrotoxicity. In addition, the newly identified genes from liver toxicity screens that are involved in the synergistic apoptosis could also play a role in TNF-α-enhanced nephrotoxicity. For example, UFD1L knockdown was shown in our screen to decrease diclofenac/TNF-α hepatotoxicity and to up-regulate A20 and IκBα levels. Since UFD1L is implicated in the closure of the nuclear envelope [102], we hypothesized that lack of UFD1L dismantled the boundary between inactive (cytoplasmic) and active (nuclear) NF-κB, allowing a rise in nuclear NF-κB presence and thereby enhancing the transcription of the negative feedback genes A20 and IκBα, as well as some of the NF-κB anti-apoptotic target genes. Such a mechanism could be preserved in kidney cells, and up-regulation of UFD1L levels by the combination nephrotoxicant/TNF-α could be a plausible mechanism involved in the synergistic apoptosis of the cells.

As functional genomics becomes more widespread, it is expected that there will be a significant increase in the use of these technologies in nephro- and hepatotoxicity studies. In addition, such technologies will certainly be useful for the development of new in vitro toxicity screening systems in the near future.

5.2.4 Conclusions

Toxicogenomics has shown its potential in identifying new mechanisms and new biomarkers of nephrotoxicity. However, several challenges need to be overcome and, once addressed, will lead to the emergence of robust screening tools and sensitive biomarkers. The integration of all the 'omics techniques and subsequent modeling approaches, the use of improved in vitro systems mimicking more the in vivo situation by integrating important components of the toxic response such as inflammation, and the use of functional genomics for detailed toxicity pathway analysis allowing the development of new in vitro screening systems will, for sure, reduce the number of drugs causing adverse reactions in the future.

References

[1] Lameire N, van Biesen W, Vanholder R. Acute renal failure. Lancet 2005;365:417–30.

[2] Marshall MR. Current status of dosing and quantification of acute renal replacement therapy. Part 2: dosing paradigms and clinical implementation. Nephrology (Carlton) 2006;11:181–91.

[3] Schiffl H. Dosing pattern of renal replacement therapy in acute renal failure: current status and future directions. Eur J Med Res 2006;11:178–82.

[4] Lieberthal W, Nigam SK. Acute renal failure II. Experimental models of acute renal failure: imperfect but indispensable. Am J Physiol Renal Physiol 2000;278:F1–F12.

[5] Thadhani R, Pascual M, Bonventre JV. Acute renal failure. N Engl J Med 1996;334:1448–60.

[6] Bush KT, Keller SH, Nigam SK. Genesis and reversal of the ischemic phenotype in epithelial cells. J Clin Invest 2000;106:621–6.

[7] Kays SE, Schnellmann RG. Regeneration of renal proximal tubule cells in primary culture following toxicant injury: response to growth factors. Toxicol Appl Pharmacol 1995;132:273–80.

[8] Gailit J, Colflesh D, Rabiner I, Simone J, Goligorsky MS. Redistribution and dysfunction of integrins in cultured renal epithelial cells exposed to oxidative stress. Am J Physiol 1993;264:F149–57.

[9] Zuk A, Bonventre JV, Brown D, Matlin KS. Polarity, integrin, and extracellular matrix dynamics in the postischemic rat kidney. Am J Physiol 1998;275:C711–31.

[10] Racusen LC, Fivush BA, Li YL, Slatnik I, Solez K. Dissociation of tubular cell detachment and tubular cell death in clinical and experimental "acute tubular necrosis." Lab Invest 1991;64:546–56.

[11] Gobe GC, Endre ZH. Cell death in toxic nephropathies. Semin Nephrol 2003;23:416–24.

[12] Padanilam BJ. Cell death induced by acute renal injury: a perspective on the contributions of apoptosis and necrosis. Am J Physiol Renal Physiol 2003;284:F608–27.
[13] Kaushal GP, Basnakian AG, Shah SV. Apoptotic pathways in ischemic acute renal failure. Kidney Int 2004;66:500–6.
[14] Toback FG. Regeneration after acute tubular necrosis. Kidney Int 1992;41:226–46.
[15] Molitoris BA, Marrs J. The role of cell adhesion molecules in ischemic acute renal failure. Am J Med 1999;106:583–92.
[16] Schnellmann RG, Kelly KJ, editors. Pathophysiology of nephrotoxic acute renal failure; 1999.
[17] Nolin TD, Himmelfarb J. Mechanisms of drug-induced nephrotoxicity. Handb Exp Pharmacol 2010;196:111–30.
[18] de Broe ME, editor. Renal injury due to environmental toxins, drugs, and contrast agents; 1999.
[19] Whelton A. Nephrotoxicity of nonsteroidal anti-inflammatory drugs: physiologic foundations and clinical implications. Am J Med 1999;106:13S–24S.
[20] Wynckel A, Ebikili B, Melin JP, Randoux C, Lavaud S, Chanard J. Long-term follow-up of acute renal failure caused by angiotensin converting enzyme inhibitors. Am J Hypertens 1998;11:1080–6.
[21] Rudnick MR, Kesselheim A, Goldfarb S. Contrast-induced nephropathy: how it develops, how to prevent it. Cleve Clin J Med 2006;73:75–80. 83–87.
[22] Burdmann EA, Andoh TF, Yu L, Bennett WM. Cyclosporine nephrotoxicity. Semin Nephrol 2003;23:465–76.
[23] Nagai J, Takano M. Molecular aspects of renal handling of aminoglycosides and strategies for preventing the nephrotoxicity. Drug Metab Pharmacokinet 2004;19:159–70.
[24] Hanigan MH, Devarajan P. Cisplatin nephrotoxicity: molecular mechanisms. Cancer Ther 2003;1:47–61.
[25] Yalavarthy R, Edelstein CL. Therapeutic and predictive targets of AKI. Clin Nephrol 2008;70:453–63.
[26] Araujo LP, Truzzi RR, Mendes GE, Luz MA, Burdmann EA, Oliani SM. Annexin A1 protein attenuates cyclosporine-induced renal hemodynamics changes and macrophage infiltration in rats. Inflamm Res 2012;61:189–96.
[27] Pabla N, Dong Z. Cisplatin nephrotoxicity: mechanisms and renoprotective strategies. Kidney Int 2008;73:994–1007.
[28] Quiros Y, Vicente-Vicente L, Morales AI, Lopez-Novoa JM, Lopez-Hernandez FJ. An integrative overview on the mechanisms underlying the renal tubular cytotoxicity of gentamicin. Toxicol Sci 2011;119:245–56.
[29] Naughton CA. Drug-induced nephrotoxicity. Am Fam Physician 2008;78:743–50.
[30] Akcay A, Nguyen Q, Edelstein CL. Mediators of inflammation in acute kidney injury. Mediators Inflamm 2009;2009:137072.
[31] Ramesh G, Brian Reeves W. Cisplatin increases TNF-alpha mRNA stability in kidney proximal tubule cells. Ren Fail 2006;28:583–92.
[32] Ramesh G, Kimball SR, Jefferson LS, Reeves WB. Endotoxin and cisplatin synergistically stimulate TNF-alpha production by renal epithelial cells. Am J Physiol Renal Physiol 2007;292:F812–9.
[33] Ramesh G, Reeves WB. TNF-alpha mediates chemokine and cytokine expression and renal injury in cisplatin nephrotoxicity. J Clin Invest 2002;110:835–42.
[34] Ramesh G, Reeves WB. TNFR2-mediated apoptosis and necrosis in cisplatin-induced acute renal failure. Am J Physiol Renal Physiol 2003;285:F610–8.
[35] Zhang B, Ramesh G, Norbury CC, Reeves WB. Cisplatin-induced nephrotoxicity is mediated by tumor necrosis factor-alpha produced by renal parenchymal cells. Kidney Int 2007;72:37–44.
[36] Dong Z, Atherton SS. Tumor necrosis factor-alpha in cisplatin nephrotoxicity: a homebred foe? Kidney Int 2007;72:5–7.
[37] Huang Q, Dunn 2nd RT, Jayadev S, DiSorbo O, Pack FD, Farr SB, et al. Assessment of cisplatin-induced nephrotoxicity by microarray technology. Toxicol Sci 2001;63:196–207.
[38] Thompson KL, Afshari CA, Amin RP, Bertram TA, Car B, Cunningham M, et al. Identification of platform-independent gene expression markers of cisplatin nephrotoxicity. Environ Health Perspect 2004;112:488–94.

[39] Ozaki N, Matheis KA, Gamber M, Feidl T, Nolte T, Kalkuhl A, et al. Identification of genes involved in gentamicin-induced nephrotoxicity in rats—a toxicogenomic investigation. Exp Toxicol Pathol 2010;62:555–66.

[40] Rokushima M, Fujisawa K, Furukawa N, Itoh F, Yanagimoto T, Fukushima R, et al. Transcriptomic analysis of nephrotoxicity induced by cephaloridine, a representative cephalosporin antibiotic. Chem Res Toxicol 2008;21:1186–96.

[41] Dieterich C, Puey A, Lin S, Swezey R, Furimsky A, Fairchild D, et al. Gene expression analysis reveals new possible mechanisms of vancomycin-induced nephrotoxicity and identifies gene marker candidates. Toxicol Sci 2009;107:258–69.

[42] Pallet N, Rabant M, Xu-Dubois YC, Lecorre D, Mucchielli MH, Imbeaud S, et al. Response of human renal tubular cells to cyclosporine and sirolimus: a toxicogenomic study. Toxicol Appl Pharmacol 2008;229:184–96.

[43] Slattery C, Campbell E, McMorrow T, Ryan MP. Cyclosporine A-induced renal fibrosis: a role for epithelial–mesenchymal transition. Am J Pathol 2005;167:395–407.

[44] Luhe A, Hildebrand H, Bach U, Dingermann T, Ahr HJ. A new approach to studying ochratoxin A (OTA)-induced nephrotoxicity: expression profiling in vivo and in vitro employing cDNA microarrays. Toxicol Sci 2003;73:315–28.

[45] Arbillaga L, Vettorazzi A, Gil AG, van Delft JH, Garcia-Jalon JA, Lopez de Cerain A. Gene expression changes induced by ochratoxin A in renal and hepatic tissues of male F344 rat after oral repeated administration. Toxicol Appl Pharmacol 2008;230:197–207.

[46] Blomme EA, Yang Y, Waring JF. Use of toxicogenomics to understand mechanisms of drug-induced hepatotoxicity during drug discovery and development. Toxicol Lett 2009;186:22–31.

[47] Amin RP, Vickers AE, Sistare F, Thompson KL, Roman RJ, Lawton M, et al. Identification of putative gene based markers of renal toxicity. Environ Health Perspect 2004;112:465–79.

[48] Fielden MR, Eynon BP, Natsoulis G, Jarnagin K, Banas D, Kolaja KL. A gene expression signature that predicts the future onset of drug-induced renal tubular toxicity. Toxicol Pathol 2005;33:675–83.

[49] Jiang Y, Gerhold DL, Holder DJ, Figueroa DJ, Bailey WJ, Guan P, et al. Diagnosis of drug-induced renal tubular toxicity using global gene expression profiles. J Transl Med 2007;5:47.

[50] Kondo C, Minowa Y, Uehara T, Okuno Y, Nakatsu N, Ono A, et al. Identification of genomic biomarkers for concurrent diagnosis of drug-induced renal tubular injury using a large-scale toxicogenomics database. Toxicology 2009;265:15–26.

[51] Minowa Y, Kondo C, Uehara T, Morikawa Y, Okuno Y, Nakatsu N, et al. Toxicogenomic multigene biomarker for predicting the future onset of proximal tubular injury in rats. Toxicology 2012;297:47–56.

[52] Suzuki H, Inoue T, Matsushita T, Kobayashi K, Horii I, Hirabayashi Y. In vitro gene expression analysis of nephrotoxic drugs in rat primary renal cortical tubular cells. J Appl Toxicol 2008;28:237–48.

[53] Thukral SK, Nordone PJ, Hu R, Sullivan L, Galambos E, Fitzpatrick VD, et al. Prediction of nephrotoxicant action and identification of candidate toxicity-related biomarkers. Toxicol Pathol 2005;33:343–55.

[54] Wang EJ, Snyder RD, Fielden MR, Smith RJ, Gu YZ. Validation of putative genomic biomarkers of nephrotoxicity in rats. Toxicology 2008;246:91–100.

[55] Vickers AE, Rose K, Fisher R, Saulnier M, Sahota P, Bentley P. Kidney slices of human and rat to characterize cisplatin-induced injury on cellular pathways and morphology. Toxicol Pathol 2004;32:577–90.

[56] Witzmann FA, Fultz CD, Grant RA, Wright LS, Kornguth SE, Siegel FL. Regional protein alterations in rat kidneys induced by lead exposure. Electrophoresis 1999;20:943–51.

[57] Charlwood J, Skehel JM, King N, Camilleri P, Lord P, Bugelski P, et al. Proteomic analysis of rat kidney cortex following treatment with gentamicin. J Proteome Res 2002;1:73–82.

[58] Kennedy S. The role of proteomics in toxicology: identification of biomarkers of toxicity by protein expression analysis. Biomarkers 2002;7:269–90.

[59] Com E, Boitier E, Marchandeau JP, Brandenburg A, Schroeder S, Hoffmann D, et al. Integrated transcriptomic and proteomic evaluation of gentamicin nephrotoxicity in rats. Toxicol Appl Pharmacol 2012;258:124–33.

[60] Meistermann H, Norris JL, Aerni HR, Cornett DS, Friedlein A, Erskine AR, et al. Biomarker discovery by imaging mass spectrometry: transthyretin is a biomarker for gentamicin-induced nephrotoxicity in rat. Mol Cell Proteomics 2006;5:1876–86.
[61] Cutler P, Bell DJ, Birrell HC, Connelly JC, Connor SC, Holmes E, et al. An integrated proteomic approach to studying glomerular nephrotoxicity. Electrophoresis 1999;20:3647–58.
[62] Aicher L, Wahl D, Arce A, Grenet O, Steiner S. New insights into cyclosporine A nephrotoxicity by proteome analysis. Electrophoresis 1998;19:1998–2003.
[63] Steiner S, Aicher L, Raymackers J, Meheus L, Esquer-Blasco R, Anderson NL, et al. Cyclosporine A decreases the protein level of the calcium-binding protein calbindin-D 28 kDa in rat kidney. Biochem Pharmacol 1996;51:253–8.
[64] Lamoureux F, Mestre E, Essig M, Sauvage FL, Marquet P, Gastinel LN. Quantitative proteomic analysis of cyclosporine-induced toxicity in a human kidney cell line and comparison with tacrolimus. J Proteomics 2011;75:677–94.
[65] Ponnudurai RP, Basak T, Ahmad S, Bhardwaj G, Chauhan RK, Singh RA, et al. Proteomic analysis of zebrafish (*Danio rerio*) embryos exposed to cyclosporine A. J Proteomics 2012;75:1004–17.
[66] Puigmulc M, Lopez-Hellin J, Sune G, Tornavaca O, Camano S, Tejedor A, et al. Differential proteomic analysis of cyclosporine A-induced toxicity in renal proximal tubule cells. Nephrol Dial Transplant 2009;24:2672–86.
[67] Zhou H, Pisitkun T, Aponte A, Yuen PS, Hoffert JD, Yasuda H, et al. Exosomal fetuin-A identified by proteomics: a novel urinary biomarker for detecting acute kidney injury. Kidney Int 2006;70:1847–57.
[68] Mischak H, Espandiari P, Sadrieh N, Hanig J. Profiling of rat urinary proteomic patterns associated with drug-induced nephrotoxicity using CE coupled with MS as a potential model for detection of drug-induced adverse effects. Proteomics Clin Appl 2009;3:1062–71.
[69] Rouse R, Siwy J, Mullen W, Mischak H, Metzger J, Hanig J. Proteomic candidate biomarkers of drug-induced nephrotoxicity in the rat. PLoS ONE 2012;7:e34606.
[70] Bandara LR, Kelly MD, Lock EA, Kennedy S. A correlation between a proteomic evaluation and conventional measurements in the assessment of renal proximal tubular toxicity. Toxicol Sci 2003;73:195–206.
[71] de Graauw M, Tijdens I, Cramer R, Corless S, Timms JF, van de Water B. Heat shock protein 27 is the major differentially phosphorylated protein involved in renal epithelial cellular stress response and controls focal adhesion organization and apoptosis. J Biol Chem 2005;280:29885–29898.
[72] de Graauw M, Le Devedec S, Tijdens I, Smeets MB, Deelder AM, van de Water B. Proteomic analysis of alternative protein tyrosine phosphorylation in 1,2-dichlorovinyl-cysteine-induced cytotoxicity in primary cultured rat renal proximal tubular cells. J Pharmacol Exp Ther 2007;322:89–100.
[73] Beger RD, Sun J, Schnackenberg LK. Metabolomics approaches for discovering biomarkers of drug-induced hepatotoxicity and nephrotoxicity. Toxicol Appl Pharmacol 2010;243:154–66.
[74] Klawitter J, Bendrick-Peart J, Rudolph B, Beckey V, Haschke M, Rivard C, et al. Urine metabolites reflect time-dependent effects of cyclosporine and sirolimus on rat kidney function. Chem Res Toxicol 2009;22:118–28.
[75] Lenz EM, Bright J, Knight R, Westwood FR, Davies D, Major H, et al. Metabonomics with 1H-NMR spectroscopy and liquid chromatography-mass spectrometry applied to the investigation of metabolic changes caused by gentamicin-induced nephrotoxicity in the rat. Biomarkers 2005;10:173–87.
[76] Serkova NJ, Christians U. Biomarkers for toxicodynamic monitoring of immunosuppressants: NMR-based quantitative metabonomics of the blood. Ther Drug Monit 2005;27:733–7.
[77] Espandiari P, Zhang J, Rosenzweig BA, Vaidya VS, Sun J, Schnackenberg L, et al. The utility of a rodent model in detecting pediatric drug-induced nephrotoxicity. Toxicol Sci 2007;99:637–48.
[78] Portilla D, Li S, Nagothu KK, Megyesi J, Kaissling B, Schnackenberg L, et al. Metabolomic study of cisplatin-induced nephrotoxicity. Kidney Int 2006;69:2194–204.
[79] Boudonck KJ, Mitchell MW, Nemet L, Keresztes L, Nyska A, Shinar D, et al. Discovery of metabolomics biomarkers for early detection of nephrotoxicity. Toxicol Pathol 2009;37:280–92.

[80] Lienemann K, Plotz T, Pestel S. NMR-based urine analysis in rats: prediction of proximal tubule kidney toxicity and phospholipidosis. J Pharmacol Toxicol Methods 2008;58:41–9.

[81] Lindon JC, Keun HC, Ebbels TM, Pearce JM, Holmes E, Nicholson JK. The Consortium for Metabonomic Toxicology (COMET): aims, activities and achievements. Pharmacogenomics 2005;6:691–9.

[82] Lindon JC, Nicholson JK, Holmes E, Antti H, Bollard ME, Keun H, et al. Contemporary issues in toxicology: the role of metabonomics in toxicology and its evaluation by the COMET project. Toxicol Appl Pharmacol 2003;187:137–46.

[83] Ebbels TM, Keun HC, Beckonert OP, Bollard ME, Lindon JC, Holmes E, et al. Prediction and classification of drug toxicity using probabilistic modeling of temporal metabolic data: the consortium on metabonomic toxicology screening approach. J Proteome Res 2007;6:4407–22.

[84] Fredriksson L, Herpers B, Benedetti G, Matadin Q, Puigvert JC, de Bont H, et al. Diclofenac inhibits tumor necrosis factor-alpha-induced nuclear factor-kappaB activation causing synergistic hepatocyte apoptosis. Hepatology 2011;53:2027–41.

[85] Sahu SC, O'Donnell Jr. MW, Sprando RL. Interactive toxicity of usnic acid and lipopolysaccharides in human liver HepG2 cells. J Appl Toxicol 2012;32:739–49.

[86] Cosgrove BD, Alexopoulos LG, Saez-Rodriguez J, Griffith LG, Lauffenburger DA. A multipathway phosphoproteomic signaling network model of idiosyncratic drug- and inflammatory-cytokine-induced toxicity in human hepatocytes. Conf Proc IEEE Eng Med Biol Soc 2009;2009:5452–5.

[87] Cosgrove BD, Alexopoulos LG, Hang TC, Hendriks BS, Sorger PK, Griffith LG, et al. Cytokine-associated drug toxicity in human hepatocytes is associated with signaling network dysregulation. Mol Biosyst 2010;6:1195–206.

[88] Matheis KA, Com E, Gautier JC, Guerreiro N, Brandenburg A, Gmuender H, et al. Cross-study and cross-omics comparisons of three nephrotoxic compounds reveal mechanistic insights and new candidate biomarkers. Toxicol Appl Pharmacol 2011;252:112–22.

[89] Xu EY, Perlina A, Vu H, Troth SP, Brennan RJ, Aslamkhan AG, et al. Integrated pathway analysis of rat urine metabolic profiles and kidney transcriptomic profiles to elucidate the systems toxicology of model nephrotoxicants. Chem Res Toxicol 2008;21:1548–61.

[90] Giaever G, Flaherty P, Kumm J, Proctor M, Nislow C, Jaramillo DF, et al. Chemogenomic profiling: identifying the functional interactions of small molecules in yeast. Proc Natl Acad Sci USA 2004;101:793–8.

[91] Parsons AB, Brost RL, Ding H, Li Z, Zhang C, Sheikh B, et al. Integration of chemical-genetic and genetic interaction data links bioactive compounds to cellular target pathways. Nat Biotechnol 2004;22:62–9.

[92] Fire A, Xu S, Montgomery MK, Kostas SA, Driver SE, Mello CC. Potent and specific genetic interference by double-stranded RNA in *Caenorhabditis elegans*. Nature 1998;391:806–11.

[93] Kiefer J, Yin HH, Que QQ, Mousses S. High-throughput siRNA screening as a method of perturbation of biological systems and identification of targeted pathways coupled with compound screening. Methods Mol Biol 2009;563:275–87.

[94] Chew J, Biswas S, Shreeram S, Humaidi M, Wong ET, Dhillion MK, et al. WIP1 phosphatase is a negative regulator of NF-kappaB signalling. Nat Cell Biol 2009;11:659–66.

[95] Halsey TA, Yang L, Walker JR, Hogenesch JB, Thomas RS. A functional map of NFkappaB signaling identifies novel modulators and multiple system controls. Genome Biol 2007;8:R104.

[96] Nickles D, Falschlehner C, Metzig M, Boutros M. A genome-wide RNA interference screen identifies caspase 4 as a factor required for tumor necrosis factor alpha signaling. Mol Cell Biol 2012;32:3372–81.

[97] Metzig M, Nickles D, Falschlehner C, Lehmann-Koch J, Straub BK, Roth W, et al. An RNAi screen identifies USP2 as a factor required for TNF-alpha-induced NF-kappaB signaling. Int J Cancer 2011;129:607–18.

[98] Wang G, Li S, Wang F, Huang S, Li X, Xiong W, et al. RNAi screen to identify protein phosphatases that regulate the NF-kappaB signaling. Front Biol 2010;5:263–71.

[99] Gewurz BE, Towfic F, Mar JC, Shinners NP, Takasaki K, Zhao B, et al. Genome-wide siRNA screen for mediators of NF-kappaB activation. Proc Natl Acad Sci USA 2012;109:2467–72.

[100] Li S, Wang L, Berman MA, Zhang Y, Dorf ME. RNAi screen in mouse astrocytes identifies phosphatases that regulate NF-kappaB signaling. Mol Cell 2006;24:497–509.
[101] Schindler A, Foley E. A functional RNAi screen identifies hexokinase 1 as a modifier of type II apoptosis. Cell Signal 2010;22:1330–40.
[102] Bays NW, Hampton RY. Cdc48-Ufd1-Npl4: stuck in the middle with Ub. Curr Biol 2002;12:R366–71.
[103] Zou W, Beggs KM, Sparkenbaugh EM, Jones AD, Younis HS, Roth RA, et al. Sulindac metabolism and synergy with tumor necrosis factor-alpha in a drug-inflammation interaction model of idiosyncratic liver injury. J Pharmacol Exp Ther 2009;331:114–21.
[104] Parsons AB, Brost RL, Ding H, Li Z, Zhang C, Sheikh B, et al. Integration of chemical–genetic and genetic interaction data links bioactive compounds to cellular target pathways. Nat Biotechnol 2004;22:62–9.
[105] Waring JF, Yang Y, Healan-Greenberg CH, Adler AL, Dickinson R, McNally T, et al. Gene expression analysis in rats treated with experimental acetyl-coenzyme A carboxylase inhibitors suggests interactions with the peroxisome proliferator-activated receptor alpha pathway. J Pharmacol Exp Ther 2008;324:507–16.

CHAPTER

'Omics in Organ Toxicity, Integrative Analysis Approaches, and Knowledge Generation

5.3

Laura Suter-Dick

University of Applied Sciences and Art, Northwestern Switzerland (FHNW), School for Life Sciences, Institute of Chemistry and Bioanalytics, Muttenz, Switzerland

5.3.1 Introduction

New technologies that are able to provide better safety assessment are necessary to advance pharmaceutical and chemical research. Particular challenges arise from the need to establish toxic liabilities inherent to new types of molecules (e.g. engineered antibodies), to understand the reasons for idiosyncratic toxicity, and to efficiently assess the safety of chemicals and new drug entities. For these purposes, the scientific community in academia and industry must apply advanced scientific knowledge and cutting-edge technologies intelligently. This is the only way to identify exposure risks accurately in a world flooded with thousands of chemicals that need safety characterization, notwithstanding growing ethical and scientific concerns that promote the diminution of animal testing. The chemical industry must ensure safety to people and the environment, the cosmetic industry likewise, and pharmaceutical companies must ensure optimal safety profiles of potential new medicines. The major challenge in this quest is to achieve an increase in predictive performance, an increase in assay throughput, and a decrease in costs, together with a reduction in animals used for safety testing. With the advent of the so-called 'omics technologies in the 1980s and 1990s, and the improvement of the bioinformatics analysis and pathway analysis tools that followed, the effects and side effects of pharmaceuticals, environmental chemicals, and other stimuli (e.g. nanoparticles) can be studied in their molecular details. From an animal welfare perspective, it was expected that modern 'omics technologies together with advanced in vitro culture systems would quickly make animal experimentation obsolete. This is not the case, as these technologies are not yet mature (and may never be) to completely supersede animal experimentation in toxicology. However, 'omics technologies should enable us to detect toxic liabilities with shorter treatment periods and/or lower doses, and thus lead to a decrease in animal usage and/or animal stress. The data generated using these advanced molecular biology tools also deliver additional knowledge on the molecular mechanisms underlying a given toxicity, providing useful information with regard to issues such as species specificity, exaggerated pharmacology, or off-target effects. In addition, profound understanding of the molecular mechanisms is extremely useful to identify new specific biomarkers that could be used to monitor possible adverse

Toxicogenomics-Based Cellular Models.

effects both in the non-clinical and the clinical setting. Thus, holistic approaches such as toxicogenomics, as well as metabolomics and proteomics, provide the means to predict toxicity based on new endpoints and to increase the understanding of the molecular events underlying a given toxicity.

An additional area of focus that can be addressed employing new technologies is interspecies translational aspects, as side effects observed in one species (e.g. the rat) are often not replicated in the non-rodent non-clinical animal model and vice versa. And both rodent and non-rodent in vivo assays have only a limited power to predict the outcome in the target population, i.e. healthy volunteers and patients. Thus, novel technologies addressing molecular endpoints and aiming to identify early safety biomarkers should enable scientists to address translational aspects of safety assessment. This is clearly reflected in a white paper (*Innovation or Stagnation*) released by the US Food and Drug Administration (FDA) in 2004, indicating that a combination of animal- or computer-based predictive models, biomarkers for safety and effectiveness, and new clinical evaluation techniques are urgently needed (http://www.fda.gov/ScienceResearch/SpecialTopics/CriticalPathInitiative/CriticalPathOpportunitiesReports/ucm077262.htm).

Regarding in vitro systems, major efforts have been invested in the field of in vitro toxicology during the last 50 years with mixed outcome. Whereas cellular assays are very useful to predict specific toxicities, such as mutagenicity or cardiac potassium (hERG)-channel blocking, the use of cell lines for the understanding of organ toxicity is still not optimal, as these simple systems greatly differ from the organs they represent in terms of functionality and cellular composition. Recently, complex organotypical in vitro systems based on three-dimensional scaffolds and co-cultures have been revisited as tools for toxicity assessment and are thought to be superior in predictive performance than simpler cell systems. This is also partly driven by the great advances in cell reprogramming achieved in the last few years, opening the possibility of generating test systems containing cells from patients with phenotypes and genotypes of interest. The incipient use of induced pluripotent stem cells (iPSC) in toxicology testing is expected to advance the field of in vitro safety assessment by generating data relevant to patients.

There are several drivers to these major technological developments applied to the field of safety assessment. On the one hand, there is a need driven by financial interests for improving the predictive accuracy of non-clinical testing by enhancing study design and measured endpoints. Classical toxicological assessment, supplemented by 'omics endpoints, has shown superiority to conventional endpoint alone in terms of sensitivity, molecular understanding, and identification of safety biomarkers. Such complex investigative studies result often in better molecular understanding of underlying pharmacological and toxicological mechanisms but also may generate large amounts of data difficult to interpret. Not only are academia and industry heavily involved in this type of research, but it is also expected that high-content, high-throughput and advanced in vitro technologies will strongly impact regulatory toxicity, mainly by combining molecular endpoints with relevant in vitro systems. In 2008, Thomas Hartung indicated that these new high-content and high-throughput technologies might be able to bring about a revolution in the field of regulatory toxicology [1]. However, regulatory acceptance of these new technologies is still at an incipient stage, despite efforts from the FDA and the European Medicines Agency (EMA) to collect and understand 'omics data. The FDA has invested largely in scientific and laboratory activities, bioinformatics, and interpretation of toxicogenomics data. It has encouraged the pharmaceutical industry to submit "mock submissions" and voluntary genomics data submissions (VXDS) to get acquainted with the analysis and interpretation processes, and some exploratory genomics data also accompany regulatory submissions. The main challenges in dealing with the data are possible misinterpretations of the results originating from uncharacterized off-target toxicological effects, as well as issues of interpretation and the large number of biologic variables that can affect results and confound gene-expression databases. Thus, some FDA officials have indicated that efforts should be placed in studying pathways related to various types of toxicity and identifying biomarkers for these pathways in order to assess the relevance of non-clinical findings to humans [2].

Also, in the large scientific field of toxicology, reflected in the literature and scientific discussions, expectations and opinions on the usefulness of molecular, 'omics, and in vitro tools are divided. While some toxicologists remain skeptical with regard to the application of new technologies for safety assessment, other scientists are convinced that understanding the intimate molecular mechanisms associated with an observed toxicity will greatly help the drug development process and advance our scientific knowledge. Others expect that these new technologies will be able to predict a toxic liability with high accuracy, within very short time frames, and possibly with a minimal usage of compound and animal testing. Yet others rely on these technologies to identify novel biomarkers of toxicity. Ideally, these newly identified biomarkers could be subsequently qualified, easily measured, and used to monitor progression and recovery of an injury in preclinical animal models and in patients. The technologies have the potential to address all these questions, but probably not simultaneously and certainly not following a one-size-fits-all solution. The matters that need to be addressed together with the applied technologies will impact the study design and the data analysis strategy. If the main question for which an answer is sought is the mechanism by which toxicity occurs, the study design needs to be appropriate and include a meaningful number of sample collection timepoints, replicates, and doses. If biomarkers are the main goal, the gold standard by which their performance is measured needs to be very clearly defined. Are we seeking a biomarker that parallels the histopathological findings or are we willing to accept prodromal biomarkers that might give us a signal before the other toxicological endpoints? Specifically for the prediction of toxicity, acceptable performance of predictive models in terms of specificity and sensitivity needs to be predefined for each experimental system. Is a prediction of hepatotoxicity with a 70% sensitivity good enough? How many false positives, i.e. what level of specificity, are we willing to accept? What are the limitations of the biological system (animal or in vitro models) and do we expect to predict specific human hepatotoxicants using rats or cell cultures? Are we also trying to address idiosyncratic events, which are, by definition, rare and unpredictable?

In any case, particularly for the pharmaceutical industry, the current research and development environment has clearly become increasingly challenging and requires sound, new, and imaginative ways to generate scientific data that will support the required characterization and differentiation of a molecule in terms of efficacy and safety, to make it a successful medicine on the market. In this chapter, some strategic considerations as well as concrete application examples regarding target organ toxicity are described.

5.3.2 Gene-expression analysis in the identification of target organ toxicity

Gene-expression analysis in toxicology (toxicogenomics) has been widely discussed and applied in predictive and mechanistic toxicology. These exploratory data have been used for internal decision making in the pharmaceutical industry and, to some extent, to support submissions to the health authorities. From a regulatory perspective, gene-expression data are generally considered exploratory and provide additional information to complement the more established endpoints (e.g. serum markers and histopathology). From a scientific research perspective, users have largely relied on pathway knowledge and gene-expression databases (DBs). The majority of DBs and models claiming good predictive performance have been generated by exposing animals (generally rats) to model compounds and generating gene-expression profiles from the target tissues, mainly liver and kidney, but 'omics approaches are amenable to any tissue type from which RNA can be extracted and data have been collected from other tissues of interest, such as heart, blood, brain, and testis.

Toxicogenomics: DBs

The amount of data generated is vast, complex, and therefore difficult to interpret statistically and biologically; therefore, it has been considered to compare data across experimental setups. Thus, several scientists, consortia, and commercial providers have generated DBs containing gene-expression fingerprints of known toxic and nontoxic substances. The specific fingerprints are generally used as comparators to predict the toxic liability of novel compounds under investigation. This approach is based in the principle that specific gene-expression changes associated with an observed pathology will be evident concomitantly to the histopathological finding or precede the histopathological lesion. Thus, a large number of compounds have been tested in an animal or cell culture models to identify their characteristic gene-expression patterns. These gene-expression patterns or fingerprints are then compared to known patterns from other compounds causing defined organ lesions or eliciting defined pharmacological responses (Figure 5.3.1). This task, although conceptually simple, is far from easily performed. Firstly, it is vital to be able to store, manage, and compute the relatively large data sets produced by microarray analysis of a large number of samples. Processing of the raw data is also pivotal to enable meaningful further analysis. There have been extensive discussions regarding the correct normalization and the appropriate statistical tests and algorithms by which relevant changes are distinguished from spurious findings due to background noise. An additional complication in the processing of the data arises from the use of different platforms to measure gene expression and from inter-laboratory variability even when using the same transcriptomics platform. Major efforts to clarify dissimilar results in different studies were performed by the MicroArray Quality Control (MAQC) project. This consortium provided experimental evidence of intra-platform consistency and inter-platform concordance regarding transcriptomics modulation. In a subsequent step, the MAQC-II project evaluated the performance of predictive models (algorithms) built using many combinations of analytical methods to analyze the same data sets. The model performance depended largely on the main endpoint analyzed and on the team proficiency, but different statistical approaches generated models of similar performance. Thus, MAQC and MAQC-II provided evidence towards establishing a framework for the use of microarrays in clinical and regulatory settings [3,4].

In addition to the technical and statistical pitfalls commonly encountered during the analysis of gene-expression data, it is mandatory for predictive toxicology that each study be carefully designed to address the underlying biological question. The selection of compounds, route of administration, dose and/or concentration, and time after exposure play a major role when trying to predict organ toxicity by comparing gene-expression changes. 'Omics approaches generate very rich data sets, but they remain a snapshot taken at a given moment of a very dynamic biological process. Therefore,

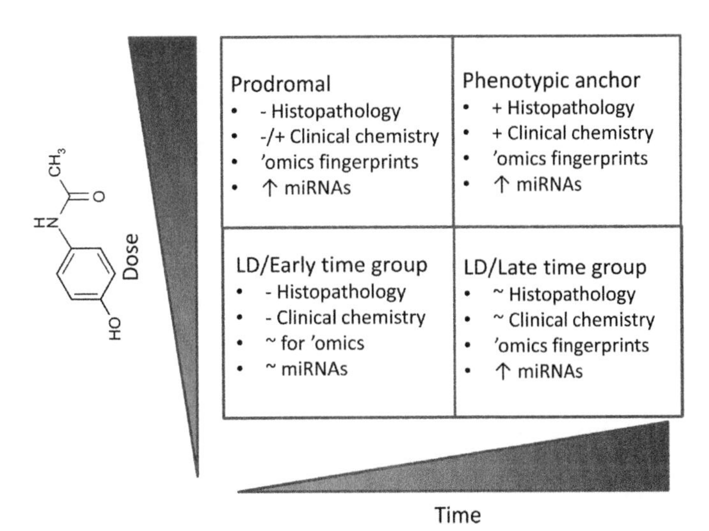

FIGURE 5.3.1 Schematic Representation of a Toxicogenomics Study Design with Two or More Doses (or Concentrations) and Two or More Time-Points.

The experiments need to be carefully designed to yield specific phenotypic anchors and capture prodromal signals. miRNAs, microRNAs.

and in order to increase the confidence in the causal or associative link between molecular endpoints (e.g. gene-expression changes) and toxicological processes, there is generally a need for a "phenotypic anchoring." These are signs other than the molecular changes that indicate that a compound or a group of compounds will ultimately lead to a lesion in a given tissue. These anchoring points may be biochemical parameters measured in serum or histopathological findings in the tissue. Carefully designed studies to populate DBs usually include at least one dose/time-point combination where such findings are observed and one dose/time-point where no findings other than molecular changes, such as transcript regulation, are seen. Thus, such studies cover the phenotypic manifestation as well as the predictive fingerprints.

Among the largest DBs are the former toxicogenomics DBs from Iconix (DrugMatrix, now available through the National Toxicogenomics Program) and GeneLogic, two commercial providers of toxicogenomics predictive models for the characterization of pharmaceuticals that became active in the late 1990s. In addition, the US Environmental Protection Agency (EPA) has applied various technologies, including toxicogenomics, to characterize chemicals in a research program entitled ToxCast, with the aims of predicting hazard, characterizing toxicity pathways, and prioritizing the toxicity testing of environmental chemicals. Another data source is the Chemical Effects in Biological Systems (CEBS) DB, a public resource hosted by the US National Institute of Environmental Health Sciences (NIEHS) that contains data of interest to environmental health scientists. CEBS houses data received from academic, industrial, and governmental laboratories and has been designed to display data in the context of biology and study design, and to permit data integration across studies. The Comparative Toxicogenomics Database (CTD) is an additional public resource that promotes understanding of the interaction of environmental chemicals with gene products, and their effects on human health. Some consortia, notably public–private partnerships such as the Predictive Safety Testing Consortium (PSTC) in the USA, the PredTox consortium in Europe, and the Toxicogenomics Project in Japan (TGP) have also generated relatively large amounts of microarray data after exposures of animals or cell cultures to toxicants [5,6]. The PSTC has an emphasis on biomarkers as it aims to identify and clinically qualify safety biomarkers for regulatory use as part of the FDA's Critical Path Initiative, a national strategy to drive innovation in the scientific processes through which medical products are developed, evaluated, and manufactured. The PredTox consortium was partly funded by the European Commission and partly by the European Federation of Pharmaceutical Industry and Associations (EFPIA) as a Framework 6 research program. Its major focus was to investigate results obtained with several 'omics technologies for the characterization of toxicity. Similarly, the TGP was funded by the Ministry of Health, Labor and Welfare, National Institute of Health Sciences (NIHS), and the working group of Japan Pharmaceutical Manufacturers Association [7].

In the majority of the DBs, genomics data is matched with conventional toxicity parameters and occasionally with other 'omics data, enabling researchers to perform meta-analysis across parameters, platforms, and data sets. In all these efforts, additional information arising from the compound structure itself has been largely ignored. Despite the vast experience in in silico approaches linking genotoxicity with specific chemical structures or chemical domains (toxicophores), this information is usually not considered when performing gene-expression analysis; hence quantitative structure–activity relationships (QSAR) modeling and toxicogenomics are typically used independently. One effort to combine these sources of information was aimed at evaluating several statistical models for predicting drug hepatotoxicity using chemical descriptors and toxicogenomics profiles of 127 drugs from the TGP DB. QSAR classification models based only on chemical descriptors gave a correct classification rate of 61%, while fingerprints based on the expression of 85 selected genes had a correct classification rate of 76%. Models combining both QSAR and gene-expression profiles were not better than genomics classifiers, but the use of both chemical and biological descriptors enriched the interpretation of the models [8]. In practical terms, and depending on the assessment method, well-performing predictive models have been described for hepatotoxicity, carcinogenicity, and nephrotoxicity. Depending on the analysis methods as well on the settings for sensitivity and specificity, most in vivo predictive models achieve a predictive performance (accuracy) greater than 80%, some even claim up to 90%. Discriminant analysis, support vector machines, and prediction analysis of microarray are some of the most commonly used machine learning algorithms used to fit models with the highly complex microarray data sets consisting of few biological replicates and thousands of variables. For example, for the liver, several models to analyze in vivo data have been published, among others that of Steiner et al. [9] and more recently that of the TGP [7]. Models also exist for proximal tubular damage in the kidney [10], and hepatocarcinogenesis [11,12].

Animal studies

As hepatotoxicity is a major concern for the pharmaceutical industry, it has been thoroughly studied using 'omics methods. There are several publications on individual substances and gene-expression DBs that are associated with hepatotoxicity. Rusty Thomas [13] published one of the first classification algorithms used for the accurate prediction of 24 hepatotoxicants from a large microarray database. Hamadeh [14] used discrimination algorithms to classify blinded samples based on a training set using high-density gene-expression profiles, and Steiner et al. [9] successfully used a support vector machine (SVM) to obtain optimal discrimination between hepatotoxic and non-hepatotoxic compounds (based on 26 compounds). A few years later, five model compounds were accurately discriminated based on gene-expression analysis measured 24 hours after a single administration [15]. These studies varied in experimental design, but all indicated the high potential of toxicogenomics for the prediction of hepatotoxicity. This approach described using model compounds has also been applied to the characterization of research compounds, although little data has been published. One very interesting, yet challenging, application of toxicogenomics is the analysis of pharmacologically related substances with different toxicological profiles. In these cases, the pharmacological activities of the compounds elicit responses that may mask their toxicological characteristics. However, it has been stated in the literature that it is possible to differentiate toxicologically relevant signals from pharmacological responses using gene-expression data and comparison with appropriately designed DBs. Two published examples describe serotonin receptor antagonists [16] and histamine-3 receptor inverse agonists [17]. These two pairs of compounds were structurally and pharmacologically similar but elicited distinct toxicity profiles in the liver. In both cases, and despite the similarities, the compounds could be clearly differentiated using hepatic genomics fingerprints in short-term in vivo studies. Moreover, the data demonstrate the predictive power of toxicogenomics, as the fingerprints obtained after a single administration were indicative of the histopathological lesions observed after 2 weeks of daily exposure.

The study of nephrotoxicity with 'omics markers focuses mainly on the more commonly observed proximal tubular damage (PTD), although some data are available on other kidney regions. This focus on one single lesion is a consequence of the anatomical complexity of the kidney; it is composed of clearly defined anatomical units with distinct cell composition and physiologies, leading to different susceptibilities to toxicants. A study to elucidate time- and dose-dependent global gene-expression changes associated with PTD in the rat, including four doses and four time-points, provided a sensitivity of 90% with a selectivity of 90%. Interestingly, 92 of the genes that drove the classification appeared to be well-known markers of kidney damage [10]. This is consistent with reports by other groups on gene-expression changes associated with proximal tubular toxicity. The identified genomic markers were mainly involved in tissue remodeling, immune response, inflammation, cell adhesion and migration, cell proliferation, and membrane transport and metabolism. Some of the genes identified using genomics approaches (e.g. kidney injury molecule 1, *KIM1*; clusterin; tissue inhibitor of metallopeptidase-1, *TIMP1*) were very consistent across experiments and have been thoroughly investigated and qualified as biomarkers for kidney damage [18–20].

The experiments investigating carcinogenesis were generally geared to generate gene-expression data after short-term (24 hours up to 2 weeks) in vivo exposures that may be able to predict the outcome of rodent life-time (2 years) bioassays. These models were generally built using gene-expression changes of selected subsets of genes, rather than the full transcriptome. The results are very promising, since gene-expression profiles from the livers of rats treated with genotoxic and non-genotoxic hepatocarcinogens show characteristic gene-expression profiles that serve for classification purposes with reported accuracy of approximately 88% [12,21]. The careful selection of biologically meaningful and statistically predictive subsets of genes makes this experimental approach amenable to evaluation using quantitative polymerase chain reaction (qPCR) [22].

In a similar fashion, single genes or small subsets of marker genes have been selected by other authors to address organ-specific toxicity. Boiling down the gene-expression fingerprints to few genes deviates from the original 'omics holistic concept but allows the employment qPCR as the technology of choice, with the advantages of qPCR being quicker, cheaper, and more quantitative than microarray analysis. By selecting genes specific for toxicity, including stress- and damage-related pathways, scientists combine a predictive approach based solely or mainly on statistical outcome and prediction accuracy with mechanistic understanding of the underlying biological processes, which are common to several organs. In some cases, a single gene has been used as a discriminator, but in general, subsets of genes comprise tens to hundreds of selected transcripts. One of the individual genes selected was *CYP1A1*, a biomarker for aryl hydrocarbon receptor (AhR) activation, a hallmark of dioxin-like toxicity that is a sensitive yet unspecific marker for AhR activation [23]. Similarly, *CYP2B10* has been identified as a genomic marker for the activation of the constitutive androstane receptor (CAR), providing supporting evidence of the species-specific differences in the observed pathophenotype [24]. More commonly, scientists select a subset of genes based on the combination of statistical and functional data. The selected genes, in general, are related to cellular processes including acute phase response, inflammation, oxidative stress, metabolic processes, protein misfolding, cell cycle regulation, apoptosis, and detoxification. For any subset of genes analyzed by qPCR, the selection of stable housekeeping genes for normalization is vital.

In vitro approaches

In addition to the large amount of genomics data generated in in vivo systems, cell culture systems to assess teratogenicity/embryotoxicity or cardiac toxicity have been published. Regarding the complex matter of teratogenesis, reproductive medicine is also entering the era of the 'omics, although its application is less widespread than for organ toxicity. It is clear that transcriptional regulation plays a major role during embryonic development and specific periods of activity of many signaling pathways have been described during fetal development. Gene-expression analysis has therefore been applied to in vitro-based developmental models, such as whole-embryo culture (WEC) and the embryonic stem cell test (EST). Genomic quantitative endpoints have been combined with the validated in vitro EST to detect teratogenicity using mouse embryonic stem cells. The developments are still at a preliminary stage, but it is expected that the use of transcriptomics endpoints to study the modulation of differentiation may be more objectively (i.e. quantitatively) and easily performed than visual assessment of the cell cultures. Preliminary data show that gene-expression signatures can be used to identify developmental toxicants. The EST has also been combined with genomics endpoints with acceptable predictive performance (63 to 83%) regarding developmental toxicity based on analysis of predefined gene sets [25]. The effects of all-trans retinoic acid (RA) on rat WEC and on rats (in vivo) across six time-points were comparable regarding the gene-expression response (directionality, significance) and functional parameters (e.g. embryonic development, cell differentiation) [26]. Away from the traditional mammalian models, toxicogenomics approaches have also been described for the study of developmental disturbances caused by ecotoxicants on zebrafish embryos, showing distinctive gene-expression profiles for the different toxicants [27].

Despite the generally good success of gene-expression analysis to predict toxicity in vivo, the implementation of predictive toxicogenomics fingerprints in in vitro systems remains difficult with regard to the detection of target organ toxicity. Cell lines differ greatly from the original organ in terms of basal status, cell biology, and response to toxic insult. Moreover, many cell lines are genetically unstable and display an abnormal karyotype. Primary cells from animals or man are considered more similar to the tissue under investigation, but are known to suffer isolation stress and to undergo relatively rapid dedifferentiation leading to death when maintained in the commonly used monolayer cultures. In particular for primary hepatocytes, it is known that they rapidly lose their metabolic ability in vitro and are thus unable to generate metabolites in pertinent amounts. A comparative transcriptomics study using several in vitro hepatic systems shows profound differences between culture systems (cell lines, primary cells in monolayer or sandwich culture, and liver slices), as well as a time-dependent deterioration. Liver slices exhibit the strongest similarity to liver tissue regarding mRNA expression, whereas cell lines are most dissimilar. Decreases in transcripts coding for cytochrome P450 are paralleled by decreased protein levels and enzymatic activity [28]. Nevertheless, some transcriptional changes observed in rat primary cell cultures (hepatocytes or renal cortical tubular cells) were considered consistent with the changes observed in vivo, but all investigators reported differences. In particular, in vitro systems fail to capture compounds that cause biliary damage and cholestasis in the liver (e.g. lithocholic acid and chlorpromazine). This is not surprising, since most in vitro systems used for toxicogenomics analysis do not recapitulate the physiological structures of the liver and fail to form bile canaliculi. Hepatotoxicants that cause direct toxicity to the hepatocyte, such as acetaminophen (APAP), lead to gene-expression changes that are similar to the in vivo findings. However, the concentrations at which these findings appear in vitro are very often much higher than the actual concentrations observed in vivo in plasma and tissue. At these concentrations, where hepatocytes are undergoing cytotoxicity that can be measured by biochemical endpoints such as enzyme leakage, reduction of tetrazolium dye (MTT test), or ATP content, the additional information gained from the complex and expensive genomics data is relatively limited. Proximal tubular kidney cells exposed to concentrations of compounds below the cytotoxic range affected a small number of transcripts. These transcripts showed modest commonalities between compounds and were mainly indicative of unspecific cellular stress, while the hallmarks of in vivo nephrotoxicity, such as *KIM1*, *TIMP1*, and clusterin, were not modulated [29]. More recently, cell systems have been employed for the detection of genotoxicity, using a transcriptomics analysis of HepG2 cells and leading to excellent results regarding predictive accuracy [30]; however, these results need to be independently reproduced and the method validated in order to be used as a possible screening tool.

Despite the modest success of large genomics in vitro databases for the prediction of organ toxicity, the effects of compounds with known pharmacological or toxicological mechanisms can be reproduced in vitro and used to elucidate mechanistic aspects of toxicity in a more controlled environment. For example, exposure of primary rat hepatocytes to peroxisome proliferator activated receptor-α (PPARα) ligands led to the induction of β-oxidation-related genes, as observed in liver tissue after in vivo exposure. However, genes related to other biological pathways that were affected in vivo (e.g. cell proliferation and apoptosis) were not modulated in the cell culture [7]. These results are in agreement with data obtained with PPARα-agonists in rodents and focusing on gene-expression changes believed to be related to the toxicology of the compounds. The gene-expression profiles after the exposure of rats (in vivo) or primary rat hepatocytes (in vitro) to non-genotoxic carcinogens (PPARα agonists), genotoxic carcinogens, and non-carcinogens differed. Two out of the three tested genotoxic carcinogens showed a similar response in vitro as in vivo with regard to DNA damage response, and were therefore correctly classified. Non-genotoxic hepatocarcinogens, on the other hand, elicited weak responses in the hepatocytes that were not concordant with the observations in animals [31]. This partial recapitulation of the in vivo effects in cell culture systems has led to the application of focused or targeted genomics approaches to differentiate compounds belonging to the same chemical or pharmacological class. Although this analysis strategy is largely similar to the general predictive algorithms, this focused approach does not strive to generate fingerprints applicable to any kind of hepatotoxicant, but rather to tweak out subtle differences in gene-expression responses that make a compound more or less likely than another one of the series to cause liver damage.

Progress in the field of generation of relevant cell culture systems has accompanied the evolution of gene expression as a means to evaluate toxic liability. Among recent developments, complex cell culture systems involving co-cultures, three-dimensional structures, and microfluidics systems are more stable, can be kept in culture for longer periods, and are believed to better recapitulate the physiology of the tissue. However, so far there are no large gene-expression databases generated in such systems. In addition, the new developments in the research on cell reprogramming and differentiation have led to the application of induced pluripotent stem cells (iPSC) differentiated into cardiomyocytes for the investigation of cardiotoxicity. These recent reports include several endpoints, such as cytotoxicity and electrophysiological characteristics, but this approach is still in its infancy with regard to the generation of substantial, toxicity-relevant genomics data [32].

5.3.3 Integration of gene-expression data with other 'omics technologies

Integration of several 'omics platforms is not trivial, as technical and biological aspects increase the degree of difficulty. Technically speaking, the number of analytes (e.g. transcripts, proteins, and endogenous metabolites) greatly differs between platforms. Also, the dynamic ranges of concentrations of said analytes are usually not comparable and the methodological sensitivities of the analysis methods are disparate. This makes the comparison of an expressed protein to a transcribed messenger RNA extremely complex. Moreover, most experimental setups only allow a "snapshot view" of the molecular environment in the tissue. This means that one can assess the level of expression of a transcript, its corresponding enzyme, and the generated metabolite at the single time-point at which the sample has been collected. With this, the kinetics of the changes that convert a transcriptional induction in the subsequent up-regulation of a protein and possibly ensuing changes in the metabolic environment can generally not be appropriately captured. To add to the complexity, exquisite feedback regulation loops may confuse matters even further, since the reduction of a given protein in cells may cause a compensatory induction of the coding transcript. Not surprisingly, many reports indicate poor concordance between gene- and protein-expression changes. However, there are ample indications that regulation of specific pathways, rather than of single genes or proteins, can be observed following pharmacological intervention or toxic insult to an organ. Some of the most prominent examples are described with the changes produced in the liver with substances leading to liver hypertrophy in the rat, such as PPARα agonists (e.g. fenofibrate) or CAR agonists (e.g. phenobarbital).

The work performed by the participants of the Framework 6 PredTox consortium highlights the usefulness of the pathway-related integration of 'omics data. In that study, rats were exposed to a series of toxicants causing liver or kidney injury. The integrated analysis of the conventional toxicological endpoints (histopathology and serum biochemistry) with liver transcriptomics data, proteomics, and metabolomics was the most informative approach for the generation of mechanistic models [6]. Novel uncharacterized compounds causing liver hypertrophy could be separated into two categories based on the molecular analysis. On the one hand, a group of chemicals elicited a marked increase in the expression of xenobiotic-metabolizing enzymes; accumulation of these proteins in the smooth endoplasmic reticulum (SER) and proliferation of the SER as the main underlying cause for the observed hepatic hypertrophy. On the other hand, two compounds caused transcriptional up-regulation of fatty acid β-oxidation, associated with the induction of genes involved in peroxisome proliferation, leading to the observed liver hypertrophy. Thus, two molecular mechanisms lead to a similar histopathological phenotype. This understanding of adverse outcome pathways can have profound implications when assessing the risk to patients. Other hepatotoxic compounds tested in the same PredTox program caused bile duct necrosis and hyperplasia, as well as hepatocyte necrosis and regeneration, together with hepatic inflammation. Gene-expression changes shared by these compounds suggested a potential sequence of molecular events preceding and accompanying the classical endpoint observations. The modulated genes were implicated in early stress responses, regenerative processes, inflammation, fibrotic processes, and cholestasis. The evaluation of metabolite profiles with liquid chromatography–linked mass spectrometry (LC-MS) showed increased levels of bile acids in response to some of these compounds, allowing a refinement in the characterization of the type of hepatobiliary toxicity. PredTox also evaluated a subgroup of nephrotoxicants by analyzing kidney tissue, as well as markers circulating in the blood. The mechanistic interpretation of the molecular processes during the proximal tubular damage observed in the kidney of the treated animals revealed transcriptional regulation of genes that had been previously associated with kidney damage, such as *KIM1*, clusterin, and *TIMP1*. In addition, several other deregulated pathways were congruent with the histopathological observations, including a specific effect on the complement system. Proteomic data obtained by two-dimensional differential gel electrophoresis (2D-DIGE) proteomics analysis of kidney tissue showed a deregulation of proteins (mostly down-regulated) involved in oxidative stress, detoxification, and energy metabolism [20,33,34].

Integration of 'omics technologies can also be applied to in vitro systems. In work performed with the cell lines HepG2 and C3A co-cultured on a biochip and exposed to the hepatotoxicant acetaminophen (APAP), the integration of the transcriptomic and proteomic analyses revealed changes in genes and proteins in the NRF2 antioxidant response pathway and fatty acid metabolism. The induction of those pathways in the biochip enhanced the metabolism of APAP when compared to monocultures. As in reported in vivo, the metabolic signature of APAP toxicity in the biochip showed changes in calcium homeostasis and lipid metabolism, as well as reorganization of the cytoskeleton at the transcriptome and proteome levels. These results exemplify the power of combining complex, sophisticated cell culture systems with integrated 'omics endpoints for the characterization of toxicity [35].

5.3.4 Systems toxicology approaches for biomarker discovery and mechanisms of toxicity

Besides the applications of 'omics for predictive toxicology, systems toxicology approaches have led to a better understanding of molecular mechanisms of toxicity and to the discovery of putative new biomarkers. Accurate biomarkers for the detection of unexpected side effects are necessary in order to be able to monitor the onset of the toxic event. Ideally, biomarkers should be noninvasive, or at least minimally invasive (e.g. blood sampling). They are most useful if they can be measured repeatedly during longitudinal studies and thus provide information about onset, progression, and reversibility of a lesion. Ideally, they should behave similarly in several species, including man, to address translational aspects and relevance to man.

Several studies aimed at discovering new biomarkers for organ damage using one or several 'omics technologies. The surface-enhanced laser desorption/ionization time-of-flight mass spectrometry (SELDI-TOF-MS) proteomics analysis in the PredTox project mentioned before led to the identification of seven plasma proteins and four liver proteins modulated in association with histopathological findings in the liver that can be considered potential safety biomarkers [36]. For one selected hepatotoxic compound, a targeted LC-MS-based proteomic analysis revealed only moderate correlation of individual protein expression with changes in mRNA expression in analysis of the same liver samples. However, both transcript and protein changes affected similar pathways, confirming also the mechanistic hypothesis that the compound acted as PPARα agonist. Hence, a panel of potential biomarkers of liver toxicity was assembled from the label-free LC-MS proteomics discovery data, the previously acquired transcriptomics data, and selected candidates identified from the literature, ultimately leading to the development of reaction monitoring assays for 48 putative hepatotoxicity markers [37]. Similarly, thiostatin and neutrophil gelatinase-associated lipocalin (NGAL), were increased in serum and urine of animals treated with hepatotoxicants in a time- and dose-dependent manner. These protein changes correlated well with mRNA expression in the target organ, and generally reflected the onset and degree of drug-induced liver injury. Statistical analysis showed that serum thiostatin is a more sensitive indicator of drug-induced hepatobiliary injury than conventional clinical chemistry parameters, i.e. alkaline phosphatase (ALP), alinine aminotransferase (ALT), and aspartate aminotransferase (AST). However, the specificity of thiostatin as a marker for hepatotoxicity needs to be established, as it is an acute-phase protein expressed in a wide range of tissues [38].

Not only the analysis of biofluids such as blood and urine serves for the identification of putative biomarkers. Sophisticated proteomics and image analysis with matrix-assisted laser desorption/ionization (MALDI) imaging and mass spectrometry (MS) can provide the spatial distribution and relative abundance of proteins in tissue, enabling differential analysis of the mass spectrum profiles. Using this approach in rats treated with gentamicin, a known nephrotoxicant, transthyretin was uniquely identified as a putative biomarker of kidney damage using a combination of tissue microextraction and fractionation by reverse-phase liquid chromatography, followed by top-down tandem MS [39]. On the other hand, the effect of gentamicin on the kidney assessed by transcriptomics and proteomics appeared to be complementary, indicating the onset of inflammatory processes as well as mitochondrial dysfunction with impairment of cellular energy production, induction of oxidative stress, and an effect on protein biosynthesis and on cellular assembly and organization. Proteomic results also provided clues for potential nephrotoxicity biomarkers such as AGAT and plasma retinoid binding protein 4 (PRBP4), which were strongly modulated in the kidney [40]. The identification and publication of putative biomarkers is, however, not sufficient to ensure their implementation in the toxicological assessment of compounds. Technical and biological validation are a necessary, yet time-consuming and difficult, step towards the qualification of newly discovered biomarkers and their acceptance in the scientific and regulatory world. An excellent example of biomarker discovery followed by qualification is represented by the kidney biomarker panel, detected originally using gene-expression, imaging, in vitro screening, and protein assays. These efforts led to a list of putative peripheral biomarkers such as the urinary glutathione S-transferase-alpha, N-acetyl-beta-d-glucosaminidase, total protein, cystatin C, beta2-microglobulin, KIM-1, lipocalin-2, and serum cystatin C. For these biomarkers, there were robust data on assay performance, generalization across several compounds, and performance (sensitivity and specificity). Also reproducibility across several experimental sites was demonstrated. These biomarkers have been qualified in the preclinical setting and are currently being qualified for clinical applications [18,41].

5.3.5 miRNAs and organ toxicity: putative biomarkers of toxicological processes

Non-coding RNAs (nc RNAs) comprise a variety of RNA species that do not encode proteins and were thought for a long time to be part of the so-called "junk DNA." Recent publications have, however, shown that several of these nc RNAs have profound effects in the regulation of transcription and translation and, therefore, play major roles in biological processes in health and disease. Non-coding RNAs include small nc RNAs that are normally less than 200 bases, such as microRNA (miRNA), short interfering RNA (siRNA), and piwi-interacting RNA (piRNA), as well as long nc RNA (lincRNA) and the also long antisense RNA. Among these macromolecules, miRNAs (approximately 22 nucleotides long) have been studied the most and have generated interest due to the pleiotropic modulation of translation that they exert. A single miRNA controls hundreds of target genes, so that miRNAs modulate a variety of biological functions, including embryogenesis, cell differentiation, tissue homeostasis, carcinogenesis, toxicity, and viral infections. Moreover, under certain conditions, miRNAs are exported from cells and remain stable in circulation, which makes them easily accessible in plasma or serum and in other body fluids. miRNAs are also easily measurable using standard molecular biology methods like quantitative reverse transcription PCR (qRT-PCR), microarrays, or sequencing, making them attractive biomarkers for pathological processes. In particular, in the toxicology field, the fact that miRNAs show tissue specificity and are relatively well conserved across species has made them prime candidates as organ-specific translational biomarkers. Several publications have studied miRNAs as biomarkers for cardiac toxicity, hepatotoxicity, muscle toxicity, and a variety of other organ toxicities, such as vasculitis. For example, plasma concentrations of miR-122, miR-133a, and miR-124 were increased in rats dosed with model toxicants that caused injuries in liver, muscle, and brain, respectively [42].

In the last few years, it has become clear that miRNAs are dominant players in different aspects of cardiac remodeling, including fibrosis, in animals and man. In mice, the overexpression of the myocardial miR-21 is related to cardiac fibrosis elicited by pressure overload. In man, the myocardial and plasma levels of miR-21 were significantly higher in aortic stenosis patients compared with controls, and correlated directly with the echocardiographic mean transvalvular gradients. The results indicate that miR-21 acts as a regulator of the fibrotic process and can be detected in the circulation [43]. Coronary artery disease and acute myocardial infarction (AMI) are currently diagnosed using serum cardiac-specific troponin levels. However, recent studies have found that circulating miRNAs are closely linked to myocardial injury and are detectable earlier than troponin [44]. Also, in the liver, miRNAs are involved in processes related to toxicity and liver injury. It has been shown that the mechanism of hepatocellular proliferation induced by PPAR-α agonists in rodents involves the down-regulation miR let-7c. This miRNA destabilizes the transcript of c-Myc, and therefore inhibits cell proliferation. Upon activation of the mouse, but not the human, PPAR-α receptor, the levels of miR let-7c are decreased, leading to an elevation of c-Myc mRNA and protein, ultimately resulting in enhanced hepatocellular proliferation [45]. One of the most abundant miRNAs in hepatic tissue is miR-122, and the relationship between levels of circulating miR-122 and liver damage due to toxic insult is currently well established. Moreover, this biomarker appears to show better tissue specificity than the traditionally measured transaminases. In addition, the data show that the sensitivity is higher and that the elevations of miR-122 occur in several species. Using APAP-induced liver injury, several groups have shown the elevation of miR-122 in mouse, rat, and man. In the mouse, APAP caused highly significant differences in the spectrum and levels of microRNAs in both liver tissues and in plasma. In particular, miR-122 and miR-192 showed dose-dependent changes in the plasma that were detectable earlier than serum aminotransferase elevations [46]. Similarly, in Sprague-Dawley rats exposed to toxic doses of APAP or the hepatotoxic herb *Dioscorea bulbifera*, miR-122 and miR-192 were among the most elevated miRNAs [47]. Underlining the translational potential of miRs as biomarkers, it has been shown that patients undergoing acute liver injury due to APAP overdose also showed elevated levels of plasma miR-122 and miR-192, whereas patients exposed to APAP but without liver injury did not show these elevated levels. In addition, patients presenting with non-APAP-induced liver injury also displayed elevated levels of those miRNAs, providing initial evidence that the effect is not due to APAP exposure but to liver injury, independent of the original cause [48]. Taken together, the results demonstrate that circulating miRNAs are suitable biomarkers of liver damage, responding prior to other circulating hepatocellular toxicity biomarkers. The large concordance between data obtained in several species supports miRNAs having a strong potential as translational biomarkers. In addition, nucleic acids are easier to measure than proteins, especially across species. Despite these very promising results, these are still early days in the use of miRNAs as biomarkers for toxicity and quite extensive additional work on qualification needs to be performed.

References

[1] Hartung T. Toward a new toxicology: evolution or revolution? Altern Lab Anim 2008;36:635–9.
[2] Jacobs A. An FDA perspective on the nonclinical use of the X-omics technologies and the safety of new drugs. Toxicol Lett 2009;186:32–5.
[3] MAQC Consortium Shi L, Reid LH, et al. The MicroArray Quality Control (MAQC) project shows inter- and intraplatform reproducibility of gene expression measurements. Nat Biotechnol 2006;24:1151–61.
[4] Shi L, Campbell G, Jones WD, Campagne F, et al. The MicroArray Quality Control (MAQC)-II study of common practices for the development and validation of microarray-based predictive models. Nat Biotechnol 2010;28:27–38.
[5] Noriyuki N, Igarashi Y, Ono A, Yamada H, Ohno Y, Urushidani T. Evaluation of DNA microarray results in the Toxicogenomics Project (TGP) consortium in Japan. J Toxicol Sci 2012;37:791–801.
[6] Suter L, Schroeder S, Meyer K, Gautier JC, Amberg A, Wendt M, et al. EU Framework 6 project: predictive Toxicology (PredTox)—overview and outcome. Toxicol Appl Pharmacol 2011;252:73–84.
[7] Uehara T, Ono A, Maruyama T, Kato I, Yamada H, Ohno Y, et al. The Japanese toxicogenomics project: application of toxicogenomics. Mol Nutr Food Res 2010;54:218–27.
[8] Low Y, Uehara T, Minowa Y, Yamada H, Ohno Y, Urushidani T, et al. Predicting drug-induced hepatotoxicity using QSAR and toxicogenomics approaches. Chem Res Toxicol 2011;24:1251–62.
[9] Steiner G, Suter L, Boess F, Gasser R, de Vera MC, Albertini S, et al. Discriminating different classes of toxicants by transcript profiling. Environ Health Perspect 2004;112:1236–48.
[10] Kondo C, Minowa Y, Uehara T, Okuno Y, Nakatsu N, Ono A, et al. Identification of genomic biomarkers for concurrent diagnosis of drug-induced renal tubular injury using a large-scale toxicogenomics database. Toxicology 2009;265:15–26.

[11] Ellinger-Ziegelbauer H, Aubrecht J, Kleinjans JC, Ahr HJ. Application of toxicogenomics to study mechanisms of genotoxicity and carcinogenicity. Toxicol Lett 2009;186:36–44.

[12] Nie AY, McMillian M, Parker JB, Leone A, Bryant S, Yieh L, et al. Predictive toxicogenomics approaches reveal underlying molecular mechanisms of nongenotoxic carcinogenicity. Mol Carcinog 2006;45:914–33.

[13] Thomas RS, Rank DR, Penn SG, Zastrow GM, Hayes KR, Pande K, et al. Identification of toxicologically predictive gene sets using cDNA microarrays. Mol Pharmacol 2001;60:1189–94.

[14] Hamadeh HK, Bushel PR, Jayadev S, DiSorbo O, Bennett L, Li L, et al. Prediction of compound signature using high density gene expression profiling. Toxicol Sci 2002;67:232–40.

[15] Zidek N, Hellmann J, Kramer PJ, Hewitt PG. Acute hepatotoxicity: a predictive model based on focused illumina microarrays. Toxicol Sci 2007;99:289–302.

[16] Suter L, Haiker M, De Vera MC, Albertini S. Effect of two 5-HT6 receptor antagonists on the rat liver: a molecular approach. Pharmacogenomics J 2003;3:320–34.

[17] Roth A, Boess F, Landes C, Steiner G, Freichel C, Plancher JM, et al. Gene-expression-based in vivo and in vitro prediction of liver toxicity allows compound selection at an early stage of drug development. J Biochem Mol Toxicol 2011;25:183–94.

[18] Dieterle F, Marrer E, Suzuki E, Grenet O, Cordier A, Vonderscher J. Monitoring kidney safety in drug development: emerging technologies and their implications. Curr Opin Drug Discov Dev 2008;11:60–71.

[19] Fielden MR, Eynon BP, Natsoulis G, Jarnagin K, Banas D, Kolaja KL. A gene expression signature that predicts the future onset of drug-induced renal tubular toxicity. Toxicol Pathol 2005;33:675–83.

[20] Matheis KA, Com E, Gautier JC, Guerreiro N, Brandenburg A, Gmuender H, et al. Cross-study and cross-omics comparisons of three nephrotoxic compounds reveal mechanistic insights and new candidate biomarkers. Toxicol Appl Pharmacol 2011;252:112–22.

[21] Ellinger-Ziegelbauer H, Gmuender H, Bandenburg A, Ahr HJ. Prediction of a carcinogenic potential of rat hepatocarcinogens using toxicogenomics analysis of short-term in vivo studies. Mutat Res 2008;637:23–39.

[22] Fielden MR, Adai A, Dunn 2nd RT, Olaharski A, Searfoss G, Sina J, et al. Development and evaluation of a genomic signature for the prediction and mechanistic assessment of nongenotoxic hepatocarcinogens in the rat. Toxicol Sci 2011;124:54–74.

[23] Hu W, Sorrentino C, Denison MS, Kolaja K, Fielden MR. Induction of cyp1a1 is a nonspecific biomarker of aryl hydrocarbon receptor activation: results of large scale screening of pharmaceuticals and toxicants in vivo and in vitro. Mol Pharmacol 2007;71:1475–86.

[24] Hoflack JC, Mueller L, Fowler S, Braendli-Baiocco A, Flint N, Kuhlmann O, et al. Monitoring Cyp2b10 mRNA expression at cessation of 2-year carcinogenesis bioassay in mouse liver provides evidence for a carcinogenic mechanism devoid of human relevance: the dalcetrapib experience. Toxicol Appl Pharmacol 2012;259:355–65.

[25] van Dartel DA, Pennings JL, de la Fonteyne LJ, Brauers KJ, Claessen S, van Delft JH, et al. Evaluation of developmental toxicant identification using gene expression profiling in embryonic stem cell differentiation cultures. Toxicol Sci 2010;119:126–34.

[26] Robinson JF, Yu X, Moreira EG, Hong S, Faustman EM. Arsenic- and cadmium-induced toxicogenomic response in mouse embryos undergoing neurulation. Toxicol Appl Pharmacol 2011;250:117–29.

[27] Sawle AD, Wit E, Whale G, Cossins AR. An information-rich alternative chemicals testing strategy using high-definition toxicogenomics and zebrafish (*Danio rerio*) embryos. Toxicol Sci 2010;118:128–39.

[28] Boess F, Kamber M, Romer S, Gasser R, Muller D, Albertini S, et al. Gene expression in two hepatic cell lines, cultured primary hepatocytes, and liver slices compared to the in vivo liver gene expression in rats: possible implications for toxicogenomics use of in vitro systems. Toxicol Sci 2003;73:386–402.

[29] Suzuki H, Inoue T, Matsushita T, Kobayashi K, Horii I, Hirabayashi Y, et al. In vitro gene expression analysis of nephrotoxic drugs in rat primary renal cortical tubular cells. J Appl Toxicol 2008;28:237–48.

[30] Magkoufopoulou C, Claessen SM, Tsamou M, Jennen DG, Kleinjans JC, van Delft JH. A transcriptomics-based in vitro assay for predicting chemical genotoxicity in vivo. Carcinogenesis 2012;33:1421–9.

[31] Doktorova TY, Ellinger-Ziegelbauer H, Vinken M, Vanhaecke T, van Delft J, Kleinjans J, et al. Comparison of hepatocarcinogen-induced gene expression profiles in conventional primary rat hepatocytes with in vivo rat liver. Arch Toxicol 2012;86:1399–411.
[32] Puppala D, Collis LP, Sun SZ, Bonato V, Chen X, Anson B, et al. Comparative gene expression profiling in human induced pluripotent stem cell derived cardiocytes and human and cynomolgus heart tissue. Toxicol Sci 2013;131:292–301.
[33] Boitier E, Amberg A, Barbié V, Blichenberg A, Brandenburg A, Gmuender H, et al. A comparative integrated transcript analysis and functional characterization of differential mechanisms for induction of liver hypertrophy in the rat. Toxicol Appl Pharmacol 2011;252:85–96.
[34] Ellinger-Ziegelbauer H, Adler M, Amberg A, Brandenburg A, Callanan JJ, Connor S, et al. The enhanced value of combining conventional and "omics" analyses in early assessment of drug-induced hepatobiliary injury. Toxicol Appl Pharmacol 2011;252:97–111.
[35] Prot JM, Briffaut AS, Letourneur F, Chafey P, Merlier F, Grandvalet Y, et al. Integrated proteomic and transcriptomic investigation of the acetaminophen toxicity in liver microfluidic biochip. PLoS ONE 2011;6:e21268.
[36] Collins BC, Miller CA, Sposny A, Hewitt P, Wells M, Gallagher WM, et al. Development of a pharmaceutical hepatotoxicity biomarker panel using a discovery to targeted proteomics approach. Mol Cell Proteomics 2012;11:394–410.
[37] Collins BC, Sposny A, McCarthy D, Brandenburg A, Woodbury R, Pennington SR, et al. Use of SELDI MS to discover and identify potential biomarkers of toxicity in InnoMed PredTox: a multi-site, multi-compound study. Proteomics 2010;10:1592–608.
[38] Adler M, Hoffmann D, Ellinger-Ziegelbauer H, Hewitt P, Matheis K, Mulrane L, et al. Assessment of candidate biomarkers of drug-induced hepatobiliary injury in preclinical toxicity studies. Toxicol Lett 2010;196:1–11.
[39] Meistermann H, Norris JL, Aerni HR, Cornett DS, Friedlein A, Erskine AR, et al. Biomarker discovery by imaging mass spectrometry: transthyretin is a biomarker for gentamicin-induced nephrotoxicity in rat. Mol Cell Proteomics 2006;5:1876–86.
[40] Com E, Boitier E, Marchandeau JP, Brandenburg A, Schroeder S, Hoffmann D, et al. Integrated transcriptomic and proteomic evaluation of gentamicin nephrotoxicity in rats. Toxicol Appl Pharmacol 2012;258:124–33.
[41] Dieterle F, Sistare F, Goodsaid F, Papaluca M, Ozer JS, Webb CP, et al. Renal biomarker qualification submission: a dialog between the FDA-EMEA and Predictive Safety Testing Consortium. Nat Biotechnol 2010;28:455–62.
[42] Laterza OF, Lim L, Garrett-Engele PW, Vlasakova K, Muniappa N, Tanaka WK, et al. Plasma microRNAs as sensitive and specific biomarkers of tissue injury. Clin Chem 2009;55:1977–83.
[43] Villar AV, Garcia R, Merino D, Llano M, Cobo M, Montalvo C, et al. Myocardial and circulating levels of microRNA-21 reflect left ventricular fibrosis in aortic stenosis patients. Int J Cardiol 2013;167:2875–81.
[44] Li C, Pei F, Zhu X, Duan DD, Zeng C. Circulating microRNAs as novel and sensitive biomarkers of acute myocardial infarction. Clin Biochem 2012;45:727–32.
[45] Gonzalez FJ, Shah YM. PPARalpha: mechanism of species differences and hepatocarcinogenesis of peroxisome proliferators. Toxicology 2008;246:2–8.
[46] Wang K, Zhang S, Marzolf B, Troisch P, Brightman A, Hu Z, et al. Circulating microRNAs, potential biomarkers for drug-induced liver injury. Proc Natl Acad Sci USA 2009;106:4402–7.
[47] Su Y-W, Chen X, Jiang Z-Z, Wang T, Wang C, Zhang Y, et al. A panel of serum microRNAs as specific biomarkers for diagnosis of compound- and herb-induced liver injury in rats. PloS ONE 2012;7:e37395.
[48] Starkey Lewis PJ, Dear J, Platt V, Simpson KJ, Craig DG, Antoine DJ, et al. Circulating microRNAs as potential markers of human drug-induced liver injury. Hepatology 2011;54:1767–76.

CHAPTER

Hepatotoxicity and the Circadian Clock: A Timely Matter

5.4

Annelieke S. de Wit[†], Romana Nijman[†], Eugin Destici[†,*], Ines Chaves[†] and Gijsbertus T.J. van der Horst[†]

[†]Department of Genetics, Erasmus University Medical Center, Rotterdam, The Netherlands,
[*]Department of Medicine, School of Medicine, University of California, San Diego, California

5.4.1 Introduction

As a direct result of the composition of our solar system, life on earth is continuously exposed to temporal changes to the environment. As well as annual, seasonal, and lunar cycles, we also experience 24-hour light/dark and temperature cycles, caused by the rotation of the earth around its own axis. To cope with these cyclic changes, organisms have acquired an internal timing mechanism with a periodicity of approximately 24 hours [1]. This circadian clock (Lat. circa = near, dies = day) is an anticipatory mechanism that allows an organism to adjust behavior, physiology, and metabolism (e.g. body temperature, sleep–wake cycle, blood pressure, and locomotor activity) to the specific needs at defined stages over the day [2,3]. The importance of circadian clocks is well illustrated by the fact that they evolved multiple times during evolution and are present in almost all life forms on earth, ranging from single cellular organisms (e.g. bacteria) to multicellular organisms (e.g. plants and animals) [4,5]. Circadian clock research has largely focused on a series of model organisms, notably the filamentous fungus *Neurospora crassa*, the plant *Arabidopsis thaliana*, the fruit fly *Drosophila melanogaster*, zebrafish, rodents, and humans [5,6].

Circadian rhythms are defined as being endogenous, self-sustained, persisting in the absence of any environmental cues (such as the light/dark cycle), and having close approximation to the period of the earth's rotation [7]. For instance, mice kept in constant darkness exhibit sleep/wake and locomotor activity cycles with a periodicity around 23.5 hours. Such rhythms are called free-running rhythms, their periodicity defined by the Greek symbol tau (τ). Likewise, humans have been demonstrated to maintain a free-running sleep/wake cycle with a periodicity slightly longer than 24 hours [8,9]. Circadian rhythms are temperature compensated and will therefore sustain the same period under different temperatures [10]. As circadian rhythms have a period of approximately, but not exactly, 24 hours, it is necessary that they are synchronized to the environment. This process is called entrainment and requires external factors, known as *zeitgebers*, of which light is the most reliable cue [4]. In its simplest form, a circadian clock can be said to be composed of three components: a

Toxicogenomics-Based Cellular Models.

central clock (an internal oscillator, generating body time), an input (keeping the clock in phase with environmental cues, notably the light–dark cycle), and an output (coupling the oscillations to biological processes). This basic scheme applies to all known circadian clocks, regardless of the species involved.

5.4.2 The mammalian circadian clock

In mammals, the central clock is located in the suprachiasmatic nucleus (SCN), a small bilateral paired structure of approximately 10,000–20,000 neurons located in the ventral hypothalamus in the brain [11–13]. These neurons are able to generate circadian rhythms in electrical, neuronal, and hormonal activities, which regulate several biological functions and output processes, including circadian patterns in behavioral activity [2,14].

To keep pace with the day/night cycle, the SCN is entrained by light. This process is called photoentrainment and is best illustrated in free-running mammals. Brief light pulses have been shown to phase-shift behavioral rhythms, and the direction and magnitude of the phase shift depend on the moment of the light pulse within the circadian cycle. For instance, a phase delay occurs when a light pulse is given during the early subjective night, while a phase advance is seen when a light pulse is given during the late subjective night. There is no phase shift when a light pulse is given during the subjective day [15].

The light information required for photoentrainment of the mammalian circadian clock is perceived by the eye and transmitted to the SCN through the retinohypothalamic tract. To this end, the inner nuclear layer of the retina is equipped with a redundant set of photoreceptor cells and pigments, notably the opsin-containing rods and cones and subset of ganglion cells that contain the photopigment melanopsin and project specifically to the SCN [16–20].

The molecular clock

At the cellular level, circadian rhythms are generated by a cell-autonomous molecular oscillator, composed of a series of clock genes and proteins (see Figure 5.4.1). The organization of the mammalian circadian oscillator is based on a complex autoregulatory transcription–translation feedback loop (TTFL) mechanism [3,21–25]. The TTFL drives transcription and translation of the clock genes, which in turn are able to rhythmically repress their own expression with a periodicity of approximately 24 hours [2].

The CLOCK and BMAL1 proteins, the positive regulators in the mammalian TTFL, are members of the basic-helix-loop-helix and period-ARNT-single-minded (bHLH/PAS) family of transcription factors. Through interactions via their PAS domains, CLOCK and BMAL1 are able to form a heterodimeric complex in the cytoplasm and, upon translocation to the nucleus, drive the transcription of genes that contain E-box enhancer elements in their promoter region [2]. Among the genes activated by the CLOCK/BMAL1 complex are the negative components of the TTFL, the period (*PER1* and *PER2*) and cryptochrome genes (*CRY1* and *CRY2*).

When PER and CRY proteins are synthesized, they are able to form PER/CRY protein complexes in the cytoplasm that shuttle between the cytoplasm and the nucleus [26,27]. When the nuclear

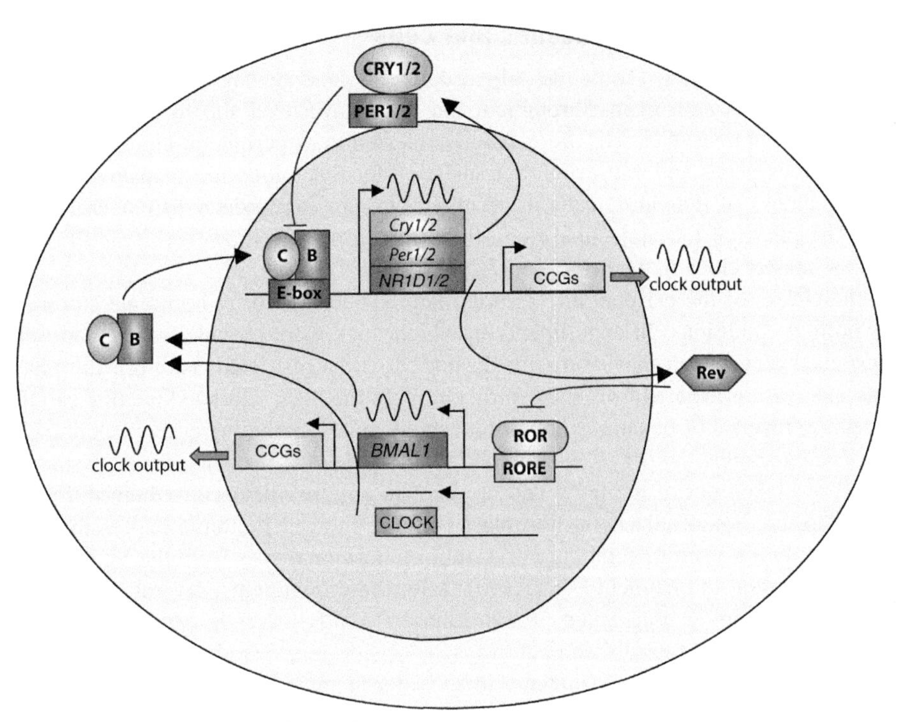

FIGURE 5.4.1 The Mammalian Circadian Oscillator.

Simplified scheme showing the mammalian circadian oscillator, composed of a negative feedback loop and one auxiliary feedback loop. Key elements in this model include CLOCK/BMAL1-driven expression of the *CRY* and *PER* genes through E-box elements in their promoters and subsequent inhibition of CLOCK/BMAL1-driven expression by CRY/PER complexes. This system is stabilized by REV-ERBα/β-driven cyclic expression of the *BMAL1* gene (the genes for REV-ERBα/β are *NR1D1/2*). This molecular oscillator is coupled to output processes through a series of E-box or RORE promoter-element-containing clock-controlled genes (CCGs). B, BMAL1; C, CLOCK.

—Reprinted with permission from Elsevier from [144].

concentration of PER/CRY complexes is sufficiently high, they inhibit the CLOCK/BMAL1 complex, thereby shutting down transcription of E-box genes, including their own transcription [28–30]. Posttranslational modifications, such as ubiquitination and phosphorylation, affect the stability of clock proteins and thereby contribute to period-length modulation [31–33].

The CLOCK/BMAL1 complex also drives the transcription of the two orphan nuclear receptor genes retinoic acid receptor related orphan receptor α (*RORA*) and nuclear receptor subfamily 1, group D, member 1 (*NR1D1* or REV-ERBα). Both RORα and REV-ERBα are involved in stabilizing the circadian oscillator by driving cyclic expression of *BMAL1*, through binding via the ROR-E elements in its promoter [2]. RORα is responsible for the transcriptional activation of *BMAL1*, while REV-ERBα represses the transcription of *BMAL1*.

Circadian clocks in peripheral tissues and cells

Initially, the SCN was considered to be the only structure to generate circadian rhythms. Today, however, virtually all tissues and organs throughout the body have been shown to contain a circadian clock [3,34]. Clocks in non-SCN cells are generally referred to as peripheral clocks and make use of the same set of clock genes and transcription–translation feedback loop mechanism [35]. The master clock keeps the clocks of individual cells in peripheral tissues synchronized by neural and humoral signals [14]. The phase of the molecular clock in peripheral tissues, however, is 3–9 hours delayed compared to the master clock in the SCN [36,37].

In vivo, the clocks of the peripheral tissues are kept synchronized by hormonal stimuli from the SCN under normal conditions. In experiments in which mice were provided with food only during the light period, during which they normally do not eat (time-restricted feeding), the phase of the peripheral clocks was dissociated from the central clock. In the liver, restricted feeding can thus result in complete phase reversal of hepatic expression of many genes, including core clock and clock-controlled genes such as *Bmal1*, *Nr1d1* (Rev-Erbalpha), *Per2* and *Dbp* [38–40].

When cells are taken out of the body and are kept in culture, they retain their ability to exhibit circadian oscillations. Immortalized rat fibroblast cultures showed rhythmic expression of the core clock genes not only after serum stimulation resembling hormonal cues from the SCN, but also after forskolin and corticosteroid treatment [34,41,150). After these treatments, several signal transduction pathways, such as the mitogen-activated protein kinase (MAPK) cascade pathway and cyclic AMP (cAMP) are activated. These transduction pathways are able to reset the circadian oscillation and thus synchronize expression of the core clock genes in a similar manner as in the SCN neurons upon treatment of animals with a clock-resetting light pulse [41,42].

5.4.3 Clock-controlled genes

The main function of the circadian system is to drive daily rhythms in behavior, physiology, and metabolism, which requires that the molecular oscillator is coupled to clock output processes. Indeed, in addition to the core clock genes, expression patterns of several other genes and proteins show a 24-hour rhythm. The circadian core oscillator regulates the transcription of these clock-controlled genes (CCGs) via clock regulatory elements, i.e. E-box, D-Box, and REV-ERB-responsive (RORE) elements [3,43]. Amongst these CCGs are transcription factor encoding genes, which further add to the complexity and dynamics of circadian gene expression by rhythmically driving output genes with a phase different from that obtained with the aforementioned promoter elements alone [44,45].

In rodents, approximately 8–10% of the SCN and peripheral tissue transcriptome is under circadian control [45–49]. The genes that oscillate vary from tissue to tissue, possibly reflecting the specific requirements for each tissue [50,51]. An important example of CCGs is the group of the basic leucine zipper (PAR-bZip) family of transcription factors, which is highly conserved in all organisms [52]. Members of this family, the albumin D-element binding protein (DBP), hepatic leukemia factor (HLF), and thyrotroph embryonic factor (TEF), are rhythmically expressed in the liver, kidney, lungs, and brain, and are known to drive rhythmic expression of various enzymes in the liver [3,53], including several P450 enzymes [54].

5.4.4 Metabolism and the circadian clock

Liver metabolism and the circadian clock

The liver is a major organ involved in metabolism. A large number of the oscillating clock-controlled genes in the liver are involved in physiological and metabolic processes [146], and many are nuclear receptors that work as transcriptional regulators capable of binding hormones and sensing concentrations of metabolites, including lipids, oxysterols, heme, and bile acids, linking the clock machinery to metabolic pathways ([55]; reviewed in detail by [56]). In turn, ligand binding to nuclear receptors regulates the recruitment of co-activators and -repressors, which modulates their transcriptional activity and thus plays an important role in the control of glucose, lipid, and mitochondrial oxidative metabolism [147].

Various enzymes in the endoplasmic reticulum membrane are involved in metabolism and xenobiotic detoxification, and some of these enzymes have been shown to play a role in many human metabolic diseases, such as obesity, insulin resistance, diabetes type 2, and atherosclerosis [148]. A critical role in the pathogenesis of chronic metabolic diseases is played by the crosstalk between the core clock machinery, nutrient sensors such as PPARs (peroxisome proliferator-activated receptors, involved in glucose and lipid metabolism), FXR (farnesoid X receptor, involved in bile acid binding), and LXR (liver X receptor, involved in cholesterol homeostasis and lipoprotein secretion) [55], and pathogen-sensing systems [57].

The clock machinery orchestrates the expression of enzymes involved in cyclic production of ligands for nuclear receptors; in turn, these receptors regulate the clock machinery through cis-regulatory elements on specific clock genes and thus play a part in the synchronization of peripheral clocks. For instance, the intracellular NAD^+/NADH ratio displays circadian oscillation [58,149] ultimately dependent on the presence of BMAL1/CLOCK. The levels of NAD^+ determine the activity of the energy sensor SIRT1 (NAD^+-dependent deacetylase sirtuin-1), which deacetylates PER2 and BMAL1 [59,60], and also LXR [61] and FXR [62], raising the possibility that the latter two are regulated through a posttranslational mechanism [151].

PPARs are key players in fatty acid metabolism and function as transcriptional regulators that sense and respond to circulating fatty acids and their metabolites. PPARα and PPARγ are under circadian control, but have also been shown to modulate transcription of core clock genes. PPARα controls the expression of genes involved in peroxisomal and mitochondrial fatty acid beta-oxidation, and regulates transcription of genes involved in lipid, cholesterol, and glucose uptake and metabolism upon binding to fatty acids [63]. Transcription of *PPARA* is induced by CLOCK and BMAL1 [22] and repressed by PER2 and CRY1 [64]. PPARγ is a regulator of adipogenesis and fat storage. It modulates genes involved in lipogenesis and storage of triglycerides, and regulates insulin sensitivity [65]. Transcription and recruitment to target promoters are inhibited by PER2 [66] and E4BP4 [65], and a reciprocal positive regulation has been shown for *PPARG* and *BMAL1*[22]. Both *PPARA* and *PPARG* are positively regulated by clock-controlled members of the PAR-bZIP family of transcription factors, including DBP, HLF, and TEF [65].

Another group of core clock genes that have been demonstrated in several studies to have direct influence on both lipid and bile acid metabolism are *NR1D1* and *2*. *NR1D1* expression is induced by the CLOCK-BMAL1 heterodimer and its gene product REV-ERBα modulates normal physiology by controlling rhythmic regulation of bile acid production via SHP and E4BP4. Likewise, REV-ERBα influences lipid metabolism by driving circadian rhythmicity in epigenetic changes, chromatin remodeling, and histone modifications of transcription factors [67]. The last is exemplified by circadian recruitment of constitutively expressed HDAC3 to promoter regions in the mouse liver, which is in phase with *Nr1d1* expression. Histone acetylation at genes regulating lipid homeostasis is inversely related to HDAC3 co-localization with REV-ERBα. Deletion of either protein or misalignment of the recruitment rhythm with fasting/feeding cycles causes perturbations in normal metabolic function and leads to hepatic steatosis [68]. The core clock genes *BMAL1* and *CLOCK* are direct target genes of both REV-ERBs [69,70]. REV-ERBα is also involved in the regulation of bile acid synthesis, including expression of P450 cytochrome CYP7a1 and sterol response element binding protein (SREBF1) [71,72]. These two are suppressed when mice are treated with synthetic REV-ERB agonists [73] and dual depletion of *Nr1d1* and *Nr1d2* genes disrupts circadian expression of both core clock and lipid homeostatic gene networks [74]. In accordance with this, enzymes involved in the glycolytic pathway and beta-oxidation of fatty acids were elevated upon treatment with REV-ERB agonists and enzymes involved in lipogenesis and cholesterologenesis were decreased, making REV-ERBs interesting targets for studying and treating various metabolic disorders [73].

Cholesterol and bile acid metabolism and homeostasis are biological mechanisms in which various nuclear receptors, transcription factors, and transporters are involved, many of which are under circadian control [75]. Bile acid metabolism is a major mechanism for elimination of cholesterol in vivo. Cholesterol has been identified as a ligand in the binding pocket of RORα. Changes in intracellular levels of cholesterol and its derivatives may modulate its transcriptional activity, suggesting a role for RORα in the regulation of cholesterol homeostasis [76]. The nuclear receptors FXR and SHR, mainly involved in controlling bile acid homeostasis, are under circadian control and thus form a hinge between key mechanisms of metabolism and circadian rhythmicity [55,77]. FXR responds to elevated bile acid levels by inducing expression of SHP, which is a potent co-repressor of the rate-limiting enzyme in bile biosynthesis, CYP7a1 [78–80]. The *CYP7A1* gene shows a distinct circadian oscillation under the control of DBP [81,82] and REV-ERBα [71]. Consistently, bile flow and biliary secretion of bile acids, cholesterol, and phospholipids exhibit daily rhythms that are in accordance with the *CYP7A1* rhythms [83]. Disruption of this coordinately regulated pathway can result in the accumulation of bile acids in the liver, a condition known as hepatic cholestasis [84,85], eventually leading to liver cell necrosis [86].

The FXR-SHP pathway controls several other genes involved in bile acid synthesis (e.g. *CYP8B1* and *CYP27*), basolateral bile acid uptake by the Na^+-taurocholate co-transporter polypeptide NTCP, which has also been shown to be under direct control of DBP [75], and apical bile acid export by the bile salt export pump BSEP [87]. These and many other bile acid transporters, which may or may not be under circadian control, have been linked to drug-induced cholestasis [88], of which posttranslational mechanisms such as localization of the transporters may be affected by immediate cholestatic effects, where the long-term maintenance of cholestasis is likely due to disruption of transcriptional mechanisms [89].

Xenobiotic metabolism and the circadian clock

Touching upon these many genes, regulators and pathways involved in liver homeostasis that are also intertwined with the core clock machinery and clock-regulated genes, it is not very surprising that various aspects of xenobiotic metabolism, a major hepatic function, are dependent on the phase of the clock as well. In mammals, the xenobiotic defense system can be divided into three functionally separate groups, referred to as phase I, II, and III enzymes.

Phase I enzymes are mainly microsomal CYPs with oxidative, reductive, and hydrolytic function adding a functional group (e.g. OH, SH, or NH_2) to the substrate, but also alcohol and aldehyde dehydrogenases and paraoxonases [90]. All CYPs require heme as a prosthetic group. Heme is also an endogenous ligand for REV-ERBα [91,92] and is reciprocally regulated by the circadian clock via the CLOCK paralog NPAS2 and PER2, and the rate-limiting heme biosynthesis enzyme aminolevulinate synthase 1 (ALAS1) [93]. CYP activity is also regulated by P450 oxidoreductase (POR) expression, which is a membrane-bound enzyme required for electron transfer to CYPs [94]. Both ALAS1 and POR are regulated by the constitutive androstane receptor (CAR), a direct target of the clock-controlled PAR-bZIP factors DBP, HLF, and TEF [52]. Moreover, some CYPs are additionally regulated by core circadian clock genes directly [95]. In general, phase I enzymes show the highest levels of expression and/or activity at times when animals are feeding and likely to encounter xenobiotics, i.e. during the night in mice [90].

Phase II consists of conjugating enzymes, comprising sulfotransferases (SULTs), UDP-glucuronosyltransferases (UGTs), NAD(P)H:quinine oxidoreductases (NQOs), epoxide hydrolases (EPHs), glutathione-S-transferases (GSHs), and N-acetyltransferases (NATs) [96]. These enzymes contribute to the conversion of lipophilic compounds to hydrophilic forms, to render them soluble enough for secretion from the liver in bile by transporters in the phase III group [96,97]. Most phase II enzymes show diurnal variation, but their pattern is not the same for the entire group; in mice, glutathione conjugation is mainly active during the early light phase, followed by glucuronidation in the late light phase and sulfation in the early dark phase [90,98].

Phase III is represented by transport proteins, like multidrug-associated proteins (MRPs) or P-glycoproteins (P-gp), which serve as barriers to limit the penetration of xenobiotics. They are involved in both import and export of compounds. On the basolateral or sinusoidal membrane, key transporters are the Na^+-taurocholate co-transporter polypeptide (NTCP or SLC10A1, major bile salt uptake transporter), organic anion transporting polypeptides (OATPs, involved in uptake of unconjugated bile salts, estrogen conjugates, and xenobiotics) and the organic cation transporter 1 (OCT1 or SLC22A1), but also conjugate export pumps (MRP1 and 3). The canalicular membrane houses many export pumps, such as multidrug resistance protein 2 (MRP2), multidrug export pump (MDR1), breast cancer resistance protein (BCRP), multidrug and toxin extrusion 1 (MATE1), and the bile salt export pump (BSEP or ABCB11) ([99,100]; reviewed in [98]). Many of the transporters on the sinusoidal and canalicular membranes of the mouse liver show circadian oscillations ([47–49,101]; reviewed by [90]).

Aryl-hydrocarbon-receptor-dependent metabolism

A well-studied xenobiotic receptor system is the aryl hydrocarbon receptor (AHR), which forms a complex with the AHR nuclear translocator (ARNT). These proteins are critical components of the xenobiotic detoxification system, in which the nuclear receptor pregnane X receptor (PXR), CAR and PPARα, various P450 enzymes, and heme-synthesizing enzymes also play a major role [56,97,102,103]. Under normal conditions, AHR resides in the cytoplasm bound to heat shock proteins, but binding to hydrophobic ligands that enter the cell through the membrane induces conformational changes causing translocation to the nucleus and its association with ARNT. In the nucleus, AHR-ARNT heterodimers induce transcriptional activation of xenobiotic metabolism pathways by binding to promoter regions of xenobiotic response elements (XREs), also called dioxin response elements (DREs) [104]. As a negative feedback mechanism, AHR is transported back to the cytoplasm and degraded by the proteasome-ubiquitin system [105], and induces its own repressor [106].

Because AHR is conserved in vertebrates and invertebrates, it is suggested that it has played an essential function through evolution [103]. AHR-ARNT complexes have been demonstrated to modulate reproductive, cardiovascular, metabolic, developmental, central nervous system, and immunological functions [98]. Organisms encounter both endogenous and exogenous ligands capable of binding AHR on a daily basis. Endogenous ligands are represented by endogenously synthesized chemicals such as indigoids, equilenin, arachidonic acid metabolites, heme metabolites, and tryptophan metabolites. Exogenous natural compounds capable of binding AHR are mainly encountered as dietary chemicals, such as polyphenols, flavonoids, carotenoids, and berberine [98]. AHR can also be translocated to the nucleus independent of ligand binding by cAMP, a core component of the circadian pacemaker, driving signaling that is significantly different from that induced by exogenous ligands [107]. It is suggested that cAMP-mediated transcription is the evolutionarily-derived primary endogenous regulator of AHR. Disrupting cAMP-mediated AHR signaling is thus considered to play a major role in the mechanisms of toxicity represented by environmental pollutants and toxic compounds [108]. Another endogenous ligand, the tryptophan photoproduct 6-formylindolo[3,2-*b*]carbazole (FICZ), is capable of influencing the circadian expression of clock genes by blocking the glutamate pathway and regulates the light-dependent circadian rhythm through AHR signaling activation [109,110]. Exogenous ligands capable of binding and translocating AHR are represented by various polycyclic aromatic hydrocarbons (PAHs), of which benzo-[a]-pyrene (B[a]P) has been studied the most, halogenated dioxins, polychlorinated biphenyls, and other compounds such as 2,3,7,8-tetrachlorodibenzo-p-dioxin (TCDD).

Many promutagens are inactive as mutagens prior to biotransformation by phase I enzymes, but oxidative metabolites can form DNA adducts, leading to toxicity and/or ultimately triggering mutagenesis and carcinogenesis. Additionally, B[a]P and TCDD induce cAMP, leading to increased levels of Ca^{2+} and activity of the Ca^{2+}/calmodulin-dependent kinase pathway that are normally modulated by melatonin, one of the principal humoral outputs for the circadian timing system. Together, this leads to increased and prolonged effects of their toxic properties [111–113]. Mice treated with TCDD and dioxin show altered behavioral circadian rhythmicity and gene expression, and reduced responsiveness of the circadian clock to changes in light/dark regimes, thus showing that xenobiotics are able to interfere with proper working of the clock machinery via activation of the AHR pathway [114–117], disrupting CLOCK-BMAL1 transcriptional activity and thereby repressing and deregulating *Per1* gene expression [117,118]. Although dimerization studies have demonstrated that AHR does not interact with many core clock protein complexes directly, shared co-mediators provide multiple interaction platforms allowing for altered influence on transcriptional machinery (reviewed in [102]). PER1 and PER2 have opposite impacts on the regulation of xenobiotic responses in the liver; PER1 functions as an inhibitory factor, whereas PER2 acts as a positive regulator in the AHR-mediated and ligand-binding-dependent activation of AHR signaling [118]. Studies performed with liver explants and cells from Per2:luciferase mice have demonstrated that activation of AHR does not cause direct changes in core clock gene expression, but the phase of the circadian clock does influence the transcriptional activity of P450 enzymes ([119]; De Wit et al., unpublished data; Nijman et al., unpublished data).

AHR signaling and circadian pathways are closely engaged in the pathophysiological basis of metabolic, immune-related, and neoplastic diseases [98]. For instance, there is evidence underlying the crosstalk between AHR activation, PPARα signaling, and altered clock machinery in the development of type 2 diabetes related to toxic exposure by altering glucose metabolism [120]. Also, epidemiological studies have found links between shift work and many forms of cancer, particularly those that have a strong link to AHR-mediated toxicity, including prostate, colorectal, and lung cancer, and non-Hodgkin lymphoma. Activation or deregulation of AHR has been associated with development of lymphomas and leukemia in mice and man (reviewed by [102]). There is substantial evidence for a link between cancer development and exposure to B[a]P and other PAHs [121].

5.4.5 DNA damage and the circadian clock

Our DNA is continuously exposed to physical and chemical genotoxic compounds of exogenous and endogenous origin (e.g. ultraviolet light, ionizing radiation, chemical pollutants, reactive byproducts of metabolism) that can damage our genome. When not repaired in time, lesions can block transcription and (depending on the DNA damage level) cause the cell to undergo apoptosis. On the other hand, replication of damaged DNA can give rise to mutations, ultimately leading to induction of cancer [122]. To counteract the deleterious effects of exposure to genotoxic agents, cells have evolved an intricate network of DNA damage-repair pathways with (partially) overlapping substrate specificity. One such repair pathway is nucleotide excision repair (NER), a multi-step cut-and-patch mechanism involving some 30 proteins [122]. Interestingly, one of the key NER factors, XPA (involved in damage recognition/verification), has been reported to be under circadian control, as evident from its rhythmic expression pattern in liver and brain [123]. The *XPA* gene exhibits two E-box elements in its promoter region, allowing transcription activation by CLOCK/BMAL, and as such is considered a CCG [123,124]. In line with the circadian expression pattern of *XPA*, repair of cisplatin-induced DNA lesions in the liver has been shown to fluctuate over the day [125].

The connection between the core clock and DNA damage appears bidirectional. It has been shown that DNA damage can reset the mammalian circadian clock in a time-dependent manner [126]. In this study, clock-synchronized rat fibroblasts were treated with a single dose of γ-radiation at a defined moment of the day, after which circadian-phase and dose-dependent phase advance of circadian rhythms was observed. This phase advance was reduced after treatment of cells with ataxia-telangiectasia-mutated (ATM) inhibitors, suggesting that resetting of the circadian clock by γ-radiation might depend on ATM [126]. Likewise, DNA damage induced by ultraviolet (UV) light, oxidative stress, and methyl methane sulfonate (MMS) was shown to induce a phase advance the circadian clock [126,127].

5.4.6 Chronotoxicity

Considering the tight link between the circadian clock and cellular metabolism, including xenobiotic metabolism and DNA damage repair, it may not come as a surprise that the (geno)toxic effect of chemical agents may depend on the time of day of exposure. This phenomenon is referred to as chronotoxicity and is well illustrated by animal exposure studies with the genotoxic agent cyclophosphamide, a chemotherapeutic and immunosuppressive agent that is widely used for treatment of several types of cancer, blood and bone marrow transplantation, and autoimmune disorders. Cyclophosphamide is a prodrug that requires metabolic activation by several cytochrome P450 enzymes [128]. Interestingly, as illustrated by animal exposure studies, cyclophosphamide sensitivity is greatly dependent on the time of drug administration.[129]. Mice treated with cyclophosphamide for three consecutive days at the beginning of the night were shown to be more tolerant to cyclophosphamide than animals treated at the beginning of the day. Likewise, other rodent studies revealed for more than 30 anticancer drugs (e.g. doxorubicin, 5-fluorourcil, celecoxib) that the animals' tolerance to the drug depended on the time of administration [129–132].

The chronotoxic properties of compounds can be very useful when treating cancer with chemotherapeutic or other anticancer agents [133–135]. Such drugs usually damage both malignant and (proliferating) healthy cells, the latter resulting in severe side effects such as nausea, vomiting, anemia, and hair loss [136,137]. Chronomodulated therapy or chronotherapy involves protocols in which chemotherapeutic agents are given at a specific moment of the day in order to minimize the side effects and cause damage predominantly to the malignant cells. As such, chronotherapy can be used to maximize efficacy, to minimize unwanted side effects, or to improve drug tolerance [134,137,138].

5.4.7 In vitro alternatives for toxicity testing

The circadian clock and in vitro alternatives for hepatotoxic risk assessment

By law, our society demands that chemical compounds be assessed for their potential to elicit (hepato)toxic and carcinogenic effects. To this end, a battery of routinely performed in vitro tests (e.g. measurement of mutations, DNA damage, chromosomal changes) are available to determine the toxicity of the agent [139,140]. Yet, as such assays are not always conclusive and many compounds are identified as false positive or negative in at least one of these tests, in vivo rodent studies are necessary in order to identify and correctly classify compounds [139]. However, apart from the high costs of animal experiments, society wants us to obey the "3Rs" principle: refinement, replacement, and reduction of laboratory animals in experimentation [141].

To decrease the number of laboratory animals used and ultimately replace their use, many ex vivo and in vitro alternatives have been and are being investigated; the liver is a major organ involved in compound metabolism and, therefore, often at the center of these alternative assays (reviewed in [142] and [143]). Several in vitro models (e.g. stable cell lines, three-dimensional tissue models) are available and frequently used to determine the (geno)toxic properties of compounds [139,141]. Nowadays, a popular approach to predict the (geno)toxic and (non)carcinogenic properties of compounds in vitro includes proteomic and transcriptomic profiling methods to establish predictive risk-marker profiles [141]. However, in such risk-assessment assays, including toxicogenomics studies, the circadian clock is usually not taken into account. Yet, the circadian clock can be predicted to prominently influence the outcome of in vivo and in vitro transcriptomics studies (and, as such, the comparison of in vivo and in vitro data) in multiple ways.

First, as xenobiotic metabolism, DNA repair, and various other toxicity-related processes are subject to circadian variation, the toxic response of the liver upon in vivo exposure to chemical compounds may depend on the time of day of exposure. Accordingly, it can be predicted that the response of the transcriptome may also depend on the circadian phase of the liver at the moment of treatment. Indeed, in a first study addressing this issue, we uncovered pronounced day–night differences in the number and category of differentially expressed genes after exposure of mice to cyclophosphamide during the day or night (Nijman et al., unpublished data). Accordingly, when interpreting (published) in vivo exposure studies, it is important to take into account at which time of the day animals have been exposed. With respect to this, it is also important to note that many rodent toxicity studies have been performed during the day, which corresponds to the inactive phase (sleep) of the animal.

Second, comparison of in vivo and in vitro exposure studies is hampered by the fact that (as discussed above) intracellular circadian clocks of cultured cells tend to desynchronize in the absence of clock-resetting signals from the master clock in the SCN because of small variations in period length; for instance, due to interactions between the core clock machinery and machinery involved in metabolism. Therefore, at the population level, the average expression level of oscillating clock-controlled genes in cultured cells may be flat, even though there are large intercellular variations between gene or protein expression level. As a consequence, in vivo/in vitro differences in gene expression levels will occur. Moreover, in a series of experiments in which we compared clock-synchronized and non-synchronized cultures, we demonstrated that in the absence of a synchronizing stimulus such as glucocorticoids (e.g. dexamethasone), cell cultures can still be partially synchronous (de Wit and Chaves, unpublished data). Indeed, culture conditions such as (but not limited to) replacing the medium (containing serum-derived clock-resetting factors), changes in temperature, or changes in glucose availability during long-term culturing, can temporarily reset the circadian clock. This implies that, in the classic experimental setup, cultures are often synchronized to some extent but the phase is unknown, making inter-experimental comparison unreliable, especially when experiments are conducted by different researchers and/or within different laboratories.

Last but not least, DNA-damaging agents are able to reset the circadian clock [126,127]. Since many in vitro systems are based on established cell lines and intracellular clocks are usually desynchronized (or partially synchronized with unknown phase) in cell cultures, exposure to DNA-damaging agents might synchronize or reset the cellular clocks, and accordingly affect the expression level of core circadian clock genes and thousands of CCGs. Such genes may accidentally be considered as (geno)toxic stress responsive genes. For example, when studying the mRNA expression level of a certain gene after exposure to a genotoxic compound, one may find the expression level of that particular gene to be increased 12 hours after exposure, and return to a lower level 24 hours after exposure (Figure 5.4.2A), which might suggest that this gene is a DNA damage responsive gene. However, when a DNA-damaging compound is able to reset the circadian clock, it might very well be that the gene of interest is not a DNA damage responsive gene, but rather a CCG and, as such, should be considered a false-positive gene (Figure 5.4.2B).

In vitro chronotoxicity assays

Given the impact of the circadian clock on liver performance (including xenobiotic metabolism and DNA repair), it is crucial that this endogenous time keeper is taken into account when designing and performing toxicity studies. We have previously proposed a new approach for in vitro toxicological and toxicogenomics studies, based on the use of clock-synchronized cell cultures [144]. In this approach (outlined in Figure 5.4.2C), intracellular clocks are synchronized prior to starting an exposure experiment. To control for clock synchronization, and to facilitate the calculation of the time of exposure, one can make use of primary cells and tissues obtained from mice with a luciferase-based circadian clock reporter gene, or alternatively introduce lentiviral- or plasmid-based clock reporter constructs in established cell lines of choice. Such in vitro liver model systems allow real-time

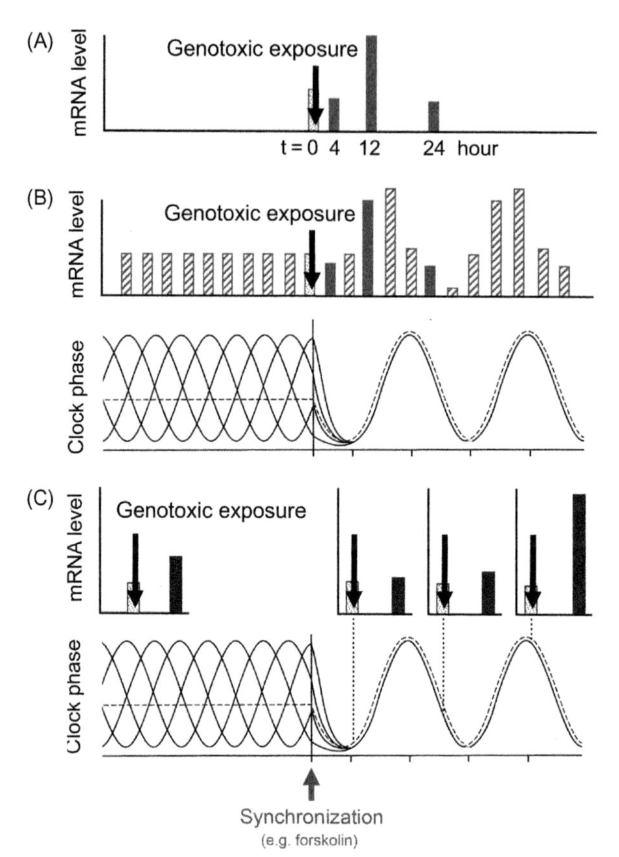

FIGURE 5.4.2 Implications of DNA-Damage-Resetting of the Circadian Clock for In Vitro Genotoxic Risk Assessment Assays.

(A) Imaginary in vitro experiment for identification of genotoxic stress markers. Cultured cells are exposed to a DNA-damaging agent. mRNA profiles are determined at various times after exposure (solid bars) and compared to those obtained from mock-treated cells (not shown) or cells harvested prior to treatment (dashed bars). The data obtained for gene X suggest that this gene is a DNA damage responsive gene. (B) The same experiment, except that mRNA levels were analyzed at 4-hour intervals (additional data represented by hatched bars). Note that exposure to the DNA-damaging agent is synchronizing the individual cellular circadian clocks, and accordingly transcription of clock (output) genes. On the basis of this experiment, gene X is likely a clock-controlled output gene. (C) Modified version of the in vitro experiment for identification of genotoxic stress markers, in which the circadian clock of the individual cells is synchronized by treatment with forskolin (or another known synchronizing compound) and cells are exposed to the DNA-damaging agent at defined phases of the circadian day. Expression profiles are determined 12 hours after treatment (solid bars) and compared to those obtained from cells harvested prior to treatment (dashed bars). Note the time-of-day differences in the response of gene Y, indicative of chronotoxic effects of the DNA-damaging agent. Also note the stronger response of gene Y in synchronized cells, as compared to non-synchronized cells.

—Reprinted with permission from Elsevier from [144].

recording of circadian clock phase. Subsequently, exposure with a chemical compound of interest (e.g. a pollutant, an industrial chemical, a pharmaceutical drug) can be initiated at defined (precalculated) phases of the circadian clock. If the tested chemical compound harbors chronotoxic properties (e.g. through clock-controlled metabolism) major differences in gene-expression profiles and biological endpoints (e.g. apoptosis, steatosis, cholestasis) can be expected, depending on the phase of the clock at the moment of exposure (see Figure 5.4.2C).

The advantage of this approach over classic liver-derived systems is that it enables us to quantify the genotoxic sensitivity of the cells in relation to circadian clock phase, thereby not only classifying an compound as being chronotoxic, but also providing novel insights into the kinetics and amplitude of (geno)hepatotoxic responses. Moreover, when the magnitude of the toxic response is dependent on circadian clock phase, the output (e.g. gene expression, proteomics, cell death) is likely to be observed at lower compound concentrations and/or shorter treatment periods than in classic toxicological experiments, thereby increasing the sensitivity of the assay. Moreover, the use of synchronized cells also eliminates the risk of nontoxic-response-related transcriptional changes originating from potential clock-resetting properties of chemical compounds, which will increase the specificity of the assay.

5.4.8 Concluding remarks

It has become evidently clear that the kinetics and dynamics of toxicological responses and side effects of drugs, poisons, or toxic substances can be influenced by the circadian clock. Over 10% of the liver transcriptome displays robust mRNA cycling, which not only well illustrates the impact of the circadian clock on various cellular key processes, but also explains why the sensitivity of tissues to environmental-toxicant- and drug-induced toxicity may depend on the time of exposure or administration (i.e. circadian clock phase). Accordingly, it is strongly recommended that the circadian clock is implemented in mechanistic research aimed at understanding (hepato)toxic responses elicited by chemical compounds, as well as in efforts to replace animal-based risk-assessment assays by validated in vitro assays.

Insight into the chronotoxic properties of chemicals will have an important societal impact. In the case of toxic industrial chemicals, knowledge of the chronotoxic properties of such compounds will contribute to the improvement of health and safety in the workplace, as employees (in addition to wearing protective clothes) can be advised to handle such chemicals only at the time of day when accidental exposure has the least impact on the body. Likewise, information on potential chronotoxic properties of anticancer drugs may help to improve therapeutic efficacy and reduce undesired side effects, and as such will improve the quality of life of the cancer patient. Rather than conventional drug administration courses in which the blood levels of drugs in patients are kept more or less constant (thereby increasing the total drug burden and contributing to toxic side effects), pharmaceutical companies should focus on intelligent drug delivery systems in which slow-release technology is combined with the patient's individual circadian rhythms [134,145].

Acknowledgments

The authors participate in The Netherlands Toxicogenomics Centre (NTC; http://www.toxicogenomics.nl) and received financial support from The Netherlands Genomics Initiative/Netherlands Organisation for Scientific Research (NGI/NWO grant no. 050-060-510).

References

[1] Edery I. Circadian rhythms in a nutshell. Physiol Genomics 2000;3(2):59–74.
[2] Lowrey PL, Takahashi JS. Mammalian circadian biology: elucidating genome-wide levels of temporal organization. Ann Rev Genomics Hum Genet 2004;5:407–41.
[3] Reppert SM, Weaver DR. Coordination of circadian timing in mammals. Nature 2002;418(6901):935–41.
[4] Bell-Pedersen D, Cassone V, Earnest D, Golden S, Hardin P, Thomas T, et al. Circadian rhythms from multiple oscillators: lessons from diverse organisms. Nat Rev Genet 2005;6(7):544–56.
[5] Dunlap JC. Molecular bases for circadian clocks. Cell 1999;96:271–90.
[6] Kantermann T, Juda M, Merrow M, Roenneberg T. The human circadian clock's seasonal adjustment is disrupted by daylight saving time. Curr Biol 2007;17(22):1996–2000.
[7] Pittendrigh C. Circadian rhythms and the circadian organization of living systems. Cold Spring Harb Symp Quant Biol 1960;25:159–84.
[8] Czeisler CA, Duffy JF, Shanahan TL, Brown EN, Mitchell JF, Rimmer DW, et al. Stability, precision, and near-24-hour period of the human circadian pacemaker. Science 1999;284(5423):2177–81.
[9] Sack R, Brandes R, Kendal A, Lewy A. Entrainment of free-running circadian rhythms by melatonin in blind people. New Engl J Med 2000;343:1070–7.
[10] Izumo M, Johnson CH, Yamazaki S. Circadian gene expression in mammalian fibroblasts revealed by real-time luminescence reporting: temperature compensation and damping. Proc Natl Acad Sci USA 2003;100(26):16089–94.
[11] Moore R. Organization and function of a central nervous system circadian oscillator: the suprachiasmatic hypothalamic nucleus. Fed Proc 1983;42(11):2783–9.
[12] Moore R, Eichler V. Loss of a circadian adrenal corticosterone rhythm following suprachiasmatic lesions in the rat. Brain Res 1972;42(1):201–6.
[13] van den Pol A. The hypothalamic suprachiasmatic nucleus of rat: intrinsic anatomy. J Comp Neurol 1980;191(4):661–702.
[14] Takahashi J, Hong H, Ko C, McDearmon E. The genetics of mammalian circadian order and disorder: implications for physiology and disease. Nat Rev Genet 2008;9(10):764–75.
[15] Pittendrigh C. The photoperiodic phenomena: seasonal modulation of the "day within." J Biol Rhythms 1988;3:173–88.
[16] Berson DM, Dunn FA, Takao M. Phototransduction by retinal ganglion cells that set the circadian clock. Science 2002;295(5557):1070–3.
[17] Gooley JJ, Lu J, Chou TC, Scammell TE, Saper CB. Melanopsin in cells of origin of the retinohypothalamic tract. Nat Neurosci 2001;4(12):1165.
[18] Hattar S, Liao HW, Takao M, Berson DM, Yau KW. Melanopsin-containing retinal ganglion cells: architecture, projections, and intrinsic photosensitivity. Science 2002;295(5557):1065–70.
[19] Provencio I, Rollag MD, Bakun A. Photoreceptive net in the mammalian retina. Nat Brief Commun 2002;415:493.
[20] Ukai H, Kobayashi TJ, Nagano M, Masumoto K, Sujino M, Kondo T, et al. Melanopsin-dependent photoperturbation reveals desynchronization underlying the singularity of mammalian circadian clocks. Nat Cell Biol 2007;9(11):1327–34.
[21] Brown SA, Kowalska E, Dallmann R. (Re)inventing the circadian feedback loop. Dev Cell 2012;22(3):477–87.
[22] Canaple L, Rambaud J, Dkhissi-Benyahya O, Rayet B, Tan NS, Michalik L, et al. Reciprocal regulation of brain and muscle Arnt-like protein 1 and peroxisome proliferator-activated receptor alpha defines a novel positive feedback loop in the rodent liver circadian clock. Mol Endocrinol 2006;20(8):1715–27.
[23] Hastings MH, Maywood ES, O'Neill JS. Cellular circadian pacemaking and the role of cytosolic rhythms. Curr Biol 2008;18(17):R805–15.

[24] Haydon MJ, Hearn TJ, Bell LJ, Hannah MA, Webb AAR. Metabolic regulation of circadian clocks. Semin Cell Dev Biol 2013;24(5):414–21.

[25] Rey G, Reddy AB. Connecting cellular metabolism to circadian clocks. Trends Cell Biol 2013;23(5):234–41.

[26] Chaves I, Yagita K, Barnhoorn S, van Der Horst GTJ, Tamanini F. Functional evolution of the photolyase/cryptochrome protein family: importance of the C terminus of mammalian CRY1 for circadian core oscillator performance. Mol Cell Biol 2006;26(5):1743–53.

[27] Yagita K, Tamanini F, Yasuda M. Nucleocytoplasmic shuttling and mCRY-dependent inhibition of ubiquitylation of the mPER2 clock protein. EMBO J 2002;21(6):1301–14.

[28] Griffin EJ, Staknis D, Weitz C. Light-independent role of CRY1 and CRY2 in the mammalian circadian clock. Science 1999;286(5440):768–71.

[29] Sato T, Yamada R, Ukai H, Baggs J, Miraglia L, Kobayashi T, et al. Feedback repression is required for mammalian circadian clock function. Nat Genet 2006;38:312–9.

[30] Shearman LP, Sriram S, Weaver DR, Maywood ES, Chaves I, Zheng B, et al. Interacting molecular loops in the mammalian circadian clock. Science 2000;288(5468):1013–9.

[31] Gallego M, Virshup D. Post-translational modifications regulate the ticking of the circadian clock. Nat Rev Mol Cell Biol 2007;8:139–48.

[32] Harms E, Kivimäe S, Young M, Saez L. Posttranscriptional and posttranslational regulation of clock genes. J Biol Rhythms 2004;19(5):361–73.

[33] Lamia K, Sachdeva U, DiTacchio L, Williams E, Alvarez J, Egan J, et al. AMPK regulates the circadian clock by cryptochrome phosphorylation and degradation. Science 2009;326(5951):437–40.

[34] Yamazaki S, Numano R, Abe M, Hida A, Takahashi R, Ueda M, et al. Resetting central and peripheral circadian oscillators in transgenic rats. Science 2000;288(5466):682–5.

[35] Yagita K, Tamanini F, van der Horst GTJ, Okamura H. Molecular mechanisms of the biological clock in cultured fibroblasts. Science 2001;292:278–81.

[36] Field MD, Maywood ES, O'Brien JA, Weaver DR, Reppert SM, Hastings MH. Analysis of clock proteins in mouse SCN demonstrates phylogenetic divergence of the circadian clockwork and resetting mechanisms. Neuron 2000;25(2):437–47.

[37] Lee C, Etchegaray J-P, Cagampang FRA, Loudon ASI, Reppert SM. Posttranslational mechanisms regulate the mammalian circadian clock. Cell 2001;107(7):855–67.

[38] Damiola F, Le Minh N, Preitner N, Kornmann B, Fleury-Olela F, Schibler U. Restricted feeding uncouples circadian oscillators in peripheral tissues from the central pacemaker in the suprachiasmatic nucleus. Genes Dev 2000;14(23):2950–61.

[39] Hara R, Wan K, Wakamatsu H, Aida R, Moriya T, Akiyama M, et al. Restricted feeding entrains liver clock without participation of the suprachiasmatic nucleus. Genes Cells 2001;6(3):269–78.

[40] Stokkan K, Yamazaki S, Tei H, Sakaki Y, Menaker M. Entrainment of the circadian clock in the liver by feeding. Science 2001;291(5503):490–3.

[41] Balsalobre A, Marcacci L, Schibler U. Multiple signaling pathways elicit circadian gene expression in cultured Rat-1 fibroblasts. Curr Biol 2000;10(20):1291–4.

[42] Akashi M, Nishida E. Involvement of the MAP kinase cascade in resetting of the mammalian circadian clock. Genes Dev 2000;14(6):645–9.

[43] Ueda H, Hayashi S, Chen W, Sano M, Machida M, Shigeyoshi Y, et al. System-level identification of transcriptional circuits underlying mammalian circadian clocks. Nat Genet 2005;37(2):187–92.

[44] Kumaki Y, Ukai-Tadenuma M, Uno KD, Nishio J, Masumoto K, Nagano M. Analysis and synthesis of high-amplitude Cis-elements. Proc Natl Acad Sci USA 2008;105(39):14946–14951.

[45] Ueda HR, Chen W, Adachi A, Wakamatsu H, Hayashi S, Takasugi T, et al. A transcription factor response element for gene expression during circadian night. Lett Nat 2002;418:534–9.

[46] Hughes ME, DiTacchio L, Hayes KR, Vollmers C, Pulivarthy S, Baggs JE, et al. Harmonics of circadian gene transcription in mammals. PLoS Genet 2009;5(4):e1000442.
[47] Miller BH, McDearmon EL, Panda S, Hayes KR, Zhang J, Andrews JL, et al. Circadian and CLOCK-controlled regulation of the mouse transcriptome and cell proliferation. Proc Natl Acad Sci USA 2007;104(9):3342–7.
[48] Panda S, Antoch MP, Miller BH, Su AI, Schook AB, Straume M, et al. Coordinated transcription of key pathways in the mouse by the circadian clock. Cell 2002;109(3):307–20.
[49] Storch K-F, Lipan O, Leykin I, Viswanathan N, Davis FC, Wong WH, et al. Extensive and divergent circadian gene expression in liver and heart. Nature 2002;417(6884):78–83.
[50] Lamia KA, Storch K-F, Weitz CJ. Physiological significance of a peripheral tissue circadian clock. Proc Natl Acad Sci U S A 2008;105(39):15172–7.
[51] Storch K-F, Paz C, Signorovitch J, Raviola E, Pawlyk B, Li T, et al. Intrinsic circadian clock of the mammalian retina: importance for retinal processing of visual information. Cell 2007;130(4):730–41.
[52] Gachon F, Olela FF, Schaad O, Descombes P, Schibler U. The circadian PAR-domain basic leucine zipper transcription factors DBP, TEF, and HLF modulate basal and inducible xenobiotic detoxification. Cell Metab 2006;4(1):25–36.
[53] Ripperger J, Shearman L, Reppert S, Schibler U. CLOCK, an essential pacemaker component, controls expression of the circadian transcription factor DBP. Genes Dev 2000;14(6):679–89.
[54] Lavery D, Lopez-Molina L, Margueron R, Fleury-Olela F, Conquet F, Schibler U, et al. Circadian expression of the steroid 15 alpha-hydroxylase (Cyp2a4) and coumarin 7-hydroxylase (Cyp2a5) genes in mouse liver is regulated by the PAR leucine zipper transcription factor DBP. Mol Cell Biol 1999;19(10):6488–99.
[55] Yang X, Downes M, Yu RT, Bookout AL, He W, Straume M, et al. Nuclear receptor expression links the circadian clock to metabolism. Cell 2006;126(4):801–10.
[56] Mazzoccoli G, Pazienza V, Vinciguerra M. Clock genes and clock-controlled genes in the regulation of metabolic rhythms. Chronobiol Int 2012;29(3):227–51.
[57] Yamamura Y, Yano I, Kudo T, Shibata S. Time-dependent inhibitory effect of lipopolysaccharide injection on Per1 and Per2 gene expression in the mouse heart and liver. Chronobiol Int 2010;27(2):213–32.
[58] Nakahata Y, Sahar S, Astarita G, Kaluzova M, Sassone-Corsi P. Circadian control of the NAD+ salvage pathway by CLOCK-SIRT1. Science 2009;324(5927):654–7.
[59] Asher G, Gatfield D, Stratmann M, Reinke H, Dibner C, Kreppel F, et al. SIRT1 regulates circadian clock gene expression through PER2 deacetylation. Cell 2008;134(2):317–28.
[60] Nakahata Y, Kaluzova M, Grimaldi B, Sahar S, Chen D, Guarente LP, et al. The NAD+ -dependent deacetylase SIRT1 modulates CLOCK-mediated chromatin remodeling and circadian control. Cell 2008;134(2):329–40.
[61] Li X, Zhang S, Blander G, Tse JG, Krieger M, Guarente L. SIRT1 deacetylates and positively regulates the nuclear receptor LXR. Mol Cell 2007;28(1):91–106.
[62] Kemper JK, Xiao Z, Ponugoti B, Miao J, Kanamaluru D, Tsang S, et al. FXR acetylation is normally dynamically regulated by p300 and SIRT1 but constitutively elevated in metabolic disease states. Cell Metab 2009;10(5):392–404.
[63] Gachon F, Leuenberger N, Claudel T, Gos P, Jouffe C, Fleury Olela F, et al. Proline- and acidic-amino-acid-rich basic leucine zipper proteins modulate peroxisome proliferator-activated receptor alpha (PPARalpha) activity. Proc Natl Acad Sci USA 2011;108(12):4794–9.
[64] Hayashida S, Kuramoto Y, Koyanagi S, Oishi K, Fujiki J, Matsunaga N, et al. Peroxisome proliferator-activated receptor-α mediates high-fat, diet-enhanced daily oscillation of plasminogen activator inhibitor-1 activity in mice. Chronobiol Int 2010;27(9–10):1735–53.
[65] Takahashi S, Inoue I, Nakajima Y, Seo M, Nakano T, Yang F. A promoter in the novel exon of hPPAR directs the circadian expression of PPAR. J Atheroscler Thromb 2010;17(1):73–83.

[66] Grimaldi B, Bellet MM, Katada S, Astarita G, Hirayama J, Amin RH, et al. PER2 controls lipid metabolism by direct regulation of PPARγ. Cell Metab 2010;12(5):509–20.
[67] Sun Z, Feng D, Everett L, Bugge A, Lazar M. Circadian epigenomic remodeling and hepatic lipogenesis: lessons from HDAC3. Cold Spring Harb Symp Quant Biol 2011;76:49–55.
[68] Feng D, Liu T, Sun Z, Bugge A, Mullican SE, Alenghat T, et al. A circadian rhythm orchestrated by histone deacetylase 3 controls hepatic lipid metabolism. Science 2011;331(6022):1315–9.
[69] Crumbley C, Burris TP. Direct regulation of CLOCK expression by REV-ERB. PLoS ONE 2011;6(3):e17290.
[70] Preitner N, Damiola F, Zakany J, Duboule D, Albrecht U, Schibler U. The orphan nuclear receptor REV-ERBalpha controls circadian transcription within the positive limb of the mammalian circadian oscillator. Cell 2002;110:251–60.
[71] Duez H, van der Veen JN, Duhem C, Pourcet B, Touvier T, Fontaine C, et al. Regulation of bile acid synthesis by the nuclear receptor Rev-erbalpha. Gastroenterology 2008;135(2):689–98.
[72] Le Martelot G, Claudel T, Gatfield D, Schaad O, Kornmann B, Lo Sasso G, et al. REV-ERBalpha participates in circadian SREBP signaling and bile acid homeostasis. PLoS Biol 2009;7(9):e1000181.
[73] Solt LA, Wang Y, Banerjee S, Hughes T, Kojetin DJ, Lundasen T, et al. Regulation of circadian behaviour and metabolism by synthetic REV-ERB agonists. Nature 2012;485(7396):62–8.
[74] Cho H, Zhao X, Hatori M, Yu RT, Barish GD, Lam MT, et al. Regulation of circadian behaviour and metabolism by REV-ERB-α and REV-ERB-β. Nature 2012;485(7396):123–7.
[75] Ma K, Xiao R, Tseng H-T, Shan L, Fu L, Moore DD. Circadian dysregulation disrupts bile acid homeostasis. PLoS ONE 2009;4(8):e6843.
[76] Wang Y, Kumar N, Solt LA, Richardson TI, Helvering LM, Crumbley C, et al. Modulation of retinoid acid receptor-regulated orphan receptor alpha and gamma activity by 7-oxygenated sterol ligands. J Biol Chem 2010;285(7):5013–25.
[77] Bookout A, Jeong Y, Downes M, Yu RT, Evans RM, Mangelsdorf DJ. Anatomical profiling of nuclear receptor expression reveals a hierarchical transcriptional network. Cell 2006;126(4):789–99.
[78] Goodwin B, Jones S, Price R, Watson M, McKee D, Moore L, et al. A regulatory cascade of the nuclear receptors FXR, SHP-1, and LRH-1 represses bile acid biosynthesis. Mol Cell 2000;6(3):517–26.
[79] Lu TT, Makishima M, Repa JJ, Schoonjans K, Kerr TA, Auwerx J, et al. Molecular basis for feedback regulation of bile acid synthesis by nuclear receptors. Mol Cell 2000;6:507–15.
[80] Moore J, Goodwin B, Willson T, Klieuwer S. Nuclear receptor regulation of genes involved in bile acid metabolism. Crit Rev Eukaryot Gene Expr 2002;12(2):119–35.
[81] Lavery DJ, Schibler U. Circadian transcription of the cholesterol 7 alpha hydroxylase gene may involve the liver-enriched bZIP protein DBP. Genes Dev 1993;7(10):1871–84.
[82] Wuarin J, Falvey E, Lavery D, Talbot D, Schmidt E, Ossipow V, et al. The role of the transcriptional activator protein DBP in circadian liver gene expression. J Cell Sci Suppl 1992;16:123–7.
[83] Nakano A, Tietz P, LaRusso N. Circadian rhythms of biliary protein and lipid excretion in rats. Am J Physiol 1990;258(5):G653–9.
[84] Alvarez L, Jara P, Sánchez-Sabaté E, Hierro L, Larrauri J, Díaz MC, et al. Reduced hepatic expression of farnesoid X receptor in hereditary cholestasis associated to mutation in ATP8B1. Hum Mol Genet 2004;13(20):2451–60.
[85] De Vree JM, Jacquemin E, Sturm E, Cresteil D, Bosma PJ, Aten J, et al. Mutations in the MDR3 gene cause progressive familial intrahepatic cholestasis. Proc Natl Acad Sci USA 1998;95(1):282–7.
[86] Woolbright BL, Jaeschke H. Novel insight into mechanisms of cholestatic liver injury. World J Gastroenterol 2012;18(36):4985–93.
[87] Kalaany NY, Gauthier KC, Zavacki AM, Mammen PPA, Kitazume T, Peterson JA, et al. LXRs regulate the balance between fat storage and oxidation. Cell Metab 2005;1(4):231–44.

[88] Wagner M, Zollner G, Trauner M. New molecular insights into the mechanisms of cholestasis. J Hepatol 2009;51(3):565–80.
[89] Trauner M, Meier PJ, Boyer JL. Molecular regulation of hepatocellular transport systems in cholestasis. J Hepatol 1999;31(1):165–78.
[90] Zhang YJ, Yeager RL, Klaassen CD. Circadian expression profiles of drug-processing genes and transcription factors in mouse liver. Drug Metab Dispos 2009;37(1):106–15.
[91] Raghuram S, Stayrook KR, Huang P, Rogers PM, Amanda K, Mcclure DB, et al. Identification of heme as the ligand for the orphan nuclear receptors REV-ERBα and REV-ERBβ. Nat Struct Mol Biol 2007;14(12):1207–13.
[92] Yin L, Wu N, Curtin JC, Qatanani M, Szwergold NR, Reid RA, et al. Rev-erbalpha, a heme sensor that coordinates metabolic and circadian pathways. Science 2007;318(5857):1786–9.
[93] Kaasik K, Lee CC. Reciprocal regulation of haem biosynthesis and the circadian clock in mammals. Lett Nat 2004;430:467–71.
[94] Gutierrez A, Grunau A, Paine M, Munro AW, Wolf CR, Roberts GCK, et al. Electron transfer in human cytochrome P450 reductase. Biochem Soc Trans 2003;31:497–501.
[95] Matsunaga N, Ikeda M, Takiguchi T, Koyanagi S, Ohdo S. The molecular mechanism regulating 24-hour rhythm of CYP2E1 expression in the mouse liver. Hepatology 2008;48(1):240–51.
[96] Xu C, Li C, Kong A. Induction of phase I, II and III drug metabolism/transport by xenobiotics. Arch Pharm Res 2005;3:249–68.
[97] Claudel T, Cretenet G, Saumet A, Gachon F. Crosstalk between xenobiotics metabolism and circadian clock. FEBS Lett 2007;581(19):3626–33.
[98] Zmrzljak UP, Rozman D. Circadian regulation of the hepatic endobiotic and xenobiotic detoxification pathways: the time matters. Chem Res Toxicol 2012;25(4):811–24.
[99] Koepsell H. Polyspecific organic cation transporters: their functions and interactions with drugs. Trends Pharmacol Sci 2004;25(7):375–81.
[100] Meier PJ, Eckhardt U, Schroeder A, Hagenbuch B, Stieger B. Special article substrate specificity of sinusoidal bile acid and organic anion uptake systems in rat and human liver. Hepatology 1997;26(6):1667–76.
[101] Akhtar RA, Reddy AB, Maywood ES, Clayton JD, King VM, Smith AG, et al. Circadian cycling of the mouse liver transcriptome, as revealed by cDNA microarray, is driven by the suprachiasmatic nucleus. Curr Biol 2002;12(7):540–50.
[102] Anderson G, Beischlag TV, Vinciguerra M, Mazzoccoli G. The circadian clock circuitry and the AHR signaling pathway in physiology and pathology. Biochem Pharmacol 2013;85(10):1405–16.
[103] Shimba S, Watabe Y. Crosstalk between the AHR signaling pathway and circadian rhythm. Biochem Pharmacol 2009;77(4):560–5.
[104] Mimura J, Fuji-Kuriyama Y. Functional role of AhR in the expression of toxic effects by TCDD. Biochim Biophys Acta 2003;1619(3):263–8.
[105] Stejskalova L, Dvorak Z, Pavek P. Endogenous and exogenous ligands of aryl hydrocarbon receptor: current state of art. Curr Drug Metab 2011;12:198–212.
[106] Zudaire E, Cuesta N, Murty V, Woodson K, Adams L, Gonzalez N, et al. The aryl hydrocarbon receptor repressor is a putative tumor suppressor gene in multiple human cancers. J Clin Invest 2008;118(2):640–50.
[107] Oesch-Bartlomowicz B, Oesch F. Role of cAMP in mediating AHR-signaling. Biochem Pharmacol 2009;77(4):627–41.
[108] Oesch-Bartlomowicz B, Huelster A, Wiss O, Antoniou-Lipfert P, Dietrich C, Arand M, et al. Aryl hydrocarbon receptor activation by cAMP vs dioxin: divergent signaling pathways. Proc Natl Acad Sci USA 2005;102(26):9218–23.
[109] Mukai M, Tischkau S. Effects of tryptophan photoproducts in the circadian timing system: searching for a physiological role for aryl hydrocarbon receptor. Toxicol Sci 2007;95(1):172–81.
[110] Rannug A, Fritsche E. The aryl hydrocarbon receptor and light. Biol Chem 2006;387(9):1149–57.

[111] Dai J, Inscho E, Yuan L, Hill S. Modulation of intracellular calcium and calmodulin by melatonin in MCF-7 human breast cancer cells. J Pineal Res 2002;32(2):112–9.
[112] Mayati A, Levoin N, Paris H, N'Diaye M, Courtois A, Uriac P, et al. Induction of intracellular calcium concentration by environmental benzo(a)pyrene involves a beta2-adrenergic receptor/adenylyl cyclase/Epac-1/inositol 1,4,5-trisphosphate pathway in endothelial cells. J Biol Chem 2012;287(6):4041–52.
[113] Monteiro P, Gilot D, Le Ferrec E, Rauch C, Lagadic-Gossmann D, Fardel O. Dioxin-mediated up-regulation of aryl hydrocarbon receptor target genes is dependent on the calcium/calmodulin/CaMKIalpha pathway. Mol Pharmacol 2008;73(3):769–77.
[114] Garrett R, Gasiewicz T. The aryl hydrocarbon receptor agonist 2,3,7,8-tetrachlorodibenzo-p-dioxin alters the circadian rhythms, quiescence, and expression of clock genes in murine hematopoietic stem and progenitor cells. Mol Pharmacol 2006;69(6):2076–83.
[115] Hogenesch JB, Chan WK, Jackiw VH, Brown RC, Gu YZ, Pray-Grant M, et al. Characterization of a subset of the basic-helix-loop-helix-PAS superfamily that interacts with components of the dioxin signaling pathway. J Biol Chem 1997;272(13):8581–93.
[116] Mukai M, Lin T, Peterson R, Cooke P, Tischkau S. Behavioral rhythmicity of mice lacking AhR and attenuation of light-induced phase shift by 2,3,7,8-tetrachlorodibenzo-p-dioxin. J Biol Rhythms 2008;23(3):200–10.
[117] Xu C-X, Krager SL, Liao D-F, Tischkau SA. Disruption of CLOCK-BMAL1 transcriptional activity is responsible for aryl hydrocarbon receptor-mediated regulation of Period1 gene. Toxicol Sci 2010;115(1):98–108.
[118] Qu X, Metz RP, Porter WW, Cassone VM, Earnest DJ. Disruption of period gene expression alters the inductive effects of dioxin on the AhR signaling pathway in the mouse liver. Toxicol Appl Pharmacol 2009;234(3):370–7.
[119] Pendergast J, Yamazaki S. The mammalian circadian system is resistant to dioxin. J Biol Rhythms 2012;27(2):156–63.
[120] Wang C, Xu C, Krager S, Bottum K, Liao D, Tischkau S. Aryl hydrocarbon receptor deficiency enhances insulin sensitivity and reduces PPAR-α pathway activity in mice. Environ Health Perspect 2011;119(12):1739–44.
[121] Krawczak M, Cooper D. p53 mutations, benzo(a)pyrene and lung cancer. Mutagenesis 1998;13:319–20.
[122] Hoeijmakers JHJ. Genome maintenance mechanisms for preventing cancer. Nature 2001;411:366–74.
[123] Kang T-H, Reardon JT, Kemp M, Sancar A. Circadian oscillation of nucleotide excision repair in mammalian brain. Proc Natl Acad Sci USA 2009;106(8):2864–7.
[124] Sancar A, Lindsey-Boltz LA, Kang T-H, Reardon JT, Lee JH, Ozturk N. Circadian clock control of the cellular response to DNA damage. FEBS Lett 2010;584(12):2618–25.
[125] Kang T-H, Lindsey-Boltz LA, Reardon JT, Sancar A. Circadian control of XPA and excision repair of cisplatin DNA damage by cryptochrome and HERC2 ubiquitin ligase. Proc Natl Acad Sci USA 2010;106(11):4890–5.
[126] Oklejewicz M, Destici E, Tamanini F, Hut RA, Janssens R, van der Horst GTJ. Phase resetting of the mammalian circadian clock by DNA damage. Curr Biol 2008;18(4):286–91.
[127] Gamsby JJ, Loros JJ, Dunlap JC. A phylogenetically conserved DNA damage response resets the circadian clock. J Biol Rhythms 2009;24(3):193–202.
[128] Gurtoo H, Marinello A, Struck R, Paul B, Dahms R. Studies on the mechanism of denaturation of cytochrome P-450 by cyclophosphamide and its metabolites. J Biol Chem 1981;256:11691–11701.
[129] Gorbacheva VY, Kondratov RV, Zhang R, Cherukuri S, Gudkov AV, Takahashi JS, et al. Circadian sensitivity to the chemotherapeutic agent cyclophosphamide depends on the functional status of the CLOCK/BMAL1 transactivation complex. Proc Natl Acad Sci USA 2005;102(9):3407–12.
[130] Blumenthal RD, Waskewich C, Goldenberg DM, Lew W, Flefleh C, Burton J. Chronotherapy and chronotoxicity of the cyclooxygenase-2 inhibitor celecoxib in athymic mice bearing human breast cancer xenografts. Clin Cancer Res 2001;7:3178–85.

[131] Granda TG, Filipski E, Attino RMD. Experimental chronotherapy of mouse mammary adenocarcinoma MA13/C with docetaxel and doxorubicin as single agents and in combination. Cancer Res 2001;61:1996–2001.
[132] Ohdo S. Chronopharmacology focused on biological clock. Drug Metab Pharmacokinet 2007;22(1):3–14.
[133] Innominato PF, Lévi FA, Bjarnason GA. Chronotherapy and the molecular clock: clinical implications in oncology. Adv Drug Deliv Rev 2010;62(9-10):979–1001.
[134] Lévi F, Okyar A. Circadian clocks and drug delivery systems: impact and opportunities in chronotherapeutics. Exp Opin Drug Deliv 2011;8(12):1535–41.
[135] Levi F, Schibler U. Circadian rhythms: mechanisms and therapeutic implications. Ann Rev Pharmacol Toxicol 2007;47:593–628.
[136] Carey MP, Burish TG. Etiology and treatment of the psychological side effects associated with cancer chemotherapy: a critical review and discussion. Psychol Bull 1988;104(3):307–25.
[137] Kondratov RV, Gorbacheva VY, Antoch MP. The role of mammalian circadian proteins in normal physiology and genotoxic stress responses. Curr Top Dev Biol 2007;78(06):173–216.
[138] Griffett K, Burris TP. The mammalian clock and chronopharmacology. Bioorg Med Chem Lett 2013;23(7):1929–34.
[139] Kirkland D, Pfuhler S, Tweats D, Aardema M, Corvi R, Darroudi F, et al. How to reduce false positive results when undertaking in vitro genotoxicity testing and thus avoid unnecessary follow-up animal tests: report of an ECVAM Workshop. Mutat Res 2007;628(1):31–55.
[140] Thybaud V, Aardema M, Clements J, Dearfield K, Galloway S, Hayashi M, et al. Strategy for genotoxicity testing: hazard identification and risk assessment in relation to in vitro testing. Mutat Res 2007;627(1):41–58.
[141] Liebsch M, Grune B, Seiler A, Butzke D, Oelgeschläger M, Pirow R, et al. Alternatives to animal testing: current status and future perspectives. Arch Toxicol 2011;85(8):841–58.
[142] Guillouzo A. Liver cell models in in vitro toxicology. Environ Health Perspect 1998;106(Suppl):511–32.
[143] Soldatow VY, Lecluyse EL, Griffith G, Rusyn I. In vitro models for liver toxicity testing. Toxicol Res 2013;2(23):23–39.
[144] Destici E, Oklejewicz M, Nijman R, Tamanini F, van der Horst GTJ. Impact of the circadian clock on in vitro genotoxic risk assessment assays. Mutat Res 2009;680(1–2):87–94.
[145] Farrow SN, Solari R, Willson TM. The importance of chronobiology to drug discovery. Exp Opin Drug Discov 2012;7(7):535–41.
[146] Asher G, Schibler U. Crosstalk between components of circadian and metabolic cycles in mammals. Cell Metab 2011;13(2):125–37.
[147] Finck BN, Kelly DP. PGC-1 coactivators: inducible regulators of energy metabolism in health and disease. J Clin Invest 2006;116(3):615–22.
[148] Hummasti S, Hotamisligil GS. Endoplasmic reticulum stress and inflammation in obesity and diabetes. Circ Res 2010;107(5):579–91.
[149] Ramsey KM, Yoshino J, Brace CS, Abrassart D, Kobayashi Y, Marcheva B, et al. Circadian clock feedback cycle through NAMPT-mediated NAD+ biosynthesis. Science 2009;324(5927):651–4.
[150] Yagita K, Okamura H. Forskolin induces circadian gene expression of rPer1, rPer2 and Dbp in mammalian rat-1 fibroblasts. FEBS Lett 2000;465(1):79–82.
[151] Yang X. A wheel of time: the circadian clock, nuclear receptors, and physiology. Genes Dev 2010;24:741–7.

SECTION

Toxicoinformatics

CHAPTER

Introduction to Toxicoinformatics

6.1

Rob H. Stierum

The Netherlands Organisation for Applied Scientific Research, Microbiology and Systems Biology, Zeist, The Netherlands

Assessing the safety of a medical drug, a food ingredient, or a chemical compound is essential to ensure human safety. Before drugs or chemicals can be administered to humans, toxicity testing is traditionally performed in animal models, where acute toxic endpoints such as mortality and behavior, and chronic toxic endpoints such as reproduction, long-term survival, and growth, can be studied. Such studies are time consuming and costly, and not in line with growing emphasis on the "3Rs" principle: the reduction, replacement, and refinement of animal use in experimentation [1]. To foster the development of alternatives to animal testing, innovative approaches have been developed. As a result, toxicology has evolved from a animal-based discipline, primarily concerned with physiological observations and histopathology, towards an interdisciplinary field, combining animal data with high-content data from different domains, including molecular biology, biochemistry, computational chemistry, in vitro cell biology, pharmacokinetic modeling, and toxicogenomics. Since 1999, when the term "toxicogenomics" was coined [2], the number of toxicogenomics papers has grown steadily to a total of nearly 1100 (Pubmed search, April 2013). Thus, since the turn of the millennium, genomics disciplines have even further increased the data content in the field of toxicology, evidence of which is provided for specific toxicological endpoints, elsewhere in this book. The purpose of the following two chapters is to provide the reader with insight into two main aspects of concern to the storage and handling of toxicogenomics data, in particular in relation to legacy toxicological data.

First, there is a need to capture, annotate, and store study-specific data and metadata into easily accessible *infrastructures*. In particular, given the heterogeneous nature of data, infrastructures are needed that are capable of storing and integrating different data types. In Chapter 6.2, Jennifer Fostel et al. provide a general overview of international database activities towards this goal, and highlight two examples: CEBS (Chemical Effects in Biological Systems), the infrastructure under development at the US National Institute of Environmental Health Sciences, and DIAMONDS, the infrastructure under development, amongst others, at the Netherlands Toxicogenomics Centre (NTC). Secondly, the development of bioinformatics *methods* towards applications in toxicology is needed. In Chapter 6.3,

Toxicogenomics-Based Cellular Models.

Kristina Hettne et al. describe bioinformatics methods for interpreting toxicogenomics data, along different lines: class discovery and separation, connectivity mapping, mechanistic analysis, and identification of early predictors of toxicity. In particular, in-depth insight is provided on the application of text-mining approaches in the analysis of toxicogenomics data.

References

[1] Russell WMS, Burch RL. The principles of humane experimental technique. London: Methuen & Co. Ltd.; 1959.

[2] Nuwaysir EF, Bittner M, Trent J, Barrett JC, Afshari CA. Microarrays and toxicology: the advent of toxicogenomics. Mol Carcinog 1999;24(3):153–9.

CHAPTER

6.2 Toxicogenomics and Systems Toxicology Databases and Resources: Chemical Effects in Biological Systems (CEBS) and Data Integration by Applying Models on Design and Safety (DIAMONDS)

Jennifer Fostel*, Eugene van Someren, Tessa Pronk†, Jeroen Pennings†, Peter Schmeits††, Jia Shao††, Dinant Kroese‡ and Rob Stierum‡‡**

**National Toxicology Program, National Institute of Environmental Health Sciences, Research Triangle Park, North Carolina, **The Netherlands Organisation for Applied Scientific Research, Microbiology, and Systems Biology, Zeist, The Netherlands, †National Institute for Public Health and the Environment (RIVM), Center for Health Protection, Bilthoven, The Netherlands, ††RIKILT—Institute of Food Safety, Wageningen University and Research Centre, BU Toxicology and Bioassays, Wageningen, The Netherlands, ‡The Netherlands Organisation for Applied Scientific Research, Risk Analysis for Products In Development, Zeist, The Netherlands, ‡‡The Netherlands Organisation for Applied Scientific Research, Microbiology and Systems Biology, Zeist, The Netherlands*

6.2.1 Introduction

In order to address questions within systems toxicology, data from a variety of disciplines are utilized. These include the toxicological phenotype, often captured in a histopathology diagnosis or through clinical pathology, in vitro assay data, such as cytotoxicity, together with a system-wide response such as transcriptomics data. These data are best integrated and understood within the context of the experiments that generated them. Thus, databases designed to capture data for systems biology capture a variety of phenotypic data, along with so-called 'omics data, and metadata describing the experimental design and execution. Databases and consortia supporting systems toxicology research are numerous, with some examples provided in Table 6.2.1.

The purpose of this chapter is to provide the reader with an in-depth overview of two data infrastructures for systems toxicology data: the Chemical Effects in Biological Systems database from the division of the National Toxicology Program (NTP) at the US National Institute of Environmental Health Sciences (NIEHS); and DIAMONDS, the infrastructure developed by The Netherlands Organisation for Applied Scientific Research (TNO), jointly with the Netherlands Toxicogenomics Center (http://www.toxicogenomics.nl). First, the structure of each of these toxicoinformatics

Toxicogenomics-Based Cellular Models.

Table 6.2.1 Overview of Relevant Databases and Consortia Supporting Systems Toxicology Research

Name	Description	URL
ArrayTrack	Tool from the FDA that allows storage of metadata and data and allows use of analysis tools	http://www.fda.gov/ScienceResearch/BioinformaticsTools/Arraytrack/default.htm
Comparative toxicogenomics database	A public website and research tool that curates scientific data describing relationships between chemicals, genes, and human diseases	http://ctdbase.org/
diXa	FP7 project to collect publicly available toxicogenomics metadata and data and perform cross-platform integrative statistical analyses, and cross-study meta-analyses, to create a systems model for predicting chemical-induced liver injury	www.dixa-fp7.eu/
NTC	Netherlands Toxicogenomics Center, aiming at the development of 'omics-based in vitro models for toxicological testing	www.toxicogenomics.nl
MGI	The Mouse Genome Informatics database, an example of a database supporting systems biology in the mouse	http://www.informatics.jax.org/
TG Gates	Repository of in vivo and in vitro transcriptomics data, clinical chemistry, and pathclogy	http://toxico.nibio.go.jp/
DrugBank	Repository of chemoinformatic information, in vivo and in vitro transcriptomics data, clinical chemistry, and pathology, with additional tools	http://www.drugbank.ca/
ASAT	Knowledge-base integrating human disease information with in vitro toxicological cata	http://asat-calc.tgx.unimaas.nl
dbXP	Nutritional Phenotype Database (dbXP) is a database environment to facilitate storage of biologically relevant preprocessed 'omics data, as well as study descriptive and phenotype data and to enable the combination of this information at different levels. Developed by TNO, NMC, NUGO, NBIC and The Hyve	https://wiki.nbic.nl/index.php/Nutritional_Phenotype_Database
HESS	HESS supports evaluation of repeated-dose toxicity by category approach	www.safe.nite.go.jp/english/kasinn/qsar/hess-e.html
tranSMART	tranSMART is a knowledge-management platform that allows investigating correlations between genetic and phenotypic data, and assessing analytical results in the context of published literature and other work	http://transmartproject.org/
TraIT	The Center for Translational Molecular Medicine's TraIT aims to develop an IT infrastructure for translational research to facilitate the collection, storage, analysis, archiving, sharing, and securing of data	http://www.ctmm.nl/en/programmas/infrastructuren/traitprojecttranslationeleresearch?set_language=en
CEBS	The Chemical Effects in Biological Systems (CEBS) database developed by NIEHS containing o.a. NTP data, in vitro and in vivo assay, and 'omics (discussed in more detail in Section 6.2.2)	http://www.niehs.nih.gov/research/resources/databases/cebs/index.cfm
DIAMONDS	Tool developed by TNO for DECO and NTC, containing toxicogenomics data (amongst others NTC TG-GATEs, DrugMatrix, ToxCast/ToxRefdb), chemoinformatic information, clin.chem, pathology, and a toolbox of analysis tools that enables meta-analysis (discussed in more detail in Section 6.2.3)	http://web-php06.tno.nl/diamonds/

FDA, US Food and Drug Administration; NIEHS, US National Institute of Environmental Health Sciences; NTP, National Toxicology Program at the NIEHS; TNO, The Netherlands Organisation for Applied Scientific Research; ASAT, Assuring Safety Without Animal Testing; FP7, Seventh framework program of the European Community for research and technological development including demonstration activities; NMC, Netherlands Metabolomics Center; NUGO, European Nutrigenomics Organization; NBIC, Netherlands Bioinformatics Center; HESS, Hazrd Evaluation Support System.

databases is explained in detail. Secondly, screenshot examples showing the setup and practical navigation through these databases are shown. Finally, the use of these databases towards integrative systems toxicology data analysis is demonstrated.

6.2.2 Chemical effects in biological systems

CEBS is designed to support systems biology and to be a resource for environmental health scientists. CEBS was initiated in 2003, when it contained approximately 25 studies [1]. By 2013, CEBS housed over 9000 studies, including the public legacy data from the National Toxicology Program (NTP), from over three decades of research into the effects of test articles in systems ranging from bacteria to rodents.

Components of a study in CEBS

The CEBS database defines a study as a self-contained scientific enquiry. That is, a study is one experiment using one type of subject, such as mice, rats, cell cultures, or bacteria. A single study has a defined start and end point, between which the study subjects are alive and being cared for, observed, and treated. The study subject is the object of the study, the organism investigated during the study. In many NTP studies, the subject is a rodent, either rat or mouse. In vitro toxicogenomics studies and genetic toxicology studies use bacterial cultures or cultured cells as the study subject. CEBS uses the term "participant" to refer to either subject or group of subjects, such as a bacterial culture or a group of animals treated as biological replicates. Specimens can be taken during the study timeline and preserved for later assays. An assay is defined as a process that takes as input a biological specimen or other tissue, and produces data. An example of an assay would be a clinical chemistry test such as the measurement of the level of alanine aminotransferase in the blood, a microscopic examination to produce a histopathological diagnosis, or a microarray experiment.

The design of CEBS is given in Figure 6.2.1. The database consists of three main sections: metadata, data, and data transformations. The study metadata provide the biological context for the study, and are described below. Connecting the study metadata to the study data is the assay input, generally a biological specimen, which is a section of tissue removed from a study subject, with all the study metadata associated with that specimen. Study data consist of all assay data from a study—histopathological observations, normalized microarray data, and so forth. Data are often summarized or conclusions are drawn from the data, and these are stored in the data transformation section of CEBS.

The CEBS database abstracts the study metadata into components: (1) the study description, listing study factors, the principal investigator, the start date, etc.; (2) groups and subjects and their attributes in the study, such as species, sex, age, strain, dose group, etc.; (3) study protocols; and (4) the study timeline, which links protocols, groups, and events. Study factors and levels are listed both as participant characteristics (e.g. the "low dose" group) and in the corresponding protocol (chemical X of concentration Y was applied by gavage). The timeline links groups, protocols, and events, and these together constitute the description of study execution.

Study protocols are divided into five classes within CEBS: care, in-life observation, stressor application, specimen preparation, and disposition. These are tailored to the type of subject and stressor. For example, an in vitro subject type would have a care protocol called culture, which includes the

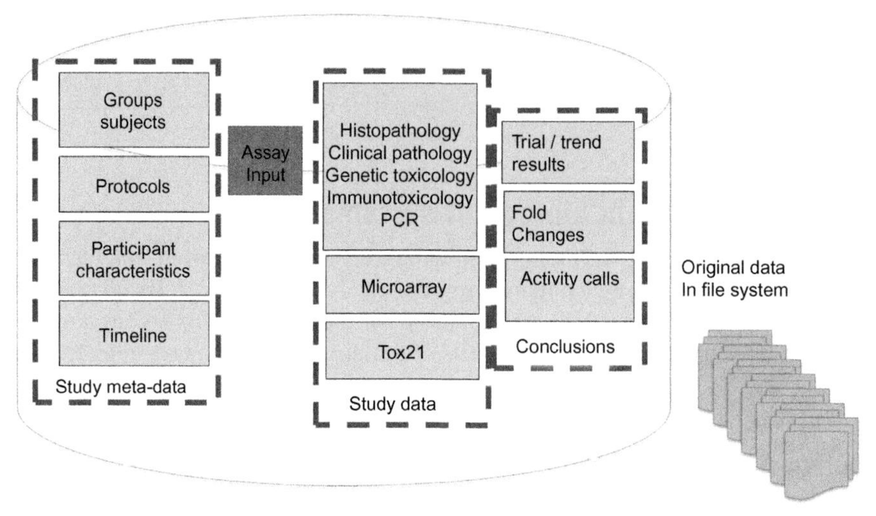

FIGURE 6.2.1 Overview of CEBS design.

The CEBS Oracle database has three major sections: study metadata, study data, and transformations of study data. All original files are also maintained in an external file system. PCR, polymerase chain reaction. Tox21 refers to the "Toxicology in the 21st Century" initiative (http://www.epa.gov/ncct/Tox21/).

type of medium, the atmosphere, and the presence of antibiotics or serum, while an in vivo subject type would have a care protocol called husbandry, which would describe protocol details including the diet, cage, and lighting schedule. A stressor protocol for a chemical stressor type would include the dose, route or administration, and vehicle, whereas a protocol for a genetic stressor such as a transgene would include the gene and type of alteration.

The timeline captures when each protocol is applied and to which group. This permits the inference of events "before" and "after" other events. An assay is any process that derives data from a specimen from the study. Assay data are linked to the subject contributing the biological material, the preparation used to obtain the specimen, and the time during the study when the specimen was taken.

Data domains and data standardization

CEBS houses data from a variety of different data domains—microarray, immunotoxicology, genetic toxicology, histopathology, clinical pathology, in-life observations, gross observations and organ weights, PCR (from polymerase chain reaction assays), enzymology (for in vitro enzyme assays), behavioral observations, and reproductive observations. All these various domains are tied to the time in the study, and the study subject, when the observation was made or the specimen collected for later generation of assay data. This permits CEBS to link all data from a given subject as well as to collect longitudinal data if a subject was observed through time.

Data are entered into CEBS using the original units used to make the measurement, and using the names for the assay that the depositor recorded. This information can be viewed in the individual study view. During the loading process, the data are converted into standard units and standard assay names within CEBS in order to support cross-study data-mining. The original assay name and units are maintained, and numerical data are copied into a standard data field using the standard unit CEBS

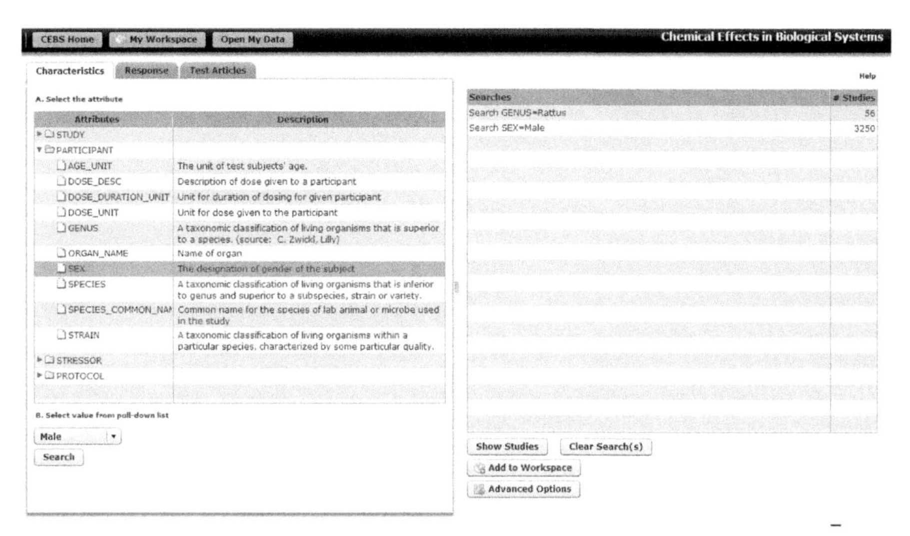

FIGURE 6.2.2 Searching within CEBS on the basis of participant (subject) characteristic.

The CEBS search screen permits the user to filter for subjects having particular characteristics (innate, such as genus and sex, and study-dependent, such as dosing and feed, for example).

lists for the given assay, using a look-up table within CEBS managed by the CEBS curation staff. Converting all the numerical data from an assay to a standard unit makes it easier to perform meta-data analysis using data from different studies in CEBS.

Data integration

In addition to the individual study view, CEBS provides the user the opportunity to select data based on the data associated with any study in CEBS. The search function has three entry points. The first entry point is the study metadata, shown in Figure 6.2.2. This page permits the user to limit the retrieved study data by diet, sex, species, chemical type, or characteristic of the participant (subject or group) or stressor. Here, two searches, one for male and the other for genus *Rattus* have been carried out. The user can also select data based on the response of the participant, as in Figure 6.2.3, which shows a search for any animal with a liver lesion.

Once a set of searches has been defined, they can be combined as shown in Figure 6.2.4, to produce, in this case, the set of studies with male rats with liver lesions. Having animals with a lesion is only half of what the user needs—generally the user also would like to retrieve the control animals as well, so that comparative data can be determined. This is done in CEBS as shown in Figure 6.2.5. The user then adds the selected subjects to the workspace section of CEBS, where the user can review the study details, retrieve a list of the test articles producing the effects, download a data matrix of all selected results, or carry out visual data-mining. In this particular search, the animals have data from clinical pathology (clinical chemistry, hematology, urinalysis), gross and in-life observations, organ weights, and tissue chemistry in addition to histopathology, and all data are available in the CEBS workspace. The user can perform visual data-mining on a clinical pathology measurement of serum albumen, shown in Figure 6.2.6, and see that one study appears to be an outlier from the others, at

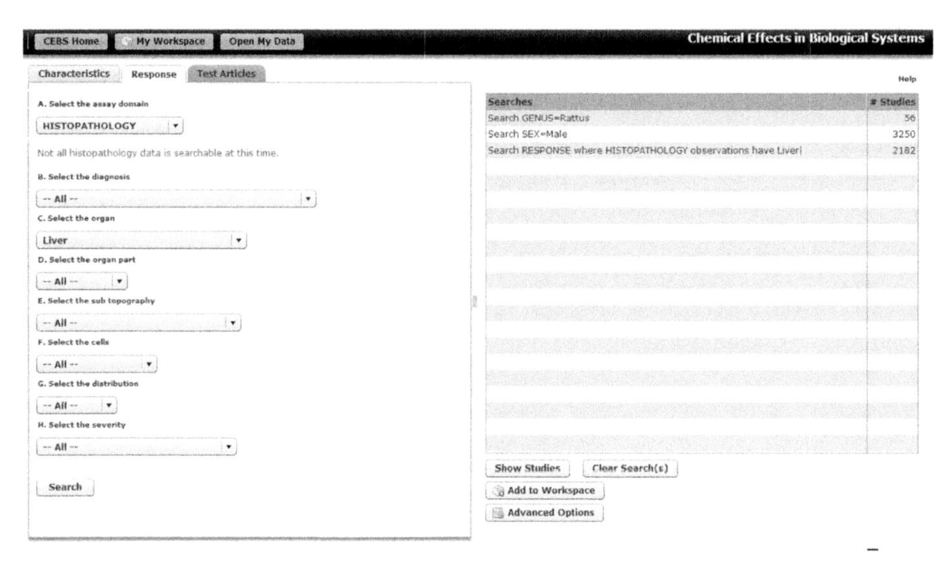

FIGURE 6.2.3 Searching in CEBS on the basis of participant response (histopathological observation).

The CEBS search screen also permits the user to search for subjects based on their response in an assay. The user selects the assay domain from a pull-down list, in this case histopathology, and then the response of interest can be selected.

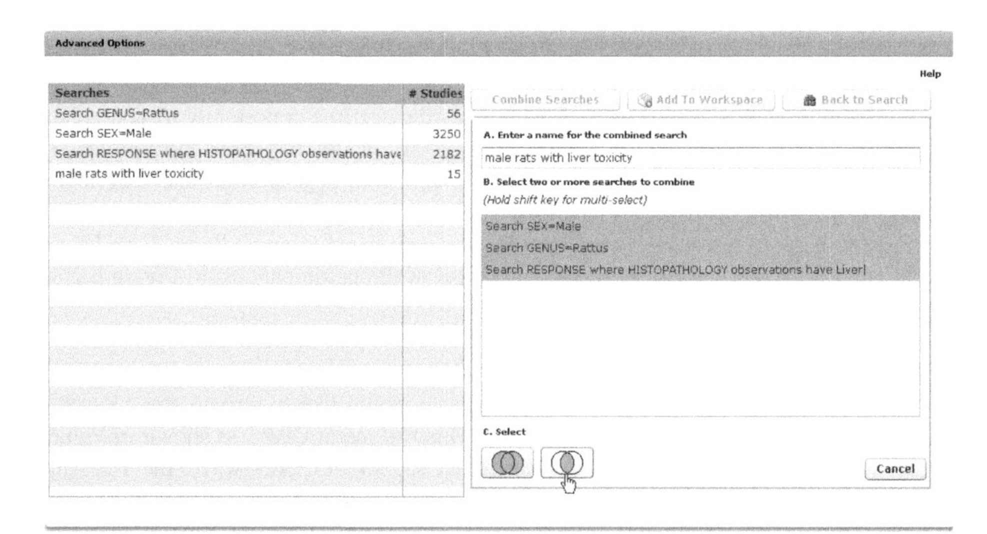

FIGURE 6.2.4 Combining searches within CEBS.

Once the user has created searches across CEBS, he may combine them using the "Combine Searches" screen in CEBS.

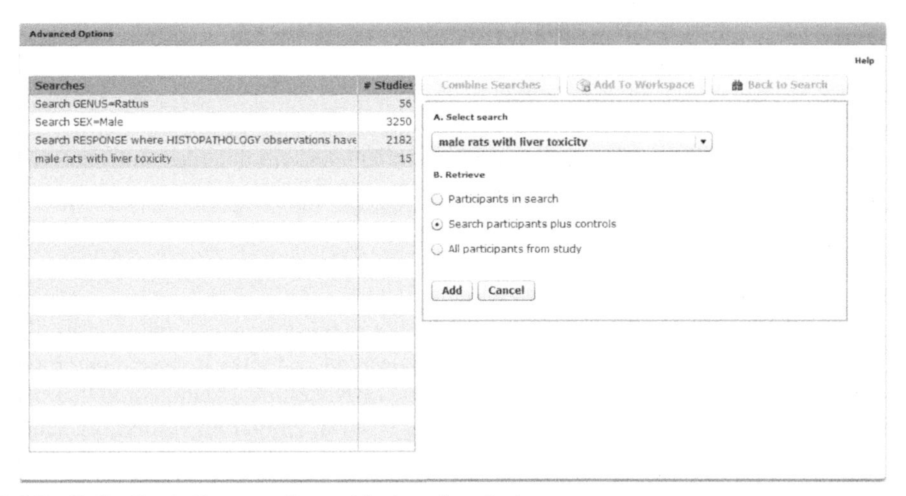

FIGURE 6.2.5 Collecting both responding subjects and controls.

The search function in CEBS permits the user to identify a set of study participants that match particular criteria. The "Send to Workspace" screen permits the user to choose to select these subjects only, or to select the matching subjects and their corresponding controls.

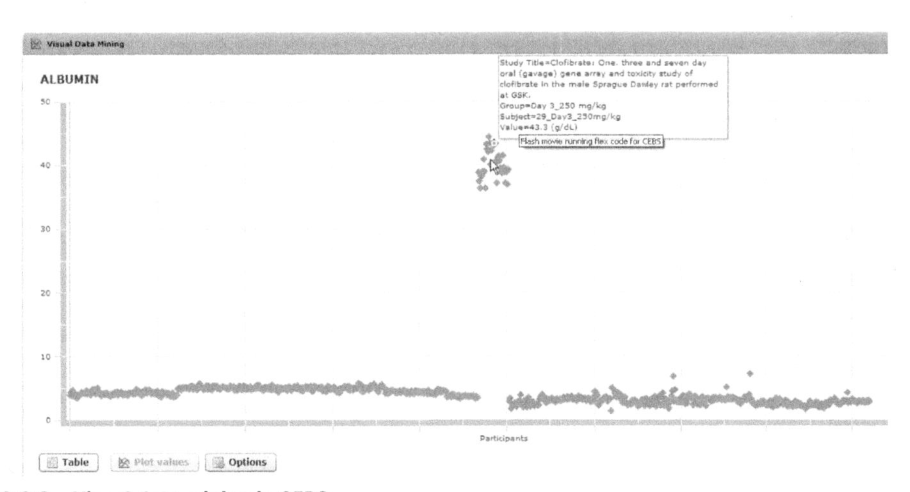

FIGURE 6.2.6 Visual data-mining in CEBS.

The user selects a particular assay with numerical output, and CEBS will display the results for the selected subjects in a scatter plot. Hovering the cursor over a point allows the user to read details of the study that produced the data.

least in the response in this assay. Hovering the cursor over the data points allows the user to see the data for the particular assay. Microarray data and other data and metadata for the same subjects can be integrated into a data matrix and downloaded.

This data integration across studies and across data domains is the feature of CEBS that supports systems toxicology, by permitting the CEBS user to combine data from subjects of interest and their respective controls, within the context of the experimental protocols and execution.

Standards

CEBS was developed after consideration of the design of public databases and data exchange formats available. The CEBS Data Dictionary reflects this design [2]. CEBS is developing tools to read data in the format defined by the Standards for Exchange of Nonclinical Data (SEND), a component of the Clinical Data Interchange Standards Consortium (CDISC, http://www.cdisc.org/send), which will allow CEBS data to be shared in this format, and for data in SEND format to be read by CEBS. CEBS aligns with the Ontology for Biomedical Investigations (OBI, http://obi-ontology.org/) and will use this format to exchange data with the semantic web. CEBS uses the MAGE-Tab format (http://www.mged.org/mage-tab/) to exchange microarray data.

The use of community standards permits facile exchange of data by using a common format where possible [3]. The SEND format is used for non-clinical data from animal studies, but does not extend to invertebrate subjects or cell culture data. For this, CEBS uses the simple investigation formatted text (SIFT) format.

Minimal and "maximal" information about a study

The metadata provided with a study in CEBS determine in part how useful those study data are for subsequent analysis. Each piece of metadata, describing the study subjects' attributes—for example, strain, age, husbandry for lab animals, and each phenotype (histopathology, behavior, and other assay data)—provides a hook for the study in other searches, allowing the data to be reused in meta-analysis. These details also permit the meta-analyst to identify causes of variability within the data set; for example, the impact of fasting status on liver gene expression [4,5], which would otherwise not be explainable.

Many attempts have been made to provide checklists and descriptions of minimal required information for various technologies (see http://mibbi.sourceforge.net/), and the ISA-Tab consortium (http://isatab.sourceforge.net/format.html) is creating tools to combine these standards into common checklists. However, these approaches begin with technology rather than biology. The age and parentage of an animal will have different importance in a reproductive toxicology study than in a general toxicology study; the presence of serum in a culture medium will have more of an impact in a test of a chemical agent in cultured cells than in a test of a physical agent, such as electromagnetic field, using the same subject. Thus, a checklist for minimal information about a toxicological study must consider the discipline, the nature of the study subject, and the nature of the test article [6]. In summary, CEBS supports the deposition of any and all metadata and data that the depositor can include, and uses a flexible, expandable methodology to collect, store, and present these data for use in data-mining and exploration in systems toxicology and environmental health science.

6.2.3 Data integration by applying models on design and safety (DIAMONDS)

Within the Netherlands Toxicogenomics Centre (NTC, http://www.toxicogenomics.nl), data from in vitro, ex vivo, and, to a limited extent, in vivo model types are generated using a variety of primary cells, cell lines, or organ samples from zebrafish and rodents. From a bioinformatics point of view, it is a challenge to build an infrastructure that can handle such diverse data types and experimental designs as well as be particularly useful for toxicogenomics analyses. This is important given the fact that NTC is in need of metadata analyses on particular toxicological problems (e.g. cholestasis), to discover which in vitro models are possible candidates for reflecting (aspects of) in vivo toxicology. One major prerequisite for such an infrastructure is to make sure the metadata are well captured. Metadata concern the description of the biological experiment from starting biological material, via treatment and 'omics platform to a data file containing the measurements. Within NTC it also entails, however, information about the project parts, people, and their roles; particularly relevant for larger multi-partner projects such as NTC. When such information is provided and regularly updated during the course of a large project, it provides a way to view the current status of the project, increases cooperation by awareness between biologists of each other's planned experiments, and allows advance knowledge of how much bioinformatics effort will be needed (as planned experiments near completion).

Two pivotal questions often raised by the biologists in the NTC project, but which hold when analyzing any 'omics data set, are:

- How do the results of **in-house-generated study data** compare to other data studies
  1. From within one's institute?
  2. From partners in participating consortia and projects?
  3. From ones that are available in the public domain?
- How can **bioinformatics tools** be easily applied to all available **data** and be **integrated** with available **biological/expert knowledge** to better understand biology and generate new knowledge?

DIAMONDS–NTC infrastructure

To answer these questions, DIAMONDS is being developed. Figure 6.2.7 illustrates the basic principles behind DIAMONDS, i.e. to build a single infrastructure that is able to integrate data, knowledge, and tools and to make comparisons between a user's own, other projects', and public data. DIAMONDS captures data from diverse domains, i.e. chemical structure information, in vitro (high-throughput screening) assay data, in vivo and in vitro 'omics data, and in vivo clinical chemistry and pathology. DIAMONDS allows the user to import public data besides private or consortium-derived data and handles access at group level. DIAMONDS also connects using application programming interfaces (APIs) to diverse knowledge databases and brings together compound structure information, pathway/gene-set information, study metadata information, and toxicological concepts. Where possible, knowledge is matched to existing ontologies and users can easily link through to external

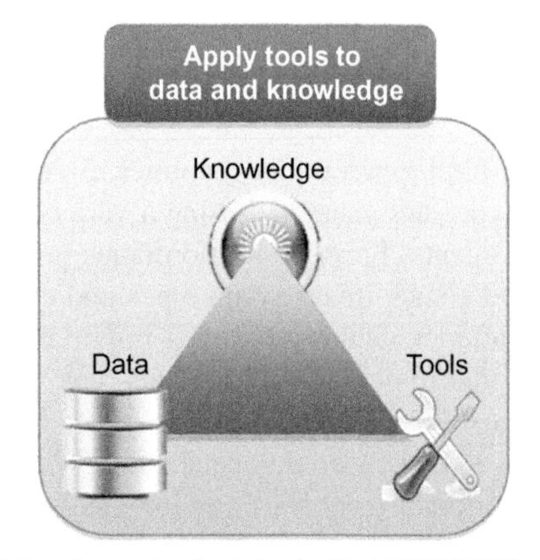

FIGURE 6.2.7 General principles behind DIAMONDS.

To integrate data, knowledge and tools into a single infrastructure (left panel) and compare own/private data with other data from projects/consortia as well as from the public domain (right panel). Data consist of in vitro/ in vivo 'omics data as well as (high-throughput) assay and toxicological phenotype data. Knowledge consists of chemoinformatics information as well as biological knowledge (pathways, interactions, and toxicological mechanisms). Tools range from basic bioinformatics tools (principal component analysis (PCA), QC) up to tools that allow pathway-based meta-analysis (network biology).

databases. The DIAMONDS infrastructure is built from a combination of a Unix file server, backup server, calculation server, MySQL database, and a Web site built in a combination of HTML, PHP, and javascript. From the Web site, the data are brought together in diverse views with main modules centered around structures, compounds, studies, meta-analysis, classification, pipelines, and users. Data can be examined using diverse visualization and analysis tools, such as quality control (QC) reports, basic multivariate data analysis such as principal components analysis, similarities, classification, and pathway scoring. In addition, DIAMONDS is designed to allow pathway-based meta-analysis by integrating data together with pathway scores generated by the integrated ToxProfiler module (http://ntc.voeding.tno.nl/toxprofiler_test/). Using the ToxProfiler algorithm [7], T-values indicating the enrichment—i.e. the under- or overexpression of gene sets relative to the total number of genes in the experiment—can be calculated for heterogeneous toxicological experiments.

Navigation through the system

Figure 6.2.8 illustrates with screenshots how researchers can visually inspect structural features of compounds and see which compounds are involved in which pathways and used in which studies. In this way, an NTC researcher knows directly which studies might be of relevance to perform a meta-analysis on. Structural similarity is based on two-dimensional (2D) or 3D fingerprints, or more structurally enriched fingerprints, commonly used in the pharmaceutical industry, such as ECFP4 and FCFP4 (Rogers and Hahn, 2010). Effectopedia (http://effectopedia.org/) is a tool developed to design adverse outcome pathways (AOPs). When viewing a compound, DIAMONDS provides information on whether that compound is used in one of the available AOPs, such that its known mechanism can be taken into account when interpreting data.

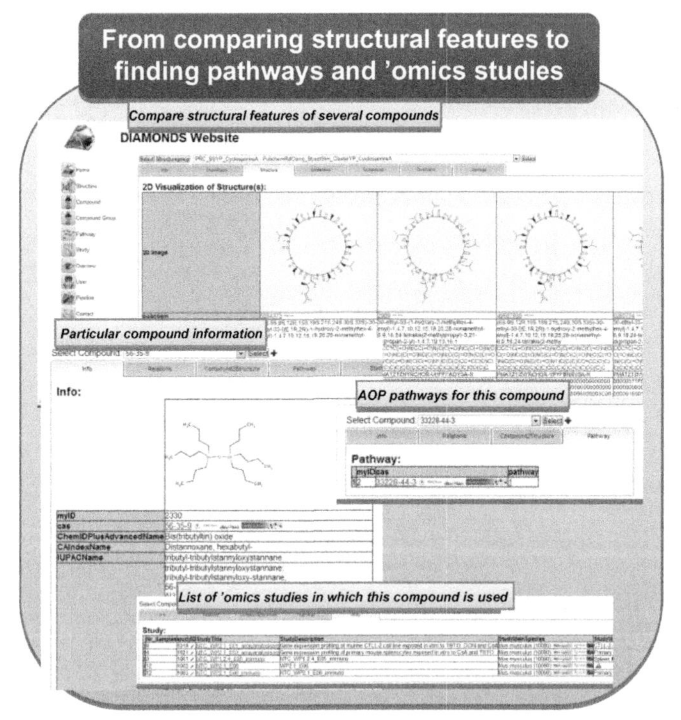

FIGURE 6.2.8 Screenshots from DIAMONDS showing how DIAMONDS links structural features of compounds to studies and pathways.

DIAMONDS allows visual and computational comparison of a group of chemical structures. Besides basic compound information, DIAMONDS also shows to which (adverse outcome) pathways a compound belongs, or in which studies it is used. Users can then jump to these pathways and/or studies or link out to external databases such as ChemIDPlusAdvanced, Pubchem, CACTVS, and Common Chemistry.

Figure 6.2.9 illustrates with screenshots how metadata are captured at study level and sample level. Where possible, controlled vocabularies and/or ontologies are being employed and link-outs to the external references, such as BioPortal, allow users to understand the terms used. However, as ontologies are never complete, users are allowed to add new terms in case no fitting term can be found (e.g. some cell lines are difficult to find in ontologies). For sustainability, DIAMONDS allows metadata to be saved as ISATAB, to be used in the ISA-Tab environment (see Minimal and "maximal" information about a study). Furthermore, an API between DIAMONDS and the Nutritional Phenotype Database (dbXP) allows studies stored in dbXP to be shown in DIAMONDS as well.

Normalization and quality control are of utmost importance when handling 'omics data. Arrayanalysis.org is a free online tool that handles normalization of transcriptomics data. DIAMONDS integrates with the output of Arrayanalysis.org and allows users to visually inspect the QC results (MA plots, heat-maps, etc). As metadata are stored, it is easy to generate a PCA to inspect the structure inside the data and change labeling and coloring according to available sample annotations.

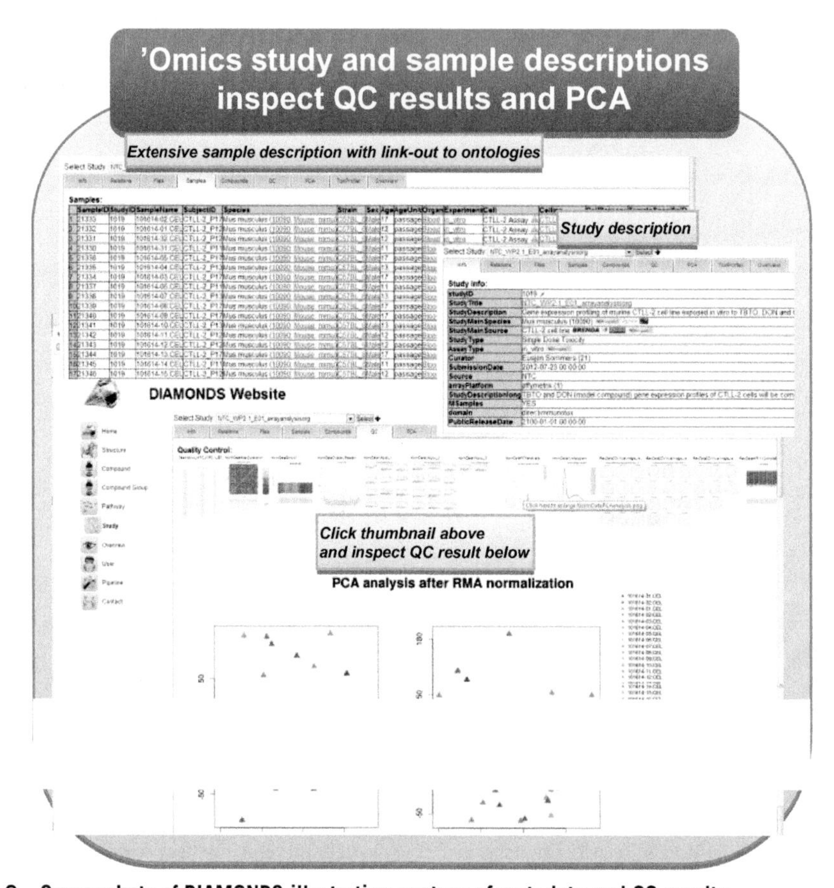

FIGURE 6.2.9 Screenshots of DIAMONDS illustrating capture of metadata and QC results.

DIAMONDS allows users to submit and retrieve metadata information about available studies. Where possible, information is matched with ontologies and users can jump to their descriptions (e.g. in BioPortal). DIAMONDS also allows visual inspection of QC plots and visualizing results using available annotations (e.g. colors in PCA plot).

DIAMONDS analysis: two examples

Figure 6.2.10 illustrates with screenshots a possible pathway-based meta-analysis. ToxProfiler is used to compute pathway scores for each sample and each pathway/gene set stored. Gene sets are based on diverse sources such as WikiPathways, KEGG, TF-binding (TFBroadC3), and tissue-specific gene sets. Users are also able to upload their own custom gene sets as well. For example, gene sets derived from in vivo toxicogenomics experiments can be employed to interrogate the toxicogenomics response obtained from various in vitro models obtained with the same compound. To perform a meta-analysis, the user first selects which studies to use (from a choice of diverse projects/sources) and subsequently which sample ratios (i.e. biological conditions) to use and whether to take

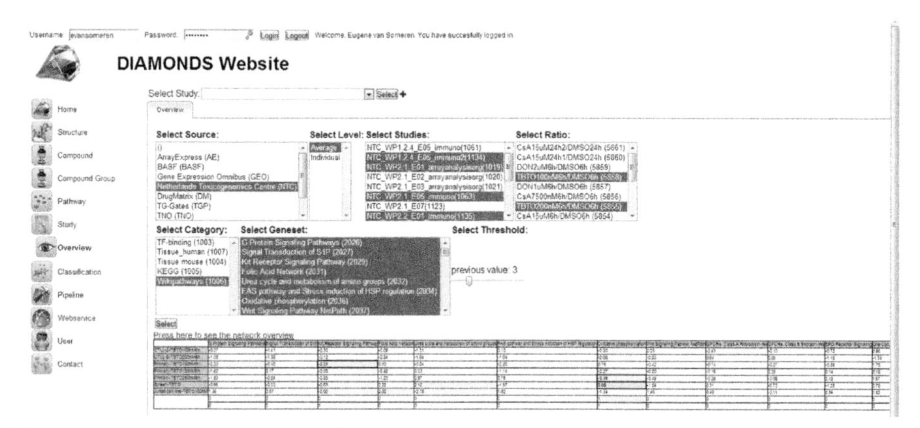

FIGURE 6.2.10 Screenshot of DIAMONDS illustrating a pathway-based meta-analysis.

A graphical sidebar on the left shows the main modules in the DIAMONDS Web site, amongst others structure/compound/groups, pathways, studies, overview (currently chosen), classification, pipelines, webservices, and user info. Choosing the overview module allows the user to select a desired combination of sample ratios (from a selection of studies and study sources) to perform a meta-analysis on. The user can also choose whether this should be done on an average (per biological condition) or on an individual ratio level. In addition, the user can indicate which combination of pathways/gene sets (from a selection of categories) should be used and which threshold to apply. Directly below, a table with the resulting pathway scores is subsequently shown to the user.

individual ratios or averages per biological condition. Then the user can choose which gene sets to analyze from diverse sources of gene sets (mentioned above) and specify a desired cut-off value for the pathway scores. Then DIAMONDS returns a fully interactive network based on Cytoscape Web [8], showing which pathways are common or specific for which samples. The user can explore this network by getting more information behind each node and search and filtering on any node property. Furthermore, users can reposition, select, and remove nodes and change their graphical properties. Figure 6.2.11 shows an example of such a meta-analysis, based on one NTC study performed by NTC partner RIKILT (Institute of Food Safety at Wageningen University), where the response of deoxynivalenol (DON) is studied in human Jurkat cells, and another NTC study by RIKILT, in which the response is studied in a murine CTLL-2 cell line at two different doses. The two doses of DON have their own specific responses, but also share several processes. There are several pathways in common to both dose levels in CTLL-2 as well as in the human Jurkat cells' response, i.e. steroid biosynthesis, bile secretion, and ribosome biogenesis in eukaryotes. The up-regulation of the process of "ribosome biogenesis in eukaryotes" that is shared between Jurkat and CTLL-2 corresponds to the classification of DON as a ribotoxic stress inducer [9].

Figure 6.2.12 shows a more extensive meta-analysis that shows how NTC data can be projected onto a pathway network generated from five diverse public toxicogenomics data sets, combining human, mouse, and rat, combining in vivo liver and lung with in vitro skin and embryonic cell lines, and combining short and late time-points. In this meta-analysis, several individual sample ratios with diverse compounds are merged into meta-nodes describing a combination of organism, organ/cell type, and known

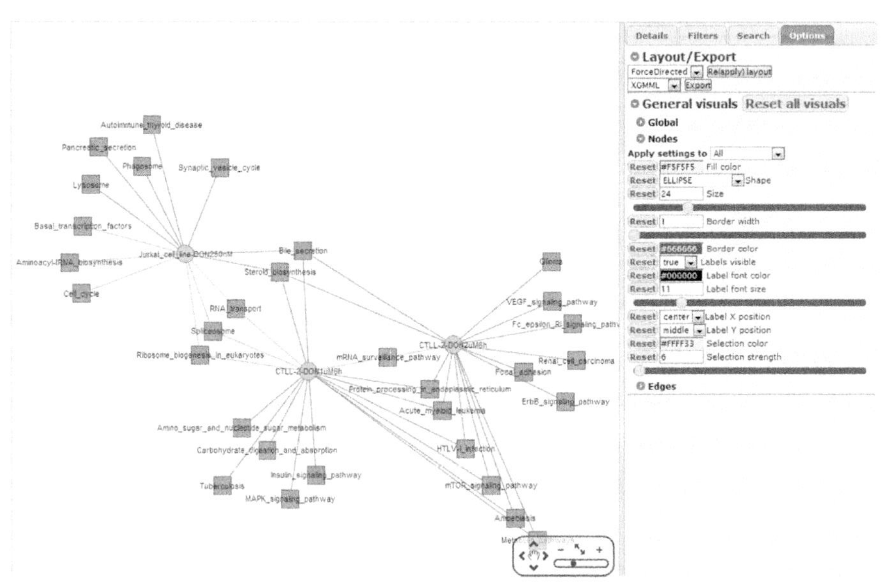

FIGURE 6.2.11 Screenshot of DIAMONDS showing the resulting network from a meta-analysis based on two datasets of the NTC.

In this meta-analysis example, the pathway response of deoxynivalenol (DON) in human Jurkat cells (one study) is compared with its response in murine CTLL-2 cells (other study). The left panel shows the resulting network, where gray circles represent biological conditions (sample ratios), purple squares represent Kyoto Encyclopedia of Genes and Genomics (KEGG) (http://www.genome.jp/kegg/) pathways, and an edge indicates that the absolute pathway score in that condition is above the user-defined threshold. The user can still interact with this network and query, filter, reposition, or hide certain nodes, get detailed information behind nodes and edges, or (as illustrated in the right panel) change the graphical properties such as color, width, text size, and shape. The user can also download the network for import into Cytoscape. The up-regulation of the process of "ribosome biogenesis in eukaryotes" that is shared between Jurkat and CTLL-2 corresponds to the classification of DON as a ribotoxic stress inducer (PMID 12773753).

toxicity, e.g. "sensitizer skin invitro human" combines several samples of sensitizing compounds to which the human HaCat cell line had been exposed. Pathways/gene sets, obtained with ToxProfiler, are connected to a meta-node if that pathway is consistently expressed in most of the samples that belong to that meta-node. Similarly, if two pathways have similar scores within samples belonging to a meta-node, these pathways are said to be co-regulated and connected in the network. This network quickly provides an overview showing which pathways are specific and/or common to different meta-nodes as well as which networks are regulated similarly. For example, it shows that there is no overlap between pathways induced by liver toxicants (separate cloud on the left middle part in Figure 6.2.12) compared with other compounds and conditions. Furthermore, carcinogens in the lung seem to have overlapping response to genotoxicants in the liver. Besides supporting mechanistic understanding, the network analysis can also be used to characterize unknown compounds. To this end, two different samples from two NTC studies that were not used to build the network were projected onto this public domain network. The first sample

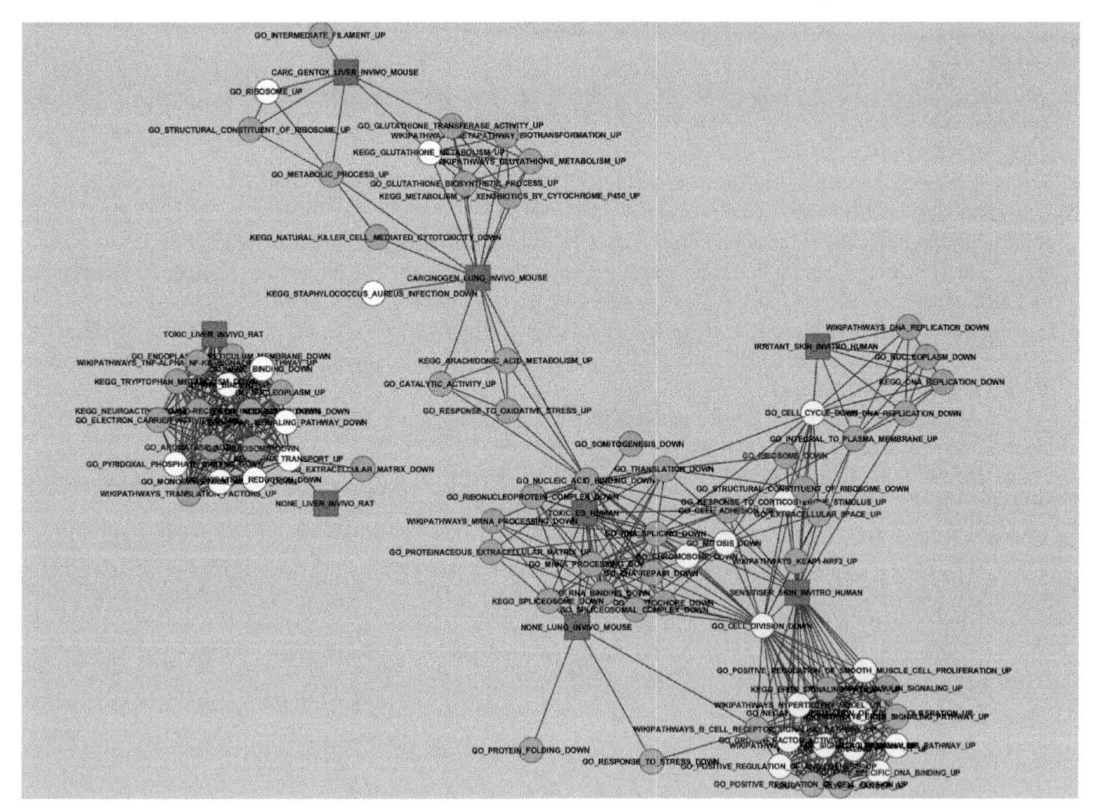

FIGURE 6.2.12 Example of a more extensive meta-analysis that integrates NTC data with public data.

The network was generated from four diverse public toxicogenomics data sets, mixing human, mouse, and rat, mixing in vivo liver and lung with in vitro skin and embryonic cell lines, and mixing short and late time-points. Square red nodes represent biological conditions (groups of sample ratios), circle nodes represent pathways/gene sets (Gene Oncology (GO) (http://www.geneontology.org/), KEGG, WikiPathways), edges between conditions and pathways indicate a significant pathway score in a majority of the underlying ratios, and edges between pathways/gene sets indicate a co-regulation within at least one condition. The pathway responses of two unseen compounds are projected onto this network by coloring the nodes green and/or yellow.

is a human HaCat cell line exposed for 4 hours to a sensitizing compound. The affected pathways are colored green in Figure 6.2.12. This compound clearly affects pathways that are mostly associated with the "sensitizer skin invitro human" meta-node, while only one pathway is associated with the "irritant skin invitro human" node, thus suggesting that this "unseen" compound behaves like a sensitizer. The second sample consists of mouse primary hepatocytes exposed for 1 day to a genotoxic liver carcinogen. The affected pathways are colored yellow in Figure 6.2.12. This compound clearly affects pathways that are mostly associated to the "toxic liver in vivo rat" meta-node, while three pathways are associated with the carcinogenic meta-nodes. This may indicate that this compound shows carcinogenic tendencies, but the response may be overshadowed by general liver toxicity.

References

[1] Waters M, Boorman G, Bushel P, Cunningham M, Irwin R, Merrick A, et al. Systems toxicology and the Chemical Effects in Biological Systems (CEBS) knowledge base. EHP Toxicogenomics 2003;111(1T):15–28.

[2] Fostel J, Choi D, Zwickl C, Morrison N, Rashid A, Hasan A, et al. Chemical effects in biological systems—data dictionary (CEBS-DD): a compendium of terms for the capture and integration of biological study design description, conventional phenotypes, and 'omics data. Toxicol Sci 2005;88(2):585–601.

[3] Fostel JM. Towards standards for data exchange and integration and their impact on a public database such as CEBS (Chemical Effects in Biological Systems). Toxicol Appl Pharmacol 2008;233(1):54–62.

[4] Boedigheimer MJ, Wolfinger RD, Bass MB, Bushel PR, Chou JW, Cooper M, et al. Sources of variation in baseline gene expression levels from toxicogenomics study control animals across multiple laboratories. BMC Genomics 2008;9:285.

[5] Corton JC, Bushel PR, Fostel J, O'Lone RB. Sources of variance in baseline gene expression in the rodent liver. Mutat Res 2012;746(2):104–12.

[6] Fostel JM, Burgoon L, Zwickl C, Lord P, Corton JC, Bushel PR, et al. Toward a checklist for exchange and interpretation of data from a toxicology study. Toxicol Sci 2007;99(1):26–34.

[7] Boorsma A, Foat BC, Vis D, Klis F, Bussemaker HJ. T-profiler: scoring the activity of predefined groups of genes using gene expression data. Nucleic Acids Res 2005;33:W592–5.

[8] Lopes CT, Franz M, Kazi F, Donaldson SL, Morris Q, Bader GD. Cytoscape Web: an interactive web-based network browser. Bioinformatics 2010;26(18):2347–8.

[9] Zhou HR, Lau AS, Pestka JJ. Role of double-stranded RNA-activated protein kinase R (PKR) in deoxynivalenol-induced ribotoxic stress response. Toxicol Sci 2003;74(2):335–44.

CHAPTER

6.3 Bioinformatics Methods for Interpreting Toxicogenomics Data: The Role of Text-Mining

Kristina M. Hettne*, Jos Kleinjans, Rob H. Stierum†, André Boorsma† and Jan A. Kors***

**Department of Medical Informatics, Biosemantics Group, Erasmus University Medical Center, Rotterdam, The Netherlands, **Department of Toxicogenomics, Maastricht University, Maastricht, The Netherlands, †The Netherlands Organisation for Applied Scientific Research, Microbiology, and Systems Biology, Zeist, The Netherlands*

6.3.1 Bioinformatics approaches to toxicogenomics data analysis

In toxicogenomics, gene and protein activity within a particular cell or tissue of an organism in response to toxic substances can be studied with the aim to predict in vivo effects from in vitro models [1]. Two important technologies in toxicogenomics are the DNA microarray—a collection of gene-specific DNA fragments attached to a solid surface—and RNA sequencing (RNA-seq), in which high-throughput sequencing technologies are applied to sequence complementary DNA (cDNA) in order to get information about a sample's RNA content. These technologies are used to measure the activity (the expression) of thousands of genes at different time-points or dose levels at once, to create a global picture of cellular function; for example, in response to a treatment. Such a global picture of gene expression is called a gene-expression profile. The assumption that similar gene-expression profiles dictate similar physiological responses underlies the use of gene-expression profiling in toxicogenomics to discern the toxicological properties of a chemical entity. It motivates the use of bioinformatics clustering approaches, where chemical entities with similar gene-expression profiles are grouped together with the aim to define chemical classes. Also, connectivity mapping, which produces a ranked list of compounds with a similar gene-expression profile to the query compound, may be employed. Gene-expression profiling is also used in mechanistic analysis, where the biology behind a toxicological endpoint at the genomics level is of interest, and in prediction analysis, where gene signatures that can act as early predictors of toxicity are sought. In the following sections, the areas of class discovery and separation, connectivity mapping, mechanistic analysis, and identification of early predictors of toxicity are described more closely.

Toxicological class discovery and separation

One of the first approaches used to analyze toxicogenomics data was the clustering of genes within or between samples based on their expression levels, with the aim to group compounds with similar toxic mechanisms [2]. However, the study design of a classic gene-expression experiment in, for

Toxicogenomics-Based Cellular Models.

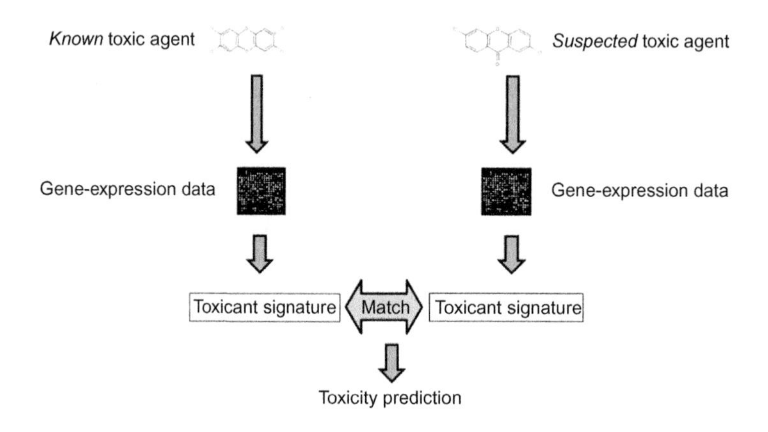

FIGURE 6.3.1

The principle of connectivity mapping.

example, cancer biology, where tumor samples are compared with corresponding normal tissue samples, is different from a typical toxicogenomics study. In a toxicogenomics study, time series and dose–response relationships play a much larger role, and the use of several compounds for comparison, spanning varying chemical spaces, further complicates the analysis. These compounds may have unique or common expression signatures and low-dose exposures that may show very small and/or early gene-expression changes difficult to distinguish from controls. Nevertheless, clustering approaches have been successfully used to separate two or more classes of samples according to a particular exposure condition or phenotype [3–5]. Other statistical methods that have been used in the area of toxicogenomics to perform class separation and/or prediction include analysis of variance (ANOVA) and linear discriminant analysis methods [6], t-test [7], linear and logistic regressions [8], principal components analysis [9], and support vector machines [10].

Connectivity mapping

Connectivity mapping is performed with the aim to infer toxicity about a new chemical entity via chemicals for which the toxicity is already known (Figure 6.3.1). A gene-expression profile is compared to reference gene-expression profiles collected in databases such as the commercial ToxExpress (from GeneLogic) [74] and the DrugMatrix database (from Entelos), now also publicly available at the US National Institute of Environmental Health Sciences [72] and the publicly available Chemical Effects in Biological Systems (CEBS) [11] database from the National Institutes of Health in the US, described above in Chapter 6.2. Pattern-matching techniques are used to produce a ranked list of chemical entities with gene-expression changes similar to the one induced by the query compound (the suspected toxic agent) (pioneered by Lamb et al. [12]).

Mechanistic analysis

Mechanistic analysis is performed to unravel the biology behind a toxicological endpoint at the genomics level, and includes investigating how the differential expression of genes affects biological pathways and processes. By describing a pathway or biological process as a group of genes and/or gene products, one can test the involvement of the particular pathway or biological process in the toxic response to a chemical. If, for example, the toxicological endpoint is hepatotoxicity (liver toxicity) in response to the drug acetaminophen, one would expect the biological process named "oxidative stress" to appear in the bioinformatic analysis results since this is one of the mechanisms by which acetaminophen causes hepatotoxicity [13]. Gene-annotation databases such as the Gene Ontology (GO) database (GeneOntology) [75] and pathway databases such as the Kyoto Encyclopedia of Genes and Genomes (KEGG) [77] and Wikipathways [14,15] are commonly used in mechanistic analyses since they contain such groupings of genes curated into biological processes and pathways, respectively. Such a group of genes can also be referred to as a gene set, and analysis methods that utilize these gene sets are called gene-set analysis (GSA) methods. An alternative to the GO and KEGG databases, and more directed at toxicogenomics, is the Comparative Toxicogenomics Database (CTD), which includes curated data from the published literature describing cross-species chemical–gene/protein interactions and chemical– and gene–disease associations [71]. GSA has become standard during recent years when analyzing gene-expression data, and a wide variety of methods have been proposed that use different statistical tests to associate gene sets with a gene-expression data set. In brief, three different types of statistical tests are available: (1) tests for the over-representation of a gene set in a list of differentially expressed genes using a hypergeometric or equivalent test; (2) methods that use the P-values of all the genes—well-known is gene-set enrichment analysis (GSEA) [16], which uses ranked P-values and tests whether the ranks of genes in a gene set differ from a uniform distribution; (3) regression analyses that use the actual expression levels of the genes in the gene set and test whether these are associated with the studied phenotype—for example, the global test [8,17]. Software programs for performing GSA or other bioinformatic analyses specifically aimed at toxicogenomics data include the commercial IPA-Tox from Ingenuity Pathway Analysis [76] and MetaDrug from GeneGo [78]. A public alternative is ToxProfiler [81], described in Chapter 6.2, from The Netherlands Organisation for Applied Scientific Research, the statistical principle of which was described by Boorsma et al. [18].

Identifying early predictors of toxicity

The identification of predictive biomarkers is another important application area for toxicogenomics. The path to finding such biomarkers by expression profiling is treacherous, but the search can be aided by bioinformatics approaches (for a recent review, see [19]). Normally, a battery of bioinformatics techniques is used, as illustrated by the two following examples. In an early study aimed at predicting nephrotoxicity, Fielden et al. [20] produced a gene signature that could predict the future development of renal tubular degeneration weeks before it appeared histologically. To derive a signature, a three-step process of data reduction, signature generation, and cross-validation was used. The signature used to predict the presence or absence of future renal tubular degeneration was derived using a sparse linear programming algorithm, which is a classification algorithm based on support vector machines. The authors noticed that many of the genes in the signature were of unknown function, and as a result the mechanism of action was difficult to infer. In another example, this time aimed at hepatotoxicity, Huang et al. [21] found that apoptosis-related genes could predict liver necrosis observed in rats exposed to a compendium of hepatotoxicants. They first grouped the liver samples by the level of necrosis exhibited in the tissue. Next, the level of necrosis was derived according to three different methods: (1) the severity scores of the injury, (2) the differentially expressed genes from an ANOVA model, and (3) the GO biological processes enrichment shared by adjacent necrosis levels. They then used a random forest classifier with feature selection to identify informative genes (a so-called "gene signature"). The gene signature was analyzed for gene function using Ingenuity Pathway Analysis for pathway enrichment and network building.

6.3.2 Text-mining and its application in toxicogenomics

The bioinformatics approaches to toxicogenomics data analysis outlined in the previous sections make use of manually curated knowledge-bases such as GO, KEGG, and CTD, either as a step in the analysis process or at the end when it is time to interpret the results. For example, gene signatures from an experiment where compounds belonging to different toxicological classes are separated based on gene-expression changes might be investigated for significantly different GO processes. However, information in manually curated knowledge-bases is not sufficient in coverage and this situation is unlikely to change in the near future [22]. Text-mining is therefore seen by many as a valuable tool to provide meaning to data [23]. Text-mining can be defined as the use of automated methods for exploiting the enormous amount of information available in the biomedical literature [24]. The following section is structured as follows: first, the basics of text-mining are introduced, followed by a description of how text-mining has been used to find explicit and implicit relationships between biomedical concepts and how it has been applied in the analysis of gene-expression data, and finally suggestions for how it can be applied in toxicogenomics are made. Box 6.3.1 explains common text-mining jargon.

BOX 6.3.1 TEXT-MINING JARGON

Concept

A concept (e.g. estrogen) is commonly defined as a cognitive unit of meaning—an abstract idea or a mental symbol sometimes defined as a "unit of knowledge." In text-mining, a concept is uniquely identifiable.

Thesaurus

A thesaurus is a list containing the preferred term referring to a concept and all its synonyms. One commonly used thesaurus in biomedical text-mining is the Unified Medical Language System (UMLS) (www.nim.nih.gov/research/umls) metathesaurus. It is organized by concept, and each concept has specific attributes (i.e. terms and definition) defining its meaning and is linked to the corresponding concept names in the various source vocabularies. For example, the terms estradiol, oestrogen, oestrogeen, oestradiol, etc. all refer to the concept estrogen.

Indexing/tagging

Indexing is the process of scanning all documents for relevant terms referring to a concept; for example, information from Medline filtered for endocrine disruption. It is also referred to as tagging. The process also usually includes storing the term statistics per document in a database.

Text corpus

A text corpus is a set of documents containing unstructured text (usually electronically stored and processed). Text corpora are used to perform information retrieval and extraction.

Precision and recall

Precision is defined as the number of relevant concepts retrieved by a search divided by the total number of concepts retrieved by that search (true positives/(true positives + false positives)), while recall is defined as the number of relevant concepts retrieved by a search divided by the total number of existing relevant concepts (true positives/(true positives + false negatives)). In broader terms: precision is the fraction of retrieved instances that are relevant, while recall is the fraction of relevant instances that are retrieved.

Concept identification in free text

A typical concept identification pipeline is shown in Figure 6.3.2. The first step in the concept identification pipeline concerns the retrieval of documents with relevant information from databases that predominantly contain textual information, a task commonly referred to as information retrieval [25]. A common resource used by the biomedical text-mining community is the MEDLINE database. A scientist seeking information usually queries MEDLINE through the Web-based interface and search engine named PubMed, which is provided by the National Library of Medicine (NLM). For larger-scale searches, programming utilities are offered by the NLM. However, owing to risks of server overload, the NLM places different limits on these services, and bioinformaticians are often faced with the need to obtain a local version of MEDLINE. A local copy also gives software developers greater control over how they use the data, and facilitates the development of customizable interfaces. A common way to store a local copy of MEDLINE is in a MySQL database, which then can be queried from the bioinformatician's programming environment; for example, in Eclipse using the Java programming language. Once the local copy is in place, an update procedure needs to be configured so that the database remains up-to-date. The local copy of the MEDLINE database can then be used for text-mining purposes. Depending on the question that is asked, different subsets of the MEDLINE

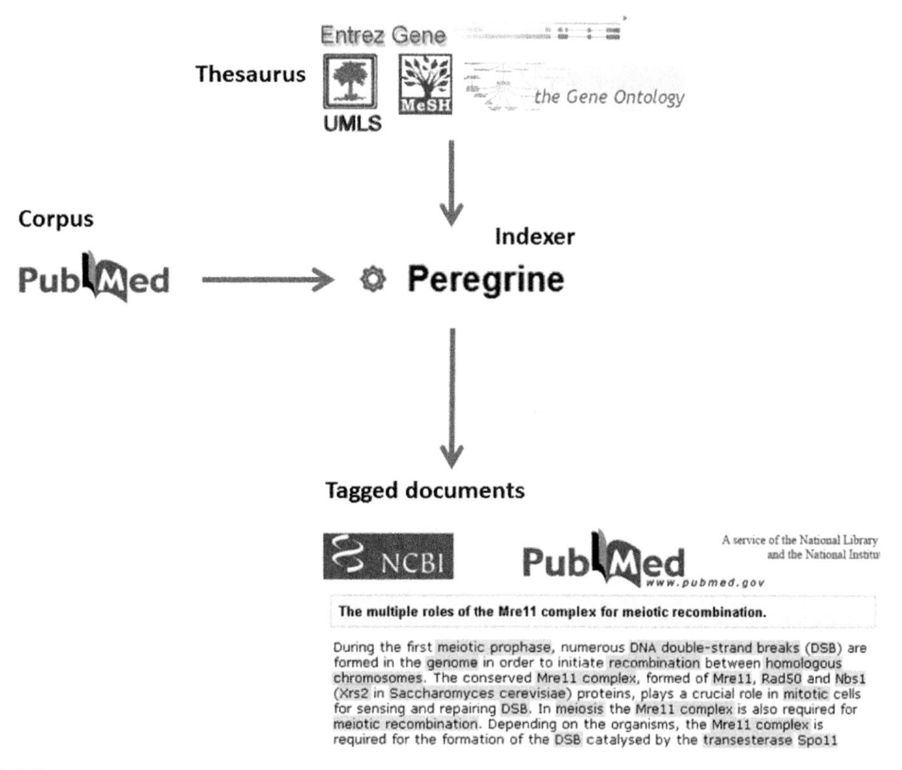

FIGURE 6.3.2

A typical concept identification pipeline.

document collection can be created. For example, if we are interested in all documents describing gene annotation in mice, we can formulate a query that will retrieve documents that only concern genes and mice. This will form our text corpus.

The next step in the concept identification pipeline concerns the identification of relevant terms in the document collection. To continue our example about mouse genes, we now would like to know which genes are mentioned in the text.

Approaches to term identification generally fall into one of three categories: thesaurus-based systems, rule-based systems, and statistics-based systems. All approaches have their disadvantages: thesaurus-based systems are dependent on fast updates, a large coverage of the underlying thesauri or dictionaries, and usually also human curation to check to what extent certain terms in text refer to one single concept; to craft the rules for a rule-based system is time consuming and requires a high level of domain knowledge; and statistics-based systems, which use machine learning techniques, need annotated corpora to train the classifiers. One important advantage of thesaurus-based systems is the possibility to perform term mapping, where an identified term is linked to a main concept and to reference data sources. A typical record in a thesaurus contains the concept listed together with its synonyms and referent data sources. A thesaurus-based system has two parts: a thesaurus and an indexing engine or tagger: a piece of software that recognizes the concepts from the thesaurus in free text. The indexer can be viewed as the integrator of the corpus text with the thesaurus to which the terms in the corpus are matched (e.g. finds estrogen, estradiol, and maps to the concept estradiol). An example of a tagger is Peregrine by Schuemie et al. [26], which has been released under an open source license, as detailed in Netherlands Bioinformatics Centre [79]. In addition to term identification, usually some form of disambiguation procedure is implemented in the tagger to map the term in the text to the correct concept. This is especially important for gene symbols and chemical name abbreviations, which are notoriously ambiguous. Disambiguation of terms is important since terms can have different meanings ("word senses") (e.g. "BAP" is a shared synonym between the two chemicals "benzo(a)pyrene" and "benzyladenine," and has an additional 58 meanings according to Acronym Finder (acronymfinder), including "blood agar plate," "BiP-associated protein," and "British Association of Psychotherapists") [70]. Peregrine [26] uses established knowledge to disambiguate terms on the fly during the indexation process.

Information extraction

Once the document corpus is in place and has been indexed by the tagger (using the thesaurus), meaningful associations between concepts can be extracted from the text, such as relationships between genes and toxicological endpoints, or chemicals and toxicological endpoints. There are currently two different approaches to this problem, one based on co-occurrence of terms and the other on natural language processing (NLP) techniques. The idea behind co-occurrence is that if two concepts are mentioned together, in, for example, the abstract of a scientific article or a sentence or phrase within that abstract, there might be a relationship between these two concepts. Since two concepts—for example, a gene and a chemical—can co-occur without there being a meaningful relationship between them, most co-occurrence-based approaches make use of algorithms that take the occurrence frequency of the concepts into account in some way [27–30]. NLP techniques focus on the extraction of precise relationships between genes and other biomedical concepts, using techniques varying from the detection of simple patterns—usually referred to as "triples," such as "chemical A–action X–gene

B" that is used by, for example, Pharmspresso [31]—to the complete parsing of whole sentences (e.g. [32]). Hybrid approaches exist as well [33–35]. As a general trend, a system-based on co-occurrence will have a higher recall and lower precision compared with a system based on NLP, meaning that more concepts are retrieved, but also containing more false positives.

Literature-based discovery

Literature-based discovery builds on the techniques for concept identification and information extraction, but adds the extra dimension of trying to discover or predict previously unknown relationships between biomedical entities. Swanson was the pioneer in this field already in the 1980s, publishing studies that were able to predict connections years before they were established in clinical trials [36,37]. He proposed a model commonly referred to as Swanson's ABC model. The model states that if "A influences B" and "B influences C," then "A may influence C." Others have built upon this model and managed to reproduce the studies by Swanson, and/or identify novel hypothetical relationships [38–48]. A few studies have been reported where new relationships have actually later been confirmed in animal models or in vitro assays: Wren et al. suggested that chlorpromazine may reduce cardiac hypertrophy, and validated this hypothesis in a rodent model [49]; Hettne et al. related the transcription factor nuclear factor kappa B to complex regional pain syndrome type I (CRPS-I) [50], a relation that was later confirmed in an animal model of CRPS-I [51]; van Haagen et al. confirmed a predicted protein–protein interaction between calpain 3 and parvalbumin B experimentally [52]; and Frijters et al. validated newly predicted relationships between compounds and cell proliferation in an in vitro cell proliferation assay [53].

To relate two concepts to each other, several authors have used the vector space model, as vectors can be compared efficiently and transparently, and the model yields a measure of the strength of the relationship [54]. An example of a tool that uses the vector space model is Anni 2.0 [55]. In Anni 2.0, a concept is represented by a concept profile. A concept profile is a list of concepts with a weight for every concept to indicate the importance of its association to the main concept, based on co-occurrence. To construct a concept profile, first PubMed records are associated to a concept using the indexing engine Peregrine equipped with a thesaurus. For all concepts except genes, the PubMed records comprise the texts in which the concept is mentioned. For genes only, a subset of PubMed records is used in order to limit the impact of ambiguous terms and distant homologs. GO terms are sometimes given as words or phrases that are infrequently found in the normal texts. To still provide broad coverage of GO terms, the PubMed records that were used as evidence for annotating genes with this GO term are added. For every concept in the thesaurus that is associated with at least five PubMed records, a concept profile is created. This concept profile is in reality a vector containing all concepts related to the main concept (direct co-occurrence), weighted by their importance using the symmetric uncertainty coefficient [55]. Figure 6.3.3 presents an overview of the concept profile generation process. Concept profiles can be matched to identify similarities between concept profiles via their shared concepts (indirect relations); for instance, to identify genes associated with similar biological processes. Any distance measure can be used for this matching, such as the inner product, cosine, angle, Euclidean distance, or Pearson's correlation. Knowledge discovery based on indirect relations might be referred to as next-generation text-mining; in comparison to first-generation text-mining, where only known relations are extracted.

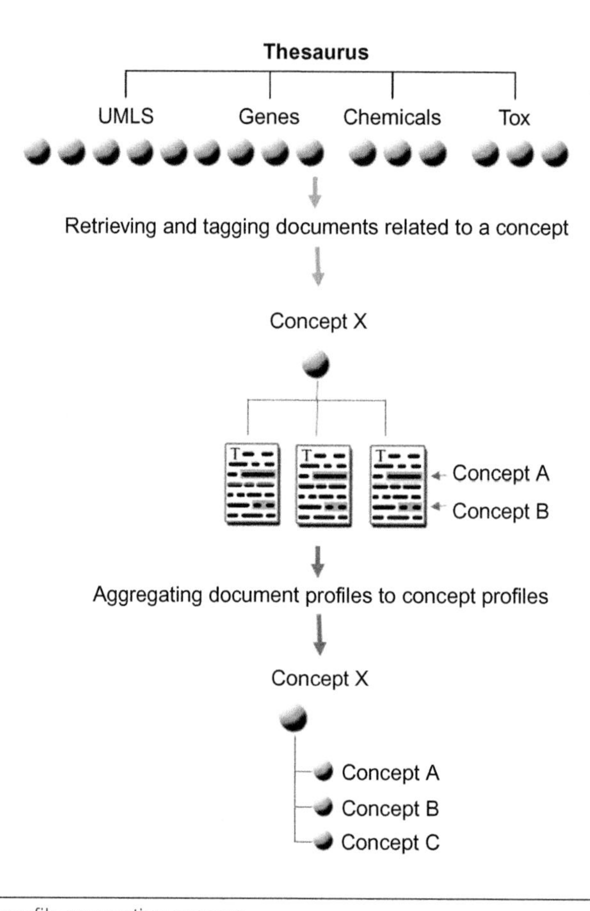

FIGURE 6.3.3

Overview of the concept profile generation process.

Assigning gene function: text-mining applied to gene-expression data

Techniques for information extraction and literature-based discovery have been applied previously to support the analysis of gene-expression data. The majority of these techniques work with gene lists. They can retrieve concepts or terms strongly associated with the selected genes and/or cluster the genes to retrieve functionally coherent subclusters [55]. The main differences between these methods concern the following two aspects: (1) usage of text words (all words in the texts or only words that map to concepts in a thesaurus) and (2) relationships considered (only direct co-occurrence or also indirect relationships). Apart from the tools that focus on retrieving gene relations from a list of genes, some methods retrieve functional associations shared between gene lists of different experiments [43,56].

The analyses performed by the mentioned text-mining-based approaches are of an exploratory nature, and do not provide a statistical evaluation for the identified associations in the context of the performed experiment. However, the algorithms can readily be combined with the three previously mentioned classes of statistical approaches for GSA. An approach for the first class of tests, for the over-representation of a gene set in a list of differentially expressed genes, could simply entail the creation of gene sets; for instance, by applying a threshold on the literature-derived association scores between genes and biomedical concepts. Sartor et al. provide literature-based gene sets in their tool ConceptGen [57], which uses Gene2MeSH [73] to identify gene and MeSH term pairs with a significantly higher number of co-occurrences than expected by chance. Frijters et al. [58] and Leong and Kipling [59] calculated biomedical term over-representation for a set of regulated genes in a similar fashion to standard class one over-representation tools.

Several text-mining approaches have been published that resemble the earlier-mentioned class two approaches that use the P-values of all the genes. Kuffner et al. [60] integrated the rank of the genes after sorting on P-value with an analysis of the literature. However, their approach was based on factorization, which complicates the interpretation of their results, and did not include formally testing retrieved associations. Minguez et al. [61] tested whether a ranked list of genes showed a significant correlation with the genes' associations to a biomedical term. These associations were based on the literature and reflect the extent to which a gene and a biomedical term that occur together in documents exceed the co-occurrence rate expected by chance.

A class three, regression-based test, which uses the actual expression levels of the genes in the gene set in combination with text-mining-derived information, is the weighted global test [62], where the weights are the concept profile matching scores. The concept profile matching scores reflect the importance of a gene for the concept of interest (for example a GO category), and provide a deeper insight into the gene-expression experiment compared with an analysis using classical GO-based gene sets. The literature-weighted global test can also be used for linking of gene-expression data to patient survival in breast cancer and to investigate the action and metabolism of drugs.

Application in toxicogenomics

Toxicogenomics could benefit from the use of text-mining through the whole process from experimental setup to the final analysis of the data. In the setup of a toxicogenomics experiment, text-mining can help answer questions regarding the selected compounds. Examples of such questions are: "which genes are known to be affected?", "which pathologies are known to be induced?", and "what other chemicals are known to have a similar effect?" Traditionally, these types of questions would be answered by searching databases such as the Toxicology Data Network "TOXNET," available from the NLM [80], PubMed, or the CTD. This usually works fine when information is sought for well-investigated and "fashionable" chemicals, such as bisphenol A, for which a search in the CTD results in 1245 manually curated gene interactions (14 March 2012), but for chemicals that are unfashionable and/or not very well investigated, the results are often disappointing. For example, a search on the fungicide flusilazole in the CTD (14 March 2012) resulted in only two manually curated genes associated with the compound. One cannot exclude the possibility that more information is locked away in the literature, confirming the non-completeness of manually curated knowledge-bases. Here, thesaurus-based text-mining can prove valuable since it can find information about chemicals in the literature, and importantly also map the chemical to its identifier (like a Chemical Abstract Service (CAS) number or InChI string) via information found in the thesaurus [63]. For example, matching the concept profile of flusilazole with all human genes in Anni 2.1 not only adds 13 genes that co-occur with flusilazole in the literature to the ones also found in the CTD, but also gives suggestions of other, probable gene interactions.

Continuing with the step after performing the toxicogenomics experiment, i.e. the data analysis, one can imagine text-mining playing a role in all the bioinformatic approaches described earlier (i.e. Toxicological class discovery and separation, Connectivity mapping, Mechanistic analysis, and Identifying early predictors of toxicity. Related to toxicological class discovery and separation and mechanistic analysis, Frijters and coworkers showed that text-mining applied to expression data from toxicogenomics experiments can separate compounds that have distinct biological activities and yield detailed insight into the mode of toxicity [64]. They created keyword profiles for compounds based on co-occurrence in MEDLINE between the mention of the differentially expressed genes in the experiments and keywords from an in-house thesaurus. The pathology thesaurus used in their study was, however, limited to liver pathologies and co-occurrences between genes and chemicals were not included.

Hettne et al. [65] succeeded in creating a text-mining-based gene set for a specific embryonic structure ("neural plate and/or tube," which corresponds to the rat postimplantation whole-embryo culture effect "posterior neuropore open"), that could discriminate triazole embryonic stem cell test (EST) gene-expression profiles from other developmental toxicant EST gene-expression profiles (carbamazepine, methotrexate, methylmercury chloride, monobutyl phthalate, monoethylhexyl phthalate, nitrofen, warfarin), and from nondevelopmental toxicant EST gene-expression profiles (monomethyl phthalate and saccharine) in a principal components analysis. In addition, by using the weighted global test to associate embryonic structure gene sets with a triazole gene-expression profile obtained through the EST, predicted effect information in the form of affected embryonic structures became readily available. The genes in the text-mining-based embryonic structure gene sets can be used to identify early predictors of toxicity. With respect to the use of text-mining to identify early predictors of toxicity and mechanistic analysis, van Dartel and coworkers used gene sets created with Anni 2.0 to show that dedicated gene sets allow for detection of cardiomyocyte-differentiation-related effects in the in embryonic stem cells in the early phase of differentiation [66], but only direct co-occurring concepts were used. Some attempts have been done to relate chemical structures to gene-expression patterns in microarray experiments using literature-derived chemical-response-specific gene sets. Concerning the application ot text-mining in connectivity mapping, Minguez et al. [61] used their tool MarmiteScan to associate chemicals with the characteristics of acute myeloid leukemia cell differentiation. However, there is no information about the size and scope of the chemical dictionary they used to mine the literature and their gene sets are not separately available, thus forcing the user to use their GSA method. There is also no possibility to test a subset of the gene sets; for example, only those relevant for evaluation of developmental toxicity. Also related to connectivity mapping, Patel and Butte [67] associated chemical-response-specific gene sets derived from the CTD with six gene-expression data sets selected based on their diversity with respect to species, chemical exposure, and cell type. Patel and Butte were able to identify chemicals applied to the samples and cell lines, and to predict known and novel environmental associations with prostate, lung, and breast cancers. Manual curation of chemical–gene interactions from publications is, however, a very time-consuming process producing high-quality information but with limited coverage, reflected by the number of chemicals for which Patel and Butte could create gene sets (1338 chemicals). In contrast, Hettne et al. [65] were able to create gene sets for 30,211 chemicals by the use of next-generation text-mining, a text-mining technique based on concept profile matching (see Literature-based discovery). Hettne and coworkers performed a head-to-head comparison of how well the CTD-based gene sets and the text-mining-based gene sets could identify chemicals applied to the same samples and cell lines as tested by Patel and Butte [67], and concluded that the CTD-based gene sets ranked higher than the text-mining-based gene sets for all data sets except one. However, Hettne and coworkers were able to predict similar and novel toxicant associations for triazoles, environmental toxicants, for which too little information is recorded in manually curated databases such as the CTD to create gene sets [65].

In conclusion, toxicogenomics is infeasible without a variety of bioinformatics analysis tools. However, the characteristics of the data from a toxicogenomics study, such as dose–response relationships and the use of multiple compounds for comparison showing very small and/or early gene-expression changes, complicate the analysis and there is a need for further developments and adjustment of bioinformatics methods in this field [68].

Text-mining is a valuable supplement to conventional bioinformatics approaches. Future directions for text-mining-based methods for toxicogenomics data interpretation include the exploitation of knowledge sources such as the Linked Open Drug Data [69] and patent databases, and the mining of more precise chemical–gene–toxicological effect relations by natural-language-processing-based methods.

References

[1] Hamadeh HK, Afshari CA, editors. Toxicogenomics: principles and applications. New York: Wiley; 2004. ISBN: 978-0-471-43417-7, p. 361.

[2] Waring JF, Ciurlionis R, Jolly RA, Heindel M, Ulrich RG. Microarray analysis of hepatotoxins in vitro reveals a correlation between gene expression profiles and mechanisms of toxicity. Toxicol Lett 2001;120(1–3):359–68.

[3] Ellinger-Ziegelbauer H, Stuart B, Wahle B, Bomann W, Ahr HJ. Comparison of the expression profiles induced by genotoxic and nongenotoxic carcinogens in rat liver. Mutat Res 2005;575(1–2):61–84.

[4] van Dartel DA, Pennings JL, Robinson JF, Kleinjans JC, Piersma AH. Discriminating classes of developmental toxicants using gene expression profiling in the embryonic stem cell test. Toxicol Lett 2011;201(2):143–51.

[5] van Delft JH, van Agen E, van Breda SG, Herwijnen MH, Staal YC, Kleinjans JC. Comparison of supervised clustering methods to discriminate genotoxic from non-genotoxic carcinogens by gene expression profiling. Mutat Res 2005;575(1–2):17–33.

[6] Bushel PR, Hamadeh HK, Bennett L, Green J, Ableson A, Misener S, et al. Computational selection of distinct class- and subclass-specific gene expression signatures. J Biomed Inform 2002;35(3):160–70.

[7] Minami K, Saito T, Narahara M, Tomita H, Kato H, Sugiyama H, et al. Relationship between hepatic gene expression profiles and hepatotoxicity in five typical hepatotoxicant-administered rats. Toxicol Sci 2005;87(1):296–305.

[8] Goeman JJ, van de Geer SA, de Kort F, van Houwelingen HC. A global test for groups of genes: testing association with a clinical outcome. Bioinformatics 2004;20(1):93–9.

[9] Hamadeh HK, Jayadev S, Gaillard ET, Huang Q, Stoll R, Blanchard K, et al. Integration of clinical and gene expression endpoints to explore furan-mediated hepatotoxicity. Mutat Res 2004;549(1–2):169–83.

[10] Steiner G, Suter L, Boess F, Gasser R, de Vera MC, Albertini S, et al. Discriminating different classes of toxicants by transcript profiling. Environ Health Perspect 2004;112(12):1236–48.

[11] Waters M, Stasiewicz S, Merrick BA, Tomer K, Bushel P, Paules R, et al. CEBS—Chemical Effects in Biological Systems: a public data repository integrating study design and toxicity data with microarray and proteomics data. Nucleic Acids Res 2008;36:D892–900.

[12] Lamb J, Crawford ED, Peck D, Modell JW, Blat IC, Wrobel MJ, et al. The Connectivity Map: using gene-expression signatures to connect small molecules, genes, and disease. Science 2006;313(5795):1929–35.

[13] Jaeschke H, McGill MR, Ramachandran A. Oxidant stress, mitochondria, and cell death mechanisms in drug-induced liver injury: lessons learned from acetaminophen hepatotoxicity. Drug Metab Rev 2012;44(1):88–106.
[14] Kelder T, van Iersel MP, Hanspers K, Kutmon M, Conklin BR, Evelo CT, et al. WikiPathways: building research communities on biological pathways. Nucleic Acids Res 2012;40:D1301–D1307.
[15] Pico AR, Kelder T, van Iersel MP, Hanspers K, Conklin BR, Evelo C. WikiPathways: pathway editing for the people. PLoS Biol 2008;6(7):e184.
[16] Subramanian A, Tamayo P, Mootha VK, Mukherjee S, Ebert BL, Gillette MA, et al. Gene set enrichment analysis: a knowledge-based approach for interpreting genome-wide expression profiles. Proc Natl Acad Sci U S A 2005;102(43):15545–50.
[17] Goeman JJ, Oosting J, Cleton-Jansen AM, Anninga JK, van Houwelingen HC. Testing association of a pathway with survival using gene expression data. Bioinformatics 2005;21(9):1950–7.
[18] Boorsma A, Foat BC, Vis D, Klis F, Bussemaker HJ. T-profiler: scoring the activity of predefined groups of genes using gene expression data. Nucleic Acids Res 2005;33:W592–5.
[19] Afshari CA, Hamadeh HK, Bushel PR. The evolution of bioinformatics in toxicology: advancing toxicogenomics. Toxicol Sci 2011;120(Suppl. 1):S225–37.
[20] Fielden MR, Eynon BP, Natsoulis G, Jarnagin K, Banas D, Kolaja KL. A gene expression signature that predicts the future onset of drug-induced renal tubular toxicity. Toxicol Pathol 2005;33(6):675–83.
[21] Huang L, Heinloth AN, Zeng ZB, Paules RS, Bushel PR. Genes related to apoptosis predict necrosis of the liver as a phenotype observed in rats exposed to a compendium of hepatotoxicants. BMC Genomics 2008;9:288-2164-9-288.
[22] Baumgartner Jr WA, Cohen KB, Fox LM, Acquaah-Mensah G, Hunter L. Manual curation is not sufficient for annotation of genomic databases. Bioinformatics 2007;23(13):i41–8.
[23] Altman RB, Bergman CM, Blake J, Blaschke C, Cohen A, Gannon F, et al. Text mining for biology—the way forward: opinions from leading scientists. Genome Biol 2008;9(Suppl. 2) S7-2008-9-s2-s7.
[24] Cohen KB, Hunter L. Getting started in text mining. PLoS Comput Biol 2008;4(1):e20.
[25] Hersh WR. Information retrieval: a health and biomedical perspective. Springer; 2009.
[26] Schuemie M, Jelier R, Kors JA. Peregrine: lightweight gene name normalization by dictionary lookup. In: Proceedings of the second biocreative challenge evaluation workshop, Madrid, Spain; 2007. pp. 131–3.
[27] Alako BT, Veldhoven A, van Baal S, Jelier R, Verhoeven S, Rullmann T, et al. CoPub Mapper: mining MEDLINE based on search term co-publication. BMC Bioinformatics 2005;6:51.
[28] Barbosa-Silva A, Soldatos TG, Magalhaes IL, Pavlopoulos GA, Fontaine JF, Andrade-Navarro MA, et al. LAITOR: literature assistant for identification of terms co-occurrences and relationships. BMC Bioinformatics 2010;11:70-2105-11-70.
[29] Hoffmann R, Valencia A. A gene network for navigating the literature. Nat Genet 2004;36(7):664.
[30] Jenssen TK, Laegreid A, Komorowski J, Hovig E. A literature network of human genes for high-throughput analysis of gene expression. Nat Genet 2001;28(1):21–8.
[31] Garten Y, Altman RB. Pharmspresso: a text mining tool for extraction of pharmacogenomic concepts and relationships from full text. BMC Bioinformatics 2009;10(Suppl. 2) S6-2105-10-S2-S6.
[32] Coulet A, Shah NH, Garten Y, Musen M, Altman RB. Using text to build semantic networks for pharmacogenomics. J Biomed Inform 2010;43(6):1009–19.
[33] Bandy J, Milward D, McQuay S. Mining protein–protein interactions from published literature using Linguamatics I2E. Methods Mol Biol 2009;563:3–13.
[34] Kemper B, Matsuzaki T, Matsuoka Y, Tsuruoka Y, Kitano H, Ananiadou S, et al. PathText: a text mining integrator for biological pathway visualizations. Bioinformatics 2010;26(12):i374–81.
[35] Krallinger M, Rodriguez-Penagos C, Tendulkar A, Valencia A. PLAN2L: a web tool for integrated text mining and literature-based bioentity relation extraction. Nucleic Acids Res 2009;37:W160–5.
[36] Swanson DR. Fish oil, Raynaud's syndrome, and undiscovered public knowledge. Perspect Biol Med 1986;30(1):7–18.

[37] Swanson DR. Migraine and magnesium: eleven neglected connections. Perspect Biol Med 1988;31(4):526–57.
[38] Chen H, Sharp BM. Content-rich biological network constructed by mining PubMed abstracts. BMC Bioinformatics 2004;5:147.
[39] Hettne KM, Weeber M, Laine ML, ten Cate H, Boyer S, Kors JA, et al. Automatic mining of the literature to generate new hypotheses for the possible link between periodontitis and atherosclerosis: lipopolysaccharide as a case study. J Clin Periodontol 2007;34(12):1016–24.
[40] Hristovski D, Peterlin B, Mitchell JA, Humphrey SM. Using literature-based discovery to identify disease candidate genes. Int J Med Inform 2005;74(2–4):289–98.
[41] Hristovski D, Friedman C, Rindflesch TC, Peterlin B. Exploiting semantic relations for literature-based discovery. AMIA Ann Symp Proc 2006;2006:349–53.
[42] Iossifov I, Rodriguez-Esteban R, Mayzus I, Millen KJ, Rzhetsky A. Looking at cerebellar malformations through text-mined interactomes of mice and humans. PLoS Comput Biol 2009;5(11):e1000559.
[43] Jelier R, 't Hoen PA, Sterrenburg E, den Dunnen JT, van Ommen GJ, Kors JA, et al. Literature-aided meta-analysis of microarray data: a compendium study on muscle development and disease. BMC Bioinformatics 2008;9:291-2105-9-291.
[44] Rzhetsky A, Iossifov I, Koike T, Krauthammer M, Kra P, Morris M, et al. GeneWays: a system for extracting, analyzing, visualizing, and integrating molecular pathway data. J Biomed Inform 2004;37(1):43–53.
[45] Schuemie M, Chichester C, Lisacek F, Coute Y, Roes PJ, Sanchez JC, et al. Assignment of protein function and discovery of novel nucleolar proteins based on automatic analysis of MEDLINE. Proteomics 2007;7(6):921–31.
[46] Tsuruoka Y, Miwa M, Hamamoto K, Tsujii J, Ananiadou S. Discovering and visualizing indirect associations between biomedical concepts. Bioinformatics 2011;27(13):i111–9.
[47] Weeber M, Vos R, Klein H, De Jong-van den Berg LT, Aronson AR, Molema G. Generating hypotheses by discovering implicit associations in the literature: a case report of a search for new potential therapeutic uses for thalidomide. J Am Med Inform Assoc 2003;10(3):252–9.
[48] Yetisgen-Yildiz M, Pratt W. Using statistical and knowledge-based approaches for literature-based discovery. J Biomed Inform 2006;39(6):600–11.
[49] Wren JD, Bekeredjian R, Stewart JA, Shohet RV, Garner HR. Knowledge discovery by automated identification and ranking of implicit relationships. Bioinformatics 2004;20(3):389–98.
[50] Hettne KM, de Mos M, de Bruijn AG, Weeber M, Boyer S, van Mulligen EM, et al. Applied information retrieval and multidisciplinary research: New mechanistic hypotheses in complex regional pain syndrome. J Biomed Discov Collab 2007;2:2.
[51] de Mos M, Laferriere A, Millecamps M, Pilkington M, Sturkenboom MC, Huygen FJ, et al. Role of NFkappaB in an animal model of complex regional pain syndrome-type I (CRPS-I). J Pain 2009;10(11):1161–9.
[52] van Haagen HH, 't Hoen PA, Botelho Bovo A, de Morree A, van Mulligen EM, Chichester C, et al. Novel protein–protein interactions inferred from literature context. PLoS ONE 2009;4(11):e7894.
[53] Frijters R, van Vugt M, Smeets R, van Schaik R, de Vlieg J, Alkema W. Literature mining for the discovery of hidden connections between drugs, genes and diseases. PLoS Comput Biol 2010;6(9) doi:10.1371/journal.pcbi.1000943.
[54] Jelier R, Schuemie MJ, Roes PJ, van Mulligen EM, Kors JA. Literature-based concept profiles for gene annotation: the issue of weighting. Int J Med Inform 2008;77(5):354–62.
[55] Jelier R, Schuemie MJ, Veldhoven A, Dorssers LC, Jenster G, Kors JA. Anni 2.0: a multipurpose text-mining tool for the life sciences. Genome Biol 2008;9(6):R96-2008-9-6-r96.
[56] Soldatos TG, O'Donoghue SI, Satagopam VP, Jensen LJ, Brown NP, Barbosa-Silva A, et al. Martini: using literature keywords to compare gene sets. Nucleic Acids Res 2010;38(1):26–38.
[57] Sartor MA, Mahavisno V, Keshamouni VG, Cavalcoli J, Wright Z, Karnovsky A, et al. ConceptGen: a gene set enrichment and gene set relation mapping tool. Bioinformatics 2010;26(4):456–63.

[58] Frijters R, Heupers B, van Beek P, Bouwhuis M, van Schaik R, de Vlieg J, et al. CoPub: a literature-based keyword enrichment tool for microarray data analysis. Nucleic Acids Res 2008;36:W406–10.
[59] Leong HS, Kipling D. Text-based over-representation analysis of microarray gene lists with annotation bias. Nucleic Acids Res 2009;37(11):e79.
[60] Kuffner R, Fundel K, Zimmer R. Expert knowledge without the expert: integrated analysis of gene expression and literature to derive active functional contexts. Bioinformatics 2005;21(Suppl. 2):ii259–ii267.
[61] Minguez P, Al-Shahrour F, Montaner D, Dopazo J. Functional profiling of microarray experiments using text-mining-derived bioentities. Bioinformatics 2007;23(22):3098–9.
[62] Jelier R, Goeman JJ, Hettne KM, Schuemie MJ, den Dunnen JT, 't Hoen PA. Literature-aided interpretation of gene expression data with the weighted global test. Brief Bioinform 2011;12(5):518–29.
[63] Hettne KM, Stierum RH, Schuemie MJ, Hendriksen PJ, Schijvenaars BJ, Mulligen EM, et al. A dictionary to identify small molecules and drugs in free text. Bioinformatics 2009;25(22):2983–91.
[64] Frijters R, Verhoeven S, Alkema W, van Schaik R, Polman J. Literature-based compound profiling: application to toxicogenomics. Pharmacogenomics 2007;8(11):1521–34.
[65] Hettne KM, Boorsma A, van Dartel DA, Goeman JJ, de Jong E, Piersma AH, et al. Next-generation text-mining mediated generation of chemical-response-specific gene sets for interpretation of gene expression data. BMC Med Genomics 2013;6:2-8794-6-2.
[66] van Dartel DA, Pennings JL, Hendriksen PJ, van Schooten FJ, Piersma AH. Early gene expression changes during embryonic stem cell differentiation into cardiomyocytes and their modulation by monobutyl phthalate. Reprod Toxicol 2009;27(2):93–102.
[67] Patel CJ, Butte AJ. Predicting environmental chemical factors associated with disease-related gene expression data. BMC Med Genomics 2010;3:17–8794-3-17.
[68] Chen M, Zhang M, Borlak J, Tong W. A decade of toxicogenomic research and its contribution to toxicological science. Toxicol Sci 2012;130(2):217–28.
[69] Samwald M, Jentzsch A, Bouton C, Kallesoe CS, Willighagen E, Hajagos J, et al. Linked open drug data for pharmaceutical research and development. J Cheminform 2011;3(1):19–2946-3-19.
[70] acronymfinder. Available at, <http://www.acronymfinder.com/>.
[71] Comparative Toxicogenomics Database. Available at, <http://ctdbase.org/>.
[72] DrugMatrix database. Available at, <https://ntp-niehs.nih.gov/drugmatrix>.
[73] Gene2MeSH. Available at, <http://gene2mesh.ncibi.org>. NCIBI.
[74] GeneLogic. Available at, <http://www.genelogic.com/knowledge-suites/toxexpress-program>.
[75] GeneOntology. Available at, <http://www.geneontology.org/>.
[76] IPA-Tox. Available at, <http://www.ingenuity.com/products/ipa-tox.html>. Ingenuity Pathway Analysis.
[77] Kyoto Encyclopedia of Genes and Genomes. Available at, <http://www.genome.jp/kegg/>.
[78] MetaDrug, Available at, <http://www.genego.com/metadrug.php>. Genego.
[79] Netherlands Bioinformatics Centre, Peregrine. Available at, <https://trac.nbic.nl/data-mining/>; 2007.
[80] TOXNET. Available at, <http://toxnet.nlm.nih.gov/>. US National Library of Medicine.
[81] Toxprofiler. Available at, <http://ntc.voeding.tno.nl/toxprofiler_test/>. Netherlands Organisation for Applied Scientific Research.

SECTION 7

Selection and Validation of Toxicogenomics Assays as Alternatives to Animal Tests

CHAPTER

Selection and Validation of Toxicogenomics Assays as Alternatives to Animal Tests

7.1

Bart van der Burg, Harrie Besselink and Bram Brouwer
BioDetection Systems b.v., Amsterdam, The Netherlands

7.1.1 Introduction: modern approaches in the development of animal alternatives

Nowadays there is increasing attention on the development of alternative, non-animal methods for safety assessment of chemicals, cosmetics, and pharmaceuticals. Increased public and political pressure has already led to a ban on using animal experiments for safety testing of cosmetic ingredients in Europe. This pressure from society coincides with new technical possibilities to generate alternative tests, which are being addressed in several large-scale initiatives, such as the US Environmental Protection Agency (EPA) ToxCast program (http://www.epa.gov/ncct/toxcast), the Netherlands Toxicogenomics Center (NTC; http://www.toxicogenomics.nl), and the EU 7th Framework Programme (FP7) projects like Chem*Screen* (http://www.chemscreen.eu). These projects all investigate the opportunities for introduction of alternative, non-animal tests in the official regulatory framework for safety assessment of chemicals. The NTC has the ambition to develop genomics-based alternative non-animal methods that can be used in safety evaluation of chemicals and drugs, for lead compound selection and screening for novel natural bioactive compounds, and for monitoring human and environmental health. In order to reach these ambitious goals, the methods developed should meet several criteria related to technological and commercial feasibility, and regulatory requirements.

The development of alternative methods to animal tests has traditionally been a slow process that is hampered by technological difficulties and regulatory obstacles due to the inherently conservative nature of regulations related to safety assessment of chemicals and pharmaceuticals. However, genomics research has opened new avenues, and led to a rapid advance in our understanding of the mechanisms of action of toxicants. New possibilities are emerging to design relatively simple cellular assays that measure defined mechanistic endpoints, i.e. activation of toxicity pathways. The use of such assays has been identified as the way forward to generate rapid and efficient toxicity screens that could replace lengthy and costly animal testing, to a large extent, in the twenty-first century [1]. Interestingly, these technical possibilities are coinciding with a great need to modernize the process of chemical risk assessment, leading to new legislation and new possibilities for alternative testing methods.

Toxicogenomics-Based Cellular Models.

Notwithstanding these exciting possibilities, an important step very often overlooked or neglected is the need for proper validation of tests being developed by the scientific community. This step is essential in any application and particularly to successfully enter the regulatory arena, a key element in the application of alternative testing methods. This chapter gives an overview of the requirements and approaches taken, focusing primarily on the use of transcriptomics methods as developed in the NTC program in the area of chemical safety evaluation.

7.1.2 Generic elements in the validation of alternative toxicity assays

Discrete processes and requirements can be distinguished in the validation of a non-animal test method [2]. First, reproducibility of the test method is key, while its relevance is the second main module of a method (Figure 7.1.1). Subdivisions can be made within these phases, with reproducibility of the method being subdivided into test definition, within-laboratory variability, transferability, and between-laboratory variability. Predictive capacity, application domain, and performance standards contribute to the assessment of the relevance of the method. These items can be seen as modules that do not need to be assessed in a predetermined order, which aims to reduce the length of the procedure [2]. This modular approach is a variation of a less flexible, classical approach, that had the aim to reduce the length of the validation process, which was and still is a problem that prevents rapid progress in the field of alternative methods.

A validated method is not necessarily a method that will be used as such, but to be implemented needs to be introduced to end-users. In the case of in-house non-regulatory use—for example, by pharmaceutical companies—the introduction and application of promising new methods can be fast. However, for regulatory use, the method generally has to be included in test guidelines designed by official regulatory bodies. There are several organizations dealing with evaluation and standardization of methods for measurement of chemical safety and for developing guidance documents. The technical specifications and requirements to be fulfilled for alternative methods can differ between these organizations, since each of these official bodies has its own specific fields of application for which it wishes to develop test methods, procedures, and guidelines [3–7].

In general, validation ultimately aims at regulatory acceptance of the method. In particular, the OECD is involved in global harmonization of test guidelines related to safety testing of chemicals, aiming at so-called mutual acceptance of data (MAD). Therefore, OECD submission, via a standard

Reproducibility	Test definition
	Within-laboratory variability
	Transferability
	Between-laboratory variability
Relevance	Predictive capacity
	Application domain
	Performance standards

FIGURE 7.1.1

The different modules in a European Union reference laboratory for alternatives to animal testing (EURL ECVAM validation process.

Adapted from [2].

project submission form (SPSF) for inclusion in the OECD test guideline development program is an important step. In this document, the background of the test method and the plan to generate lacking validation data are described. To facilitate the validation process, sponsors can be approached, like the former European Centre for Validation of Alternative Methods (ECVAM), now part of the European Union Reference Laboratory for alternatives to animal testing (EURL ECVAM). These sponsors can be approached using standard test submission forms. As a part of the validation process, a validation management group (VMG) is involved in management of the process [2]. This can be organized by the sponsor or the regulatory agency itself. In any case, an independent peer review is needed as the final step before the method is accepted. In all this, it should be noted, however, that while validation of a test method is one step away from acceptance by a regulatory authority, regulatory authorities like the US Food and Drug Administration (FDA) and the EPA do not necessarily require a test method to be formally validated to accept data produced by these methods, while data produced by a validated method still may need supplementation with data from other tests, including animal data [30].

7.1.3 Stages in the process of development of validated tests

In most research programs, like the NTC program, tests developed have only been used for scientific purposes and have not yet been (extensively) validated. They have not gone through technological scrutiny and have not been evaluated from the viewpoints of robustness, transferability, turnaround time, cost efficiency, or options for developing a high-throughput mode, or options to meet the regulatory requirements. Therefore, it is necessary to put guided efforts into those aspects in order to be able to translate the predominantly scientific tools into validated test methods, meeting international requirements. The criteria necessary to meet for the various methods depends on their stage of development and their field of application. i.e. the criteria are less stringent for scientific applications and for lead compound screening purposes, while they will be much more stringent for official purpose testing, such as alternative tests for the safety assessment of drugs and chemicals. A complicating factor is that the regulatory framework for application of alternative, non-animal methods and interpretation of data from such alternative methods for, e.g. safety evaluation of chemicals, is still largely in its infancy, in particular with respect to the possibilities to incorporate modern technologies.

In principle, the process of generation of accepted bioassays or test methods follows a similar sequence of development:

1. Scientific phase:
   - Investigation into target/biomarker identification.
   - Understanding of the mechanistic base.
   - Development of an in vitro bioassay or test method based on target/biomarker.

   Output is a scientific tool.
2. Validation phase:
   - Pre-validation of in vitro bioassay or test method.
     - Draft standard operational procedures (SOPs).
     - Reliability of test.
     - Predictability of test for specific endpoint.
     - Transferability of test.

- Development of final SOPs.
- Validation of in vitro bioassay, or test method.
    - Idem pre-validation but with more stringent inter-laboratory validation and controlled testing procedures.

Output is a validated test method.

3. Implementation phase:
    - Acceptance of test method and guideline by (inter)national bodies and users (i.e. chemical and pharmaceutical industries).
    - Implementation/dissemination strategy.

Output is a fully accepted test method ready for implementation.

7.1.4 Method validation in relation to its intended use

The extent of validation, proof of performance, predictability, and development of guidelines is dependent on the intended use, or application of the respective method or bioassay.

Scientific research purposes

For scientific research purposes, the use of a biochemical target or of a potential biomarker to study qualitatively the effects exerted by certain chemicals on this target, or of biomarker response, does not require an extensive period of validation and proof of performance from a reliability point of view. Neither is this required a priori from a predictability point of view, since this is mostly part of the research question addressed, i.e. to find novel targets and/or mechanisms of action.

Estimated time frame for validation/implementation: 1–6 months.

Monitoring purposes

For monitoring purposes (e.g. environmental, food, or clinical applications) the use of bioassays based on a biochemical target or biomarker of response do require extensive validation, which is further enhanced if the bioassay is developed for the purpose of a quantitative method instead of a screening tool. The performance characteristics to be met are dependent on the specific guidelines that may exist already for chemical-analytical methods in the field of food, human, and environmental monitoring. There is a need to develop updated chemical-analytical guidelines specifically to adjust for inclusion of bio-based methods. The methods used for monitoring purposes do not need to undergo an extensive validation study with respect to predictability of effects, because they are usually used as a surrogate for quantifying certain "analytes," e.g. contaminants or additives. If the bioassay is also used for measurements for official purposes, i.e. for fulfillment of food quality guidelines for export purposes, it is not enough to just have a proper validation of the method in place. In this case, it is also a requirement to be accredited (e.g. ISO 17025 [5]) as a laboratory to perform those tests—for example, for food safety purposes.

Estimated time frame for validation/implementation of bioassays for monitoring purposes: 6 months (screening) up to 2 years (accredited quantitative assay).

Alternative, non-animal tests

For alternative, non-animal tests, the use of bioassays, using a biochemical target, or a biomarker of response should be based on sound scientific evidence indicating its key role in the mode of action of a certain group of chemicals or of a certain type of toxic endpoint. This could involve either a "single-target" type bioassay, or a "multiple-target" assay (e.g. expression array). It is imperative for a bioassay with an intended use as an alternative, non-animal method that the predictability of the biomarker response for the toxicological endpoint is scientifically well founded. Such a validation for predictability is essential for each component in an array of expressions (e.g. DNA expression fingerprint). This part of the validation requires an extraordinary effort in terms of resources and time with involvement of one or several in vitro and in vivo study comparisons. Once the predictability aspects are clear and supported by a sound scientific understanding, the aspects of reliability, performance characteristics, and inter-laboratory comparison have to be fulfilled as well. Official bodies such as the OECD should then be addressed to initiate further steps for inclusion in the OECD test guideline program. This may involve intermediate steps to sponsor validation efforts, e.g. through ECVAM.

Estimated time frame for development of a fully validated and approved test and test guidelines: 5–10 years.

7.1.5 Generic bottlenecks in the validation process

Obviously, one of the main problems in validating assays is the length of the procedures, and consequently the large amount of investment needed. This makes it unattractive for academic scientists to be involved in these processes, and the large investments needed without a clear market perspective scares away commercial interest. Even with sponsoring by organizations like EURL ECVAM to fund formal validation efforts of promising assays, major hurdles still remain. First of all, there is a substantial investment to be made even before tests are considered for validation. Secondly, in the past, the route to application of validated tests has been difficult because this validation does not equal regulatory acceptance. Because of this, nowadays ECVAM actively engages in regulatory submissions of validated tests. Validation processes usually are much longer than needed for the actual practical work to be done. Much time is spent on preparation, communication, and review of data. This calls for a more practical and straightforward approach to validation, in which performance characteristics and standards, rather than expert opinions, play a key role. OECD is now engaged in setting up performance-based test guidelines for tests that address similar endpoints as already validated tests [8]. This can be expected to reduce the length of so-called "me-too" validation processes considerably. This process can also accommodate technical changes of a method measuring the same toxicity pathway.

A second major problem in developing alternative testing methods is that risk assessment using experimental animals is hampered by uncertainties, as reflected by uncertainty factors when extrapolating data from animals to humans. On the one hand, one may argue that it is relatively easy to improve or replace an imperfect system. However, the reliance on this system by regulators is high and because of this animal studies are used as a gold standard for alternative methods. This poses a problem since a surrogate model is now being used to validate new models, and false predictions in the surrogate model will affect the predictive capacity of the new models. This situation is difficult to change since there are rarely any more directly relevant data—i.e. from human exposures to toxicants at realistic exposure levels that can be linked to disease outcomes [9]. Only for relatively acutely acting chemicals are such data available, making validation of new test methods much more straightforward.

An interesting recent issue that has raised some controversy is that, very often, according to the valorization/dissemination requirements of research programs, one should seek patent protection for a scientific tools whenever possible and appropriate. One issue of concern is, however, that the OECD and other official bodies by default prefer not to use patented methods to enter the guideline evaluation and approval scheme, in order to prevent monopoly positions in regulatory screening for patent holders.

7.1.6 Feasibility: a practical approach to application

Typically, the procedures leading to a validated test are lengthy and the path to regulatory acceptance even longer (i.e. 5–10 years). For practical reasons, therefore, in the context of a research program with limited lifespan, choices have to be made in the approach to validate assays, since otherwise these procedures may not be ended before the end of the lifespan of the consortium. The advantage of the modern approach of validating tests being designed by ECVAM is that it is modular and accepts a stepwise filling of the different modules [2]. However, when considering other applications as well, as is being done for tests in large consortia like the NTC, organizations like ECVAM and OECD are not the only target to focus on in validating tests. The different regulatory bodies, however, all share in common that the tests and guidelines developed should be robust and preferably globally applicable allowing mutual acceptance of data [3–7]. Therefore, when choices have to be made in validation of a test method, a focus on reproducibility and transferability is advisable. In general, several aspects of the predictivity and the mechanistic base of the method have been established by the scientific group developing the method, but very often these less scientifically appealing steps have received limited attention, and their importance and resource intensity are usually largely underestimated.

Other issues are important as well when aiming at the transfer of a scientific method to a validated and accepted test method. First of all, as discussed above, patenting of a method is generally not helpful in regulatory acceptance, and should also be considered in the very long time to market of these methods, which reduces the time in which the investments in a patent can be earned back. However, this does not exclude a more commercial approach in this area. In fact, to guarantee the sustainability of availability and distribution of key tools, training, service, and support it is essential to aim at a self-supporting (i.e. commercial) activity, particularly at the implementation stage. This should be offered at a reasonable cost for end-users. Without such a strategy, the often relatively complex validated alternative methods may end up not being used. Therefore, it is important to set up criteria to evaluate whether candidate methods are likely to succeed going through a validation process without to many delays and complications. The next section aims to provide guidance to this process.

7.1.7 Evaluation criteria for prioritization of scientific tools to enter a pre-validation process

There are several aspects to consider when prioritizing scientific tools for entering programs to convert them into officially approved, fully validated, and technologically and economically feasible test methods. Prioritization is necessary because of the considerable investment in time and resources required to complete this process. Furthermore, it is necessary to select the most advanced scientific tools and the intended fields of application requiring least intensive validation, such that one can harvest "quick wins" that set the stage for the validation of more complex assays. To facilitate this process, we have defined a number of criteria.

Criterion 1: scientific basis for predictability

For any target (protein, gene, receptor) or for any biomarker identified that one wishes to use as the basis for a novel bioassay or alternate test method, it is not enough to just show a good correlation between an in vitro bioassay (including the intended target) and an in vivo toxic endpoint. It is also required to establish a sound scientific basis for its role in the mode of toxic action, or for its key position in toxic response pathways. This scientific understanding is very helpful—for example, for the regulator for interpretation of the in vitro responses observed and for discriminating adaptive responses from real toxicological effects.

This is an absolute prerequisite for any scientific tool to pass the translational research phase successfully and be converted into an officially accepted bioassay or test method for regulatory purposes. It is, however, not clear from the existing guidelines for method development how extensive this scientific underscoring should be.

Criterion 2: single- vs multiple-target-type assays

Single-target-type assays are bioassays and tests based on the binding to and response from one single target molecule by a toxicant, usually a key protein or gene (receptor, specific enzyme, target gene). The relevance of this single gene or protein and its key role in the mode of action should be established as indicated before. However, the genomics revolution creates mostly multi-target-type assays, i.e. where a gene-expression profile of several key genes is involved in the overall response to a toxicant. The validation in terms of predictability for multi-target-type assays is much more complicated and, in essence, requires a validation data set and scientific basis for each of the individual target genes in the expression profile. One approach may be the phenotypic anchoring of gene-expression changes to identify molecular mechanisms and candidate markers of toxicity. A second approach involves the identification and validation of predictive gene-expression signatures of toxicity. Nowadays, official regulatory bodies do not allow a toxicogenomics-based profile to be used as a definite test, screening test, or a replacement test. However, regulatory bodies will allow toxicogenomics data to be added as toxicity supportive information in a qualitative sense.

Therefore, from a practical point of view, bioassays and test methods based on a single target are more likely to pass the translational research phase and are more readily accepted than multi-target methods. This should not discourage researchers from continuing with their research and development of expression profiles and other multi-target-type assays, because these may be most valid in the future.

Criterion 3: (pre-)validation status of tools

In general, the more (pre-)validation data are available for a certain bioassay or test method, the better chances there are for fast passage through the translational research funnel and realization of a "quick win." However, before starting a pre-validation or validation phase, one should first identify for which purpose or application one wishes to develop the bioassay or test method: for scientific purposes, for monitoring purposes, or as an alternative test method for regulation of chemical safety.

The extent of (pre-)validation and the time frame required differ greatly for these different fields of application. Furthermore, it is advisable that consideration is also given to technological and commercial feasibility aspects.

Criterion 4: technological and commercial feasibility aspects

For parties interested in commercial application of the bioassays or test methods and thereby securing sustainability of the method, it is essential to also consider aspects of a practical nature: that the methods are affordable and easily transferable from one laboratory to another.

The developed methods should be commercially attractive, practical, affordable, and easily transferable.

7.1.8 Validation of toxicogenomics assays

The validation of toxicogenomics assays is not essentially different from that of any other type of assay. However, because of the complexity of the methods, variability due to this complexity, and ongoing technological improvements, validation of such methods is complicated. These aspects are often challenging to the methods of validation themselves. To be able to accommodate these novel toxicogenomics methods, it therefore may even be needed to adapt the process of validation; of course, without changing its basic principles of being able to establish and safeguard the relevance and reproducibility of a test method.

Initial discussions of toxicogenomics methods center around the variability of the assays due to technical imperfections of these methods and the rapid changing of the technology, making them essentially unsuitable to enter a validation process. Most of these technical challenges have been overcome and even cross-platform comparison is now good, particularly for highly expressed genes [10].

What remains, however, is that modern technologies change more rapidly than an experimental animal, and validation methods should be adapted to allow such variations without going through the whole procedure again. It should be kept in mind that while experimental animals do not change obviously over time, an astonishing variability is accepted in guideline studies, in which the animal species rather than the strain is defined. Very often, even different but closely related species can be chosen, like the rat and mouse, even in very recent guidelines (see, for example, the test guideline on repeated dose toxicity: [11]). This is remarkable, since it has been shown repeatedly and consistently that not only species but also strain differences considerably influence experimental outcome in toxicity studies and, thus, risk assessment [12–16]. It seems that confidence in experimental animals as guideline models is overestimated, and that reluctance to accept alternatives by regulators not only resides in scientific arguments but also includes hesitation to go for the unknown. In that respect, it is important to note that many of the animal models that are currently used have not undergone the prior validation that now is being stressed to be essential for alternative methods. Instead, confidence in those methods has grown during their use. Validation, however, can and will provide an important element in this building of confidence. The initial use of toxicogenomics and molecular screening methods in a regulatory context, therefore, very likely will be to provide additional information in combination with classical animal tests. It is very likely that when those methods, after prolonged use, turn out to be reliable and informative, their weight in regulatory decisions will increase. Obviously, in this process, a test with a clear mechanistic base that is easy to understand and interpret will have a clear advantage over a black-box approach.

Typically, however, toxicogenomics methods still are relatively black-box approaches seeking to find signature profiles of gene-expression patterns that can predict the response of a cell or organism to a toxic insult, e.g. by a chemical. Initially, the selection of these gene-expression profiles has been based on purely statistical methods, involving a training set of chemicals to generate the signature followed by a validation set to test the robustness of the signature [17]. This robustness has often turned out to be disappointing for statistical reasons; while the initial fitting of a multi-parameter array to a limited set of profiles usually results in consensus signatures without much problem, additional testing can lead to disappointing results. This is not essentially different from other methods, which usually perform better with a training set than with the validation set. However, a problem with a classical, purely statistical, toxicogenomics approach is that the mechanistic base of the gene signature is often lacking. In designing alternative methods, it is important to consider this mechanistic base [18,19] and, in fact, description of it is a part of the EURL ECVAM test submission form. This will also facilitate acceptance by regulatory agencies (see above). The OECD has adopted the concept of designing adverse outcome pathways (AOPs; [20]) for chemical risk assessment, using molecular screening assays as the in vitro starting point. With this approach, the in vitro mechanistic effect is coupled to a predicted adverse outcome in an organism. While an important concept, it should be noted that many mechanistic effects in vitro can lead to a plethora of adverse outcomes in experimental animals and humans, making these pathways complicated networks rather than linear pathways. For example, activation of the retinoic acid receptor can lead to a multitude of effects that depend on the chemical dose, life stage, species, and timing [21]. This obviously complicates the AOP concept. However, it does not mean that the screening of retinoid-like effects of chemicals does not provide valuable information. In contrast, it shows that the results of one mechanistic screening test can explain a host of effects in animal experiments, and through this such a test can replace many experimental animals. It also shows that a strong mechanistic base, in the case of toxicogenomics and molecular screening methods through systems biology approaches [22], is an important way forward in regulatory acceptance.

7.1.9 Perspectives

Clearly, when proper choices are made to increase confidence of regulators, toxicogenomics-based methods have an important future role in replacing animal experimentation. In particular, toxicogenomics and related molecular approaches have led to the identification of toxic modes of action of toxicants. Based on these insights, many tests have been generated that assess specific modes of action, e.g. activation of specific receptors or signal transduction pathways in cells. Very often, such relatively simple and straightforward mechanistic tests are based on genetically modified cells in which activation or repression of such pathways can be assessed very specifically and in a relatively cost-effective manner. An example of this type of assay is reporter gene assays measuring receptor activation, such as the above-mentioned retinoic acid receptor or other important receptors like dioxin- and steroid receptors. Panels of such assays have been generated [23], which can be automated efficiently [29] and used in high-throughput screening methods to predict chemical toxicities [28]. Such screening methods can very likely provide a way forward in chemical risk assessment, and may form an important basis for integrated testing strategies (ITS) as are being used in REACH (the Registration, Evaluation, Authorisation and Restriction of Chemicals regulation of the EU) [24]. As an important step in creating confidence, tests that are part of the screening panel should be validated to determine their predictivity in comparison with animal experiments. Possibly, expansion of the battery of tests may not require a full validation. Rather, the validated tests may form the core of the screening panel, and similar methods for other toxicological endpoints may undergo a partial validation only. This can be done using performance standards and by focusing on the reproducibility of the test in addition to its mechanistic base. The predictive capacity should focus on the entire battery of tests (or genes) rather than the individual assays. This can all be part of a stepwise approach to promote the increased use of non-animal methods for chemical risk assessment.

References

[1] National Research Council. Toxicity testing in the twenty-first century: a vision and a strategy. : Committee on Toxicity and Assessment of Environmental Agents; 2007.

[2] Hartung T, Bremer S, Casati S, Coecke S, Corvi R, Fontaner S, et al. A modular Approach to the ECVAM principles on test validity. ATLA 2004;32:467–72.

[3] European Medicines Agency (EMEA). Guideline to bioanalytical method validation; 2011.

[4] Interagency Coordinating Committee on the Validation of Alternative Methods (ICCVAM). Validation and regulatory acceptance of toxicological test methods. NIH publication 97-3981; 1997.

[5] International Organization for Standardization (ISO)/International Electrotechnical Commission (IEC). Standard ISO/IEC 17025: general requirements for the competence of testing and calibration laboratories; 2005.

[6] International Union of Pure and Applied Chemistry (IUPAC). Technical report: harmonized guidelines for single laboratory validation of methods of analysis. Pure Appl Chem 2002;74:835–55.

[7] OECD. OECD series on testing and assessment, No.34 Guidance on the validation and international acceptance of new or updated test methods for hazard assessment. Paris: OECD; 2005.

[8] OECD. Guidelines for the testing of chemicals Performance-based test guideline for stably transfected in vitro assays to detect estrogen receptor agonists (TG455). Paris: OECD; 2012.

[9] Leist M, Hasiwa N, Daneshian M, Hartung T. Validation and quality control of replacement alternatives—current status and future challenges. Toxicol Res 2012;1:8–22.

[10] Liu F, Kuo WP, Jenssen TK, Hovig E. Performance comparison of multiple microarray platforms for gene expression profiling. Methods Mol Biol 2012;802:141–55.

[11] OECD. Guidelines for the testing of chemicals Repeated dose 28-day oral toxicity study in rodents. (TG 407). Paris: OECD; 2008.

[12] Bannasch P. Strain and species differences in susceptibility to liver tumour induction. IARC Sci Publ 1983; 51:9–38.

[13] Hayes AW, Dayan AD, Hall WC, Kodell RL, Williams GM, Waddell WD, et al. A review of mammalian carcinogenicity study design and potential effects of alternate test procedures on the safety evaluation of food ingredients. Regul Toxicol Pharmacol 2011;60(1 Suppl):S1–S34.

[14] Kacew S, Ruben Z, McConnell RF. Strain as a determinant factor in the differential responsiveness of rats to chemicals. Toxicol Pathol 1995;23:701–14.

[15] Miller MS, Gressani KM, Leone-Kabler S, Townsend AJ, Malkinson AM, O'Sullivan MG. Differential sensitivity to lung tumorigenesis following transplacenta exposure of mice to polycyclic hydrocarbons, heterocyclic amines, and lung tumor promoters. Exp Lung Res 2000;26:709–30.

[16] Spearow JL, Doemeny P, Sera R, Leffler R, Barkley M. Genetic variation in susceptibility to endocrine disruption by estrogen in mice. Science 1999;285:1259–61.
[17] Salter AH, Nilsson KC. Informatics and multivariate analysis of toxicogenomics data. Curr Opin Drug Discov Dev 2003;6:117–22.
[18] Corvi R, Ahr HJ, Albertini S, Blakey DH, Clerici L, Coecke S, et al. Validation of toxicogenomics-based test systems: ECVAM-ICCVAM/NICEATM considerations for regulatory use. Environ Health Perspect 2006;114:420–9.
[19] Hartung T. From alternative methods to a new toxicology. Eur J Pharm Biopharm 2011;77:338–49.
[20] Ankley GT, Bennett RS, Erickson RJ, Hoff DJ, Hornung MW, Johnson RD, et al. Adverse outcome pathways: a conceptual framework to support ecotoxicology research and risk assessment. Environ Toxicol Chem 2010;29:730–41.
[21] Collins MD, Mao GE. Teratology of retinoids. Annu Rev Pharmacol Toxicol 1999;39:399–430.
[22] Cummings A, Kavlock R. A systems biology approach to developmental toxicity. Reprod Toxicol 2005; 19:281–90.
[23] Sonneveld E, Jansen HJ, Riteco JAC, Brouwer A, van der Burg B. Development of androgen- and estrogen-responsive bioassays, members of a panel of human cell-line-based highly selective steroid responsive bioassays. Toxicol Sci 2005;83:136–48.
[24] Grindon C, Combes R, Cronin MT, Roberts DW, Garrod JF. Integrated testing strategies for use in the EU REACH system. ATLA 2006;34:407–27.
[25] ECVAM. General guidelines for submitting a proposal to ECVAM for the evaluation of the readiness of a test method to enter the ECVAM prevalidation and/or validation process. Available at: <https://evcam.jrc.it>; 2005.
[26] Festing MF. Inbred strains should replace outbred stocks in toxicology, safety testing, and drug development. Toxicol Pathol 2010;38:681–90.
[27] Food and Drug Administration (FDA). Guidance for industry: bioanalytical method validation; 2001.
[28] Piersma AH, Schulpen SHW, Uibel F, van Vugt-Lussenburg B, Bosgra S, Hermsen SAB, et al. Evaluation of an alternative in vitro test battery for detecting reproductive toxicants. Reprod Toxicol 2013;38:53–64.
[29] van der Burg B, van der Linden SC, Man HY, Winter R, Jonker L, van Vugt-Lussenburg B, et al. A panel of quantitative CALUX® reporter gene assays for reliable high-throughput toxicity screening of chemicals and complex mixtures. In: Steinberg P, editor. High-throughput screening methods in toxicity testing. New York: John Wiley and Sons. ISBN: 9781118065631.
[30] Wilcox N, Goldberg A. Food for thought … on validation: a puzzle or a mystery—an approach founded on new science. ALTEX 2010;28:3–8.

SECTION

Toxicogenomics Implementation Strategies

8

CHAPTER

Toxicogenomics Implementation Strategies 8.1

Wouter T.M. Jansen and Jan Hendrik R.H.M. Schretlen
PricewaterhouseCoopers Advisory N. V., The Hague, The Netherlands

8.1.1 Introduction

This chapter contains some operational advice for starting toxicogenomics (TGX) product/service providers. This advice is based on desk-research, an internal interview round among members of The Netherlands Toxicogenomics Center, and rigorous market analysis based on in-depth strategic interviews with TGX experts and decision makers within different industries and markets. These experts include key representatives of industry, academia, and the regulatory field from across the globe, who expressed their expectations with regard to the market for TGX for pharmaceutical, chemical, and cosmetic products.

TGX is a technology that enables toxicity testing without using animals. Hence, TGX represents a potentially interesting market. The European market, however, is still in an early latent phase, because TGX technology is still developing. As with all innovations, the market dimensions and dynamics become fully apparent after the technology and derived products and services are established. Ready-to-use TGX products and services are scarce and the European TGX market is substantial. This, with the market latency, complicates a solid quantitative market analysis. This chapter mainly relies on qualitative strategic information at board-room level from leading industries. Quantitative data were not obtained, and this chapter will, therefore, not provide an exhaustive quantitative market overview.

The TGX market is dispersed and complex. A clear focus and stepwise approach is needed to enter this market. The main goal of this chapter is to provide technical and business arguments to guide the focus and strategy of novel TGX product/service providers. The objective is to highlight some of the opportunities and roadblocks, and to translate these into high-level guidance on effective strategies for the launch and growth of TGX product/service providers.

The relevance of TGX for toxicity testing depends on the availability of robust predictive assays, mechanistic assays, and software tools. A TGX-based **predictive assay** can be used to *predict* whether a given substance is toxic, whereas a TGX-based **mechanistic assay** can be used to *explain* substance toxicity by providing information on the toxicological mode of action of the substance. Given the enormous amount of raw data produced by TGX and TGX-derived assays, good **software tools** to *analyze and interpret* TGX data are indispensable. Before such products can enter the market, a validation

Toxicogenomics-Based Cellular Models.

process should take place, in which performance, robustness, and relevance of the products is evaluated in different test laboratories.

8.1.2 The TGX market is driven by regulations

Toxicology (from the Greek words τοξικός—toxicos—meaning poisonous, and logos) is the study of the adverse effects of chemicals on living organisms. As stated in the European legislation, all pharmaceutical, chemical, cosmetic, and other products present in, or to be approved for, the European market must be safe for the consumer. This requires intensive safety testing and risk assessment before and after the launch of the product. For example, it has been estimated that 100,000 deaths occur annually in the US due to adverse drug reactions (ADRs), that 10% of all inpatients experience a serious ADR and that 5% of all hospital admissions are due to ADRs.

Whereas, until recently, the tools to determine the toxicological profile of products mainly consisted of animal experiments, their use has now been narrowed down by economic, political, and legal issues. The focus market sectors—the pharma, cosmetics, and chemical industries—are driven by societal and regulatory expectations. Pharmaceutical companies face risks of drug failure in late clinical or even post-marketing phases, and hence are looking for tools to speed up the drug development process, reduce time to market, and adequately address safety issues during all developmental phases. The 7th Amendment to the Cosmetics Directive requires the phasing out of animal experiments by 2009 and 2013 for, among other things, reproductive toxicity, which has a profound impact on the development and use of animal-free toxicology tests. In addition, since cosmetic ingredients are chemicals that fall under Directive 67/548/EC and Regulation (EC) no. 1907/2006, the Cosmetics Directive is closely linked to the chemicals legislation. The chemical industry aims to ensure their license to market by means of REACH (EU regulation of Registration, Evaluation, Authorisation and Restriction of Chemicals) compliance. Finally, the economic downturn forces industries to focus on cost-saving strategies. Figure 8.1.1 illustrates the legislative and program time frames that push the business case of TGX product/service providers. TGX-based safety testing holds the promise of answering all of these altered societal and regulatory expectations, by reducing, refining, or replacing (the "3Rs") animal testing in a cost-saving manner. TGX market opportunities have been described in the Cambridge Health Tech report [1]. The Cambridge Health Tech report estimated that the total possible TGX revenues would range from 78 million Euros in 2007 to 235

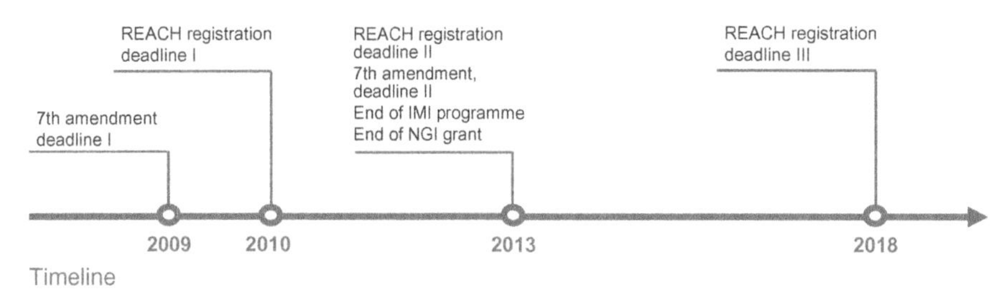

FIGURE 8.1.1
Legislation and program time frames that push the TGX product/service providers' business case.

million Euros in 2012. However, as regulation is a major driver of the TGX market, this has a profound effect on the strategy that TGX product/service providers should follow to successfully enter this complex market.

8.1.3 The European TGX market is still latent

In spite of the promise of TGX as an alternative, cost-effective in vitro tool for animal-based safety testing, the European TGX market is still in a latent phase. Although industries face substantial challenges and tight deadlines with respect to alternatives for animal testing, the TGX technology and acceptance is not fully there yet to fill the gaps. The TGX market is, therefore, not a fast-growing market, as hypothesized by the Cambridge Health Tech report, but rather appears to be an upcoming market. The figures described in the Cambridge Health Tech report represent the potential market share for TGX product/service providers, once all prerequisites have been fulfilled. This especially concerns the need for rigorous (regulatory) validation of to-be-commercialized products, as outlined below. TGX efforts within the different markets and industries are largely confined to public–private partnerships, grant-based research, and in-house programs conducted within large pharma. Commercial toxicogenomic applications are scarce. USA-based providers of TGX databases and services (Gene Logic and Iconix) have failed, showing that this is not yet a viable business model. The main reason for this is that the added value of the databases and services was considered to be relatively poor. Moreover, pharmaceutical companies were reluctant to share data with these TGX companies, which prevented further improvement the databases and derived services. As a result, the players in the market are all somewhat hesitative about the merits of TGX in the short term, and require well-performing, well-validated products and services. The lesson to learn for TGX product/service providers is to be prudent in entering this market. In close collaboration with launching customers, products and services need to be well developed and validated before market entrance can be successful. The main products and services that are currently on the market refer to generic genomics technologies and support tools, such as data-analysis software. However, given the amount of competition in this field and the presence of open-source software, this is not a very lucrative business.

8.1.4 The TGX market develops towards mechanistic understanding of the toxicology mode of action

The most significant trends in the TGX market are the more realistic perception of TGX merits and limitations, and—in fact closely linked to this—the trend towards mechanistic understanding of toxicity profiles. As with all new promising technologies, after initial over-expectations, the bubble bursts, and finally settles in a more realistic view on the merits and limitations of the new technology. History teaches that whereas initial expectations are often too optimistic, long-term expectations are often too pessimistic. The same seems to hold true for TGX. After unbridled enthusiasm in the early days of this technology, currently TGX faces some skepticism. Although several hurdles are still to be taken, this pessimism may reflect disappointment over the initial overstated expectations of the technology. So, is there an interesting market for TGX? Or is the market still immature and complex? Actually, both questions can be answered with yes.

The main issue for TGX is to strengthen the link with human risk assessment. Predictive toxicology should be accompanied by mechanistic understanding of toxicology signatures to increase the value of its relationship with safety testing and human risk assessment. Vice versa, once the toxicology mode of action of a given compound is elucidated, this may provide valuable predictive information on its adverse effects in animals and/or humans. As such, predictive and mechanistic TGX are intertwined.

8.1.5 The best market segments for TGX product/service providers are pharmaceutical and cosmetics companies

Safety testing is important in, among other areas, the food industry, chemical sector, cosmetics sector, and pharmaceutical sector. Nutritionals and chemicals, however, are far from ideal markets.

Nutritionals

Although food regulatory authorities are aware of the potential of TGX-based safety assessment methods in food safety testing, regulations are still vague as they are heavily influenced by political interests. As the food industry possesses substantial political influence, food safety decisions have direct political implications. Furthermore, given the relatively small profit margin associated with nutritionals, any commercial safety assessment methodology has to be offered at very low cost.

Chemicals

Our findings indicate that the chemicals market is not suitable in the short term for TGX product/service providers either. However, it is recommended to keep monitoring this REACH-driven market, as it can be swiftly influenced by alterations in policy making.

Pharma

Within pharmaceutical companies, the so-called 'omics area, in combination with novel tools and platforms (i.e. combinatorial chemistry and high-throughput screening), has led to a vast increase in early lead compounds in the developmental process. All of these compounds need to be tested for potential therapeutic and adverse effects. Many lead compounds fail owing to toxicity problems in later developmental phases. This means a gigantic waste of money. It is thought that the costs associated with the development of a novel drug—over $800 million—are largely related to failure of all the other lead compounds for the same drug. In addition, the discovery and development of new compound classes for more complex diseases, in combination with their swift entry in the market upon drug approval, lead to increasing post-marketing toxicology problems. Withdrawal of a blockbuster drug due to adverse effects is disastrous for a company's reputation and revenues. Early toxicity testing and, hence, the availability of in vitro toxicology assays are thus of utmost importance.

However, failures of TGX service providers GeneLogic, Epidaurus, and Iconix indicate that this is not an easy market. The US Food and Drug Administration (FDA) and the European Medicines Agency (EMEA) are facing political pressure to act proactively in the acceptance and uptake of alternatives to classical, animal-based toxicology testing. However, the Vioxx scandal and the uncertainty with novel assays render regulators prudent. On the one hand, the general public and the market request more and better safety reporting, whereas, on the other hand, regulators raise a barrier towards entering this market, as they require fully validated assays that are reliable, robust, AND provide insight in to why and how a given substance is or is not toxic.

Pharmaceutical companies are interested in commercial products and services when they bring additional unique knowledge, with less attention to cost-saving or capacity-related solutions. TGX product/service providers need to deliver packages of highly dedicated, validated products and services, also when providing custom solutions to specific requests from the customer. More details on market volumes and prerequisites are given below. Companies tend to cooperate in public–private collaborations in a joint effort to use TGX for mechanistic research and biomarker development.

Cosmetics

The cosmetics market is driven by the 7th Amendment, that requires the gradual ban of animal safety testing for cosmetics. The cosmetics sector is highly aware of the urgency to develop and implement alternative in vitro safety-testing tools. Several cosmetic companies are to some extent engaged in TGX research through in-house programs and/or participation in public–private partnerships under the European Cosmetics Association COLIPA. Most respondents, however, are rather unaware of the possibilities (and limitations) of TGX. At the moment, they are tending to stick to current in vitro assays and the use of known components that are considered to be safe in the development of novel cosmetic products.

The cosmetics sector is mainly interested in predictive and mechanistic toxicology assays. When a toxicology problem is encountered for which in-house expertise is absent and that literature studies and available assays cannot solve, cosmetics industries are willing to outsource or buy in TGX-based products and services. Simple assays will be bought in, whereas for complex assays (e.g. zebrafish models) outsourcing is the preferred method. These assays should perform similarly to or better than current gold standards and need to be rigorously validated. Cost issues are of somewhat less relevance. Only when there are no alternatives does the cosmetics sector consider assays commercially attractive that are solely scientifically validated, and not by the regulatory authorities.

In spite of the total ban on animal testing, the cosmetics sector is still focused on animal-based in vitro testing, meaning that novel in vitro safety assays must be compared to animal data as the gold standard. The demand for regulatory validation puts the focus towards animals rather than humans, as more animal data are available, and human toxicity data are scarce. As one respondent stated:

> *The focus should remain on animal-based testing as the golden standard as human data are too scarce. In fact, a human-based skin irritation assay developed by MatTek was declined by ECVAM [the European Centre for the Validation of Alternative Methods], as it should be revised to compare the data to animal data.*

Other markets and industries

Although not subject of the market survey performed as part of this chapter, Clinical Research Organizations and diagnostic firms may represent an interesting market as well. Strategic alliances with diagnostic firms may allow them to integrate TGX-based products and services and allow TGX product/service providers to profit from their existing customer network (including small pharmaceutical companies and other small and medium enterprises. However, diagnostic firms may not be eager to integrate TGX into their portfolios, and rigorous validation of TGX products seems important for this sector as well. For example, a provider of regulatory toxicity studies for pharmaceutical companies stated that their clients are mainly interested in transcriptomics and polymerase chain reaction (PCR)-based testing technologies. In addition, public and semi-public organizations such as universities and hospitals represent other potential markets. TGX product/service providers may develop toxicological data-analysis and interpretation tools that are not stand-alone, but can be integrated into existing data-analysis software. As such, data-analysis providers may also represent a market for TGX product/service providers. Finally, novel types of substances (i.e. nutriceuticals within the cosmetic sector and small molecules and biologics in the pharmaceutical sector) may boost the TGX market as well.

8.1.6 Validated predictive and mechanistic toxicology assays and data-analysis/interpretation services

Within these markets, validated predictive and mechanistic toxicology assays and data-analysis/interpretation services are commercially most relevant.

Predictive assays

Predictive assays are of interest to all three different markets: pharma, cosmetics, and chemicals. However, there is skepticism over whether predictive toxicity assays can predict long-term biological endpoints, i.e. over 4 weeks. Predictive toxicology assays are statistical tools that determine a toxic signature based on a set of test and reference compounds. Their robustness relies on the size (and quality) of these test and reference compounds used to "train" the model. This is an inherent weakness in technical optimization, validation, and commercialization of predictive toxicology assays. Even though high sensitivity and specificity can be reached for a given compound domain, extrapolation to other test compound domains may be difficult as these models are often over-fit to a given set of compounds and test conditions. Furthermore, without mechanistic information on the biological relevance of toxicology signatures, predictive toxicology assays are a rather "black-box" approach, which hampers their regulatory acceptance and uptake. In principle, commercial predictive toxicology assays could be of use both for in-house testing (e.g. drug development) and for regulatory approval (e.g. FDA drug approval). However, rigorous validation is crucial for both. Some respondents indicated that well-validated assays would be bought in on a case-by-case basis, whereas other respondents indicated that such assays may be integrated into the safety program. Predictive toxicology assays are mainly used in early stages of developmental processes, e.g. for lead optimization and prioritization. Predictive toxicology assays that are suitable for regulatory purposes may take another 5–10 years. However, once these assays are properly validated, "the sky is the limit," as one of the respondents stated.

Currently, big pharmaceutical companies often have developed organ-specific predictive toxicology assays—especially liver toxicology assays—that are used for in-house toxicology programs. Given the limitations of predictive toxicology assays as mentioned above, pharmaceutical companies would only buy in commercial predictive toxicology assays when they are robust and well validated, both for in-house developmental programs and for regulatory use. Although assays that are not validated by regulatory authorities can be used in regulatory processes—e.g. for grouping of related compounds, data triangulation, and increasing the weight of evidence—it is unlikely that this can be done on a regular commercial basis. Regulatory validation is not strictly necessary for in-house testing purposes, but certainly boosts the commercial value for predictive toxicology assays for in-house testing programs. Some quotes from respondents are:

- *Cosmetics sector:* "Only when we encounter a big toxicology problem and there are no alternatives, assays are already commercially attractive to us when technically validated only."
- *Pharmaceutical companies sector:* "We would never let go of a lead-compound based on assays that are not fully validated."
- *Pharmaceutical companies sector:* "Within our company *internal* acceptance and uptake of TGX assays is already one of the biggest hurdles. Therefore, it is highly unlikely that we will buy-in predictive toxicology assays that are not formally validated."
- *Chemical sector:* "We do not normally do mechanistic work, and TGX has little utility in REACH guideline testing studies as long as it is not formally taken-up in the REACH protocol."

Carcinogenicity predictive assays

Carcinogenicity predictive assays are considered to be highly relevant by most respondents across the different markets. They are, in particular, needed for non-genotoxic compounds. For genotoxic compounds, several in vitro models exist, and currently in the USA in silico models are being developed that perform reasonably well. In vivo carcinogenicity testing in rodents is costly, time consuming, and lacks specificity. However, development and short-term commercialization of these assays is difficult. The assay under development is based on a selection of 100 genes. It predicts carcinogenicity in the presence or absence of genotoxicity, thus categorizing substances in four different classes. When this assay can be successfully developed and validated, long-term revenues should be substantial.

Teratogenicity and immunogenicity predictive assays

Teratogenicity and immunogenicity predictive assays are also considered highly relevant to respondents, although development and short-term commercialization of these assays is difficult. The main difficulty for reprotoxicity is the presence of multiple endpoints.

Mechanistic research

Mechanistic research is of interest to pharmaceutical and cosmetics companies and, to a lesser extent, to the chemical sector. Mechanistic research fits the general market trend towards mechanistic understanding of toxicology mode of action, rather then relying on predictive black-box approaches alone. Mechanistic research is considered to be relevant, for example, to find multilevel or translational biomarkers. However, mechanistic research is currently conducted in joint public–private research partnerships, rather than on a commercial basis. Pharmaceutical companies are reluctant to outsource mechanistic research to commercial providers. First, it is difficult to validate and, hence, value commercial mechanistic research. One of the pharmaceutical companies respondents mentioned that "there is nothing to validate here." Second, outsourcing of mechanistic research is regarded as too risky, as it is considered a quite undirected approach that anything may come out of. Third, pharmaceutical and cosmetics companies would be unwilling to share compounds and/or data with third parties.

Mechanistic assays

Mechanistic assays are of special interest to pharmaceutical and cosmetics companies. What is meant here by mechanistic assays is cell-based reporter gene assays. For in-house purposes, they would be largely used retrospectively, i.e. to understand in vivo toxicological phenotypes. They are used for lead characterization and can be used for regulatory purposes as well, to underpin or explain a certain mode of action or to accumulate the weight of evidence. Since mechanistic assays are used retrospectively, they are used on a case-by-case basis, rather than being incorporated in developmental in-house programs. Thus, they may occupy a somewhat smaller and/or more unpredictable market niche than predictive toxicology assays.

Commercial data analysis and interpretation

Commercial data analysis and interpretation can be attractive. Respondents indicated that there is little interest in data-analysis software per se. There are multiple providers on the market and there is open-source software available. Most respondents stated that they have their own software providers, to whom they tend to stick. This means that TGX product/service providers should either provide competitive software at very low cost, or provide high-quality data-analysis and interpretation packages. The latter is the best option, as pharmaceutical companies are especially interested in the combination of toxicological data analysis and toxicological expertise to interpret this data. Data-analysis tools and toxicological expertise may thus represent a commercially attractive combination. These data-analysis and interpretation tools include data-mining, integration, statistics, analysis, and data interpretation by toxicological experts. Data-analysis and consulting services can be combined through the establishment of a commercial knowledge portal. Here is a typical response from one of the pharmaceutical company respondents:

> *We work with GeneGo and Ingenuity. Main drivers for selection of other service providers are cost and technological expertise to interpret the data. Costs should be below 25,000 Euros per analysis. Otherwise it becomes more cost-effective to hire an employee to do the job.*

Data analysis and consulting services can be exploited on a relatively short-term basis. However, whereas, for products, marketability depends on proof of concept and formal acceptance, for services, the provider's reputation and low costs are crucial.

8.1.7 Competitors

As the European market is rather immature, launching a TGX company is certainly not without risks, but also holds the promise of becoming European market leader in a field without yet much competition. In fact, the main competition will arise from the TGX product/service provider's customers themselves. Customers are also potential competitors as, in particular, pharmaceutical companies are developing assays as well. Some companies already have their own knowledge portal. With respect to data analysis, competition is fierce. There are many software providers in the marketplace as well as open-source software. However, in combination with toxicological expertise to interpret these data, there is little competition. Within the USA, the TGX market is more mature, but the same is true there for the development of TGX products and services. The observed protectionism in the US market hampers TGX product/service providers entering this market.

8.1.8 Investments

Performance and validation up to the regulatory level are indispensable for predictive and mechanistic toxicity assays to become commercially attractive to the market. Validation is the most crucial and costly element in the commercialization of TGX products and services. Therefore, the validation costs and timeline for TGX product/service providers' mechanistic assays, predictive assays, and software tools are outlined in more detail below.

Mechanistic assays

For both mechanistic and predictive toxicology assays, the in-house pre-validation phase will, on average, take 2 years, and cost approximately €450,000 per assay. The main costs are associated with the external validation process. A typical external validation process lasts 3 years and costs about €300,000 per assay. Subsequent regulatory implementation typically lasts 2–7 years. The length of this period depends mainly on the amount and quality of available reference substances and animal data, the need for further optimization of the assay, the duration of financial and administration procedures, and the process to reach consensus on the regulatory uptake of the assay. The external validation process requires participation of three different test laboratories. The time frame and costs are largely dependent on the complexity of the assay. When one of the reference laboratories is an ECVAM laboratory, costs can be reduced. This, however, is dependent on the availability of reference data and animal data and the capacities of the ECVAM laboratory. ECVAM is able to subsidize some of the external validation costs. External validation can also be performed in consortia to share the costs. This, however, occurs mainly on a noncommercial basis. It is questionable whether collaboration in public–private consortia to validate TGX product/service providers' products and services represents a commercially attractive business model. In order to reduce the validation costs, several assays may be validated simultaneously. In-house experience may further increase the efficiency of the validation process and hence reduce validation costs. A typical example of a validation process on in vitro safety testing is for a skin irritation assay. External validation costs were €800,000 in total for three different test methods.

Predictive assays

As compared with mechanistic assays, which mostly focus on a single toxicological pathway, predictive assays rely on the presence of many components (genes or other biomarkers) that together predict the toxic potential of a given substance in a statistical model. Understanding the (potential) biological role of each component is crucial to value the relevance of the assay results. Hence, for predictive assays, apart from the costs mentioned above for mechanistic assays, it is essential for the individual components (genes) to be biologically validated. Depending on the target, this biological validation

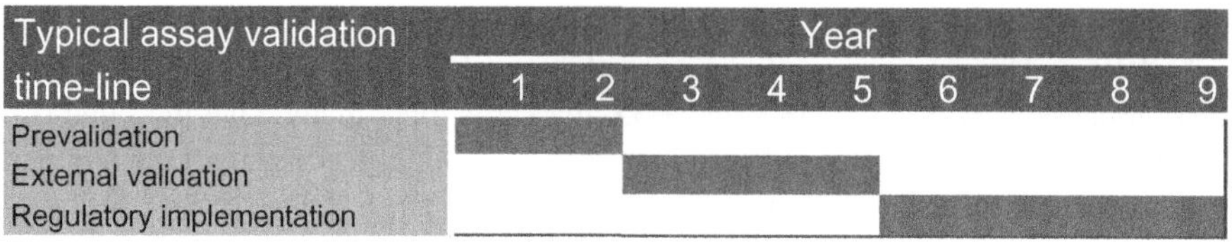

Typical assay validation time-line	Year								
	1	2	3	4	5	6	7	8	9
Prevalidation									
External validation									
Regulatory implementation									

FIGURE 8.1.2

Validation timeline.

may amount to €500,000 per gene. However, in an efficient biological validation process it should be possible to validate an average number of 50 genes for €3 to 4 million in total. Given these huge costs, the Organisation for Economic Co-operation and Development (OECD) is considering revising the regulations on this point. A typical validation timeline for predictive and mechanistic assays is given in Figure 8.1.2.

Data analysis and interpretation

Validation and regulatory acceptance of a software package or module is dependent on early involvement of regulatory offices during development/validation. Three successful examples are:

1. GeneLogic, which developed pathway-analysis software together with the FDA in a 5-year period. Although the software development was a success, the sustainability of this company was not, for reasons mentioned before.
2. The Q-SAR toolbox that was developed by the OECD in close collaboration with regulatory offices using public means.
3. PSAS, a software module for statistical analysis of clinical data. Submission of clinical data to the FDA for statistical analysis has to be presented in PSAS format. Ideally software development and regulation need to be closely linked.

TGX product/service providers can provide added value mainly within the field of data interpretation, in providing toxicological expertise to currently available software tools, including consulting services and toxicological data-interpretation tools.

1. *Consulting services:* TGX service providers can provide data interpretation services on a commercial basis to translate the generic data provided by clients in toxicological relevance. As this will be based on expert opinion rather than statistical models, these services will be explorative (i.e. hypothesis generation on type and severity of toxicities) rather than conclusive. This does not require substantial investment for TGX service providers.
2. *Providing toxicological data-interpretation tools:* There are plenty generic software providers on the market to service 'omics data analysis and integration. However, the relation with specific toxicological endpoints is lacking. There is a need to integrate toxicological knowledge into these data-analysis tools. In this way, the translation of generic data to toxicological relevance can be facilitated. An example is GeneGo, a leading provider of data-mining and analysis solutions in systems biology. GeneGo's MetaCore is an integrated knowledge database and software tool for pathway analysis of experimental data and gene lists. GeneGo's ToxHunter is a toxicological add-on to the generic data-analysis tool MetaCore.

8.1.9 Revenues

This chapter is based on a strategic rather than a quantitative market analysis. Hence, only a semi-quantitative analysis of possible investments and revenues is provided.

Two semi-quantitative approaches are pursued, to provide estimates on TGX product/service providers' revenues. As data obtained by these approaches are in part complementary to each other, each can be used to substantiate the data obtained in the other.

1. **Individual market volumes.** Respondents have been asked to provide absolute market volumes for TGX products and services within their company and their industry/market sector in general. In this way, the market opportunities for such products and services can be assessed at the level of individual companies. Whereas the Cambridge Health Tech report envisaged a fast-growing market, currently the market dynamics have slowed down. Most respondents regard the market as relatively stable in anticipation of new (validated) products and services to enter the market. As one respondent stated, "The TGX market has the potential to grow. The tests of today, however, will need a long time to get established and accepted, also by the regulators." Therefore, in a conservative estimation, it is assumed (at the time of writing, in 2013) that the TGX market will not grow further over the next decade.
2. **Cost savings.** As the main economic driver for TGX-based assays is the refinement/reduction/replacement of costly animal-based safety testing, cost savings upon the introduction of TGX-based assays by TGX product/service providers can be determined by estimating their potential to reduce the number of animals needed for safety testing. These total cost savings represent the rough market volume that can be occupied by TGX-based toxicity tests.

Individual market volumes

Mechanistic assays

As mechanistic tests are already in use, mechanistic assays seem to be products that are within reach. The downside of these types of assay is that they are very specific, so they can only be used for addressing specific toxicological questions. For individual top-10 pharmaceutical companies, market volume for mechanistic assays varies between several hundred thousand and two million Euros per year. One thousand Euros per assay is considered reasonable. Depending on the number of lead compounds, 20–50 compounds per project could be tested, leading to €20,000 to €50,000 per project. For one big pharmaceutical company, uptake of such a mechanistic assay in all developmental programs would lead to an annual revenue for a TGX product/service provider of approximately two million Euros.

Predictive assays

Market volumes for predictive assays are highly dependent on their performance and regulatory uptake, and may vary per company between a hundred thousand and two million Euros as well. This greatly depends on whether predictive assays are, like mechanistic assays, added primarily "on top" of the regular testing program, to obtain additional (early) information on substance toxicities, or when the number of animal-based safety tests is significantly reduced upon the introduction of these assays.

Data analysis and interpretation

For the chemical and pharmaceutical markets, half a million Euros for a corporate license on software for data analysis and interpretation is considered acceptable. Given the fierce competition in generic software tools, and the absence of toxicity-specific data-analysis and interpretation tools, TGX product/service providers should have good opportunities to offer toxicological-specific data-analysis and interpretation services. It will, however, take quite some time for companies to switch to a novel data-analysis provider. First, a new software program should be tested within companies for a while to determine its added value. It should have clear advantages compared with the old program, as companies consider adapting to a new software program to be a high-impact decision. Criteria for changing to a novel data-analysis provider are its reliability to translate raw data into applied data (e.g. predicting cell physiology), user-friendliness, inclusion of new tools (enabling analysis of pathways, for example), compatibility with other platforms and tools (e.g. Affimetrix), and the inclusion of toxicology-specific data with toxicology-related endpoints. Finally, a good track record of the provider is considered important, also from the regulatory standpoint. The market and revenues for data-analysis and interpretation tools depend on whether these tools can be sold as stand-alone products to the end-users within the pharma, cosmetics, and chemical sectors. As there is little commercial interest in data-analysis software per se, TGX product/service providers should aim at developing toxicity-specific models that can be integrated as an add-on to current data-analysis tools. Depending on the compatibility and ease of combining novel tools with current data-analysis tools available on the market, TGX product/service providers should aim at selling these products to software providers, or directly to the end-users (or both).

Cost savings

Predictive toxicology assays are hard to develop and validate, but once this has been achieved their market potential is superior compared with mechanistic toxicity assays. Obviously, this especially holds true when this assay reduces the number of animals needed for safety testing. As one respondent stated:

> *Predictive tests are more difficult to develop than mechanistic assays, as you have to cross species boundaries in order to have a good product. If you crack that problem and you can replace animal-based testing with these tests, the sky is the limit. The expectations are that you can ultimately reach 50% of the total safety budget, but it's a long way to go. The current 28-day in vivo toxicology assay is hard to beat with predictive toxicology assays. You first have to run them in parallel for a few years, so it will take you at least 5–10 years to get a good product.*

When predictive assays indeed hold the promise of replacing such costly animal-based safety tests, a significant increase in market volume can be envisaged. For example, REACH imposes huge costs on the chemical industry, especially due to the costly endpoints in chronic carcinogenicity studies, skin sensitization, and developmental and reproductive toxicity studies (see Table 8.1.1). When a well-performing and regulation-implemented predictive assay is able to replace costly in vivo experiments to achieve up to 10% cost reduction, such an assay would have a potential market volume ranging from roughly 5 to 50 million Euros per year for the chemical sector.

Breakdown of the Cambridge Health Tech report data into total market volumes for mechanistic assays (€10 million) and predictive assays (€8 million) do not take into account the toxicity types covered by these assays. When TGX product/service providers focus on carcinogenicity only, approximately 15–25% of overall market share as described by the Cambridge Health Tech report can be reached, as outlined below. The relative market shares for carcinogenicity testing as compared to other

Table 8.1.1 Potential of Predictive Assays to Reduce Animal-Based Safety-Testing

REACH: Additional Costs for Chemical Industry*		Potential Market Volume of Predictive Toxicology Assays when Enabling 10% Cost Reduction in Endpoint Tests	
Endpoint	**M €/Year**	**Predictive Assay**	**M €/Year**
In vivo mutagenicity	129	Genotoxicity	13
Long-term repeated-dose (carcinogenicity) study	52	Carcinogenicity	5
Skin sensitization	40	Skin sensitization	4
Developmental toxicity study	476	Teratogenicity	48
Two-generation reproduction toxicity	376	Reproduction	38

*Source: EC report on assessment of additional testing needs under REACH (2003).

Table 8.1.2 Number and Percentage of Animals Used for Different Types of Toxicity Testing

Toxicity Types	Number of Animals	Percentual Use
Carcinogenicity and mutagenicity	47,663	6*
Reproduction and developmental	120,575	15
Skin sensitization and irritation	47,546	6
Other	566,082	72
TOTAL	**781,866**	**100**

*6% for carcinogenicity and mutagenicity testing is an underestimation, as part of these tests are cataloged under chronic toxicity testing (category "other").

toxicity testing are around 10–15% for pharma, 15–20% for the chemical sector, and 5% for the cosmetics sector. When these percentages are weighted against the market volumes within these different market, a total market volume of approximately two million Euros is reachable for a mechanistic carcinogenicity assay. For predictive assays, respondents indicated similar absolute market volumes, adding up to a total of six million Euros for three predictive toxicity assays. For pharma, the relatively low percentages for carcinogenicity research are related to the fact that biologics (antibodies), cancer drugs, and short-term treatments do not require testing for carcinogenicity purposes. Pfizer was mentioned as a company who might spend more on these types of assays. These data are roughly confirmed by the third report of the European Commission to the council and European Parliament on animal use within the EU (2003) [2]. This report calculated the use of animals for different toxicity purposes (see Table 8.1.2).

As animal-based safety testing is the most expensive part of safety assessments, and TGX-based assays are anticipated to refine/reduce/replace these animal-based experiments, the relative use of animals for different toxicity types is useful as a rough estimate of the relative importance of these different toxicity types.

In contrast to mechanistic and predictive assays, which each cover a specific toxicity domain, data-analysis and interpretation services cover a broader spectrum of toxicities. Hence, there is a substantial market volume for these services, although it has to be realized that competition is fierce in this field and, hence, to acquire substantial market share such tools and services would require clear competitive advantages over current data-analysis and interpretation tools.

8.1.10 Portfolio management

Portfolio management may be based on feasibility, mechanistic focus, and core expertise. The complexity of the technology and the market requires a well-defined focus and proactive portfolio management. Although all abovementioned products and services are of commercial interest to the market, the TGX market is dispersed in the sense that different companies in different markets require different products and services, often on a case-by-case basis. In other words, there is no single blockbuster in TGX-derived products and services. However, the general trend is clearly towards mechanistic in contrast to purely predictive toxicology. In addition, entrance hurdles to the market are substantial in terms of building a solid reputation and collaborations to (co-)develop well-validated products and services.

Hence, TGX product/service providers could build their portfolio management around three criteria:

Focus on feasibility. Which products require relatively straightforward validation efforts and realistic timeframes to become commercially attractive?
Focus on mechanistic toxicology. Which products best reflect the market trend towards mechanistic understanding of toxicology profiles?
Focus on core-expertise. Which products and services best reflect the core expertise of the (novel) TGX product/service provider?

These criteria can be translated to a clearly defined product and service portfolio to be offered by TGX product/service providers, as outlined below.

The validation process is crucial to enter the market

TGX represents a novel market with substantial volume. There are, however, several obstacles that have to be overcome to open up and exploit this market. These obstacles include, to some extent, the immaturity of TGX technology itself, but mostly refer to validation aspects. Since several TGX companies in the field have gone bankrupt, the initial unbridled enthusiasm for TGX as an alternative to toxicology testing in animals has somewhat waned. This is a step forward to the field in the sense that lessons have been learned and the possibilities and limitations of TGX are more sharply defined. Nevertheless, the real success stories for TGX are still to come. There is, on the one hand, a clear need to replace, reduce, and refine toxicology testing in animals but, on the other hand, there is still some skepticism on the potential of TGX to fill in the gaps. Moreover, regulators are conservative and tend to go for certainty. All these developments underscore the importance of well-validated assays.

Human vs rodent TGX assays

Which is the preferred species to base TGX assays on: human or animal? This is a matter of choice for the ideal versus the realistic approach. Initially, TGX product/service providers are advised to focus on the development and marketing of animal-based TGX assays (i.e. TGX assays that are biologically validated through animal data as the gold standard). Although all respondents agree that, in the end, human-based TGX assays are preferred, this is considered not feasible in the next 5 to 10 years. The market is conservative, still considers animal experimentation as gold standard, and requires assays well validated against this perceived gold standard. Validation is very difficult for human-based TGX assays. Human toxicology data are largely lacking, as lead compounds and drugs are abandoned upon adverse effects observed in humans. Human TGX data can be obtained via micro-dosing experiments in human volunteers and by sharing toxicology data on failed drugs. When TGX product/service providers are closely collaborating with pharmaceutical companies on the development and validation of novel products and services, human toxicology data may also be obtained from clinical trials, under strict confidential disclosure agreements.

Cell-based reporter gene assays

Cell-based reporter gene assays are of value for both predictive and mechanistic toxicology. In vitro assays are needed to predict and explain toxicity profiles of novel substances. Prediction without mechanism is not well accepted. Both aspects are, in fact, needed, and they are to some extent intertwined. There is a clear market trend towards mechanistic understanding of toxicology profiles and hence the development of mechanistic toxicology assays. Cell-based carcinogenicity assays have the capacity to prosper in this trend and to combine mechanistic and predictive toxicology aspects. Cell-based reporter gene assays take profit from the discovery of novel potential hubs by transcriptomics-based predictive toxicology. The other way around, cell-based reporter gene assays can support the biological (functional) validation of genes to be taken up in predictive toxicology assays. Finally, cell-based assays can serve predictive purposes themselves as well. With the incremental addition of novel hubs and cell types, multi-well reporter assays can be generated. In combination with machine learning, these multi-well assays can be trained for predictive toxicology purposes. The multi-well approach is also considered to be important for mechanistic purposes, as the correlation between single endpoints or single cellular phenotypes and animal and human pathology may be low.

Cell-based mechanistic assays

Cell-based mechanistic assays require a different validation process than predictive toxicology assays and mechanistic research, which may speed up the time to market.

Predictive toxicology assays are considered interesting once validated to the regulatory level. This is very cumbersome, and several respondents are skeptical about the feasibility of commercial predictive toxicology assays. Mechanistic toxicology research is considered interesting on a noncommercial basis only, as it is costly, validation is impossible, there is no certainty about the outcome, and the structure of outsourcing these services to a consortium is not attractive to industries. Cell-based mechanistic toxicology assays, however, perfectly reflect the trend towards mechanistic understanding of toxicity profiles, and require a less comprehensive validation process than predictive toxicology assays.

Cell-based assays facilitate bridging the cross-species barrier

For mechanistic assays, the animal–human cross-species barrier is not as black and white as for predictive toxicology assays. Focusing on animal-based in vitro assays (that facilitate the crossing of the in vitro–in vivo barrier since validation in an animal model is possible) or human-based in vitro assays (to tackle the cross-species barrier) is not the main issue. Knowledge from both in vitro and in vivo animal and human data is needed to extract crucial toxicological pathways that determine toxicological endpoints, and to distinguish them from pharmacological pathways that mediate the desired pharmacological effect of the substance. The basic idea here is that identification of the pathways that determine toxicology endpoints would have the best predictive value when applied in a bioassay and, thus, may enable omitting the actual measurement of these endpoints in animal models. Cell-based assays can then be developed based on these key pathways and processes. It is not crucial initially to base these assays on human cells, although it may be perceived as important by regulators and other stakeholders.

8.1.11 Conclusion

Within the top 10 pharma, cosmetics, and chemicals markets, the best market segments for TGX products and services are pharma and cosmetics. Given the technical and validation hurdles for entering the market, TGX product/service providers need to pursue a clear focus, based on realistic investments over realistic timeframes. In general, investments to develop and validate these assays are substantial and assays need to be customized to the needs of launching customers. TGX product/service providers should build firm partnerships with industry leaders to fine-tune portfolio management, facilitate product development, and customize and validate products towards specific industry needs.

References

[1] Acton G. Toxicogenomics and predictive toxicology: market and business outlook. Cambridge Healthtech Advisors; 2004.

[2] Pedersen F, de Bruijn J, Munn S, van Leeuwen K. Assessment of additional testing needs under REACH—Effects of Q(SARS), risk based testing and voluntary industry initiatives. European Commission, Directorate General, Joint Research Center, Institute for Health and Consumer Protection. September 2003. <http://ec.europa.eu/enterprise/sectors/chemicals/files/reach/testing_needs-2003_10_29_en.pdf>.

Index

Note: Page numbers followed by *"f"* and *"t"* refer to figures, and tables, respectively.

A

B

D

E

F

G

H

Q

R

U

V

W

X

Y

Z

Printed and bound by CPI Group (UK) Ltd, Croydon, CR0 4YY
08/05/2025
01864979-0003